# CAMBRIDGE LIBRARY COLLECTION

*Books of enduring scholarly value*

## Technology

The focus of this series is engineering, broadly construed. It covers technological innovation from a range of periods and cultures, but centres on the technological achievements of the industrial era in the West, particularly in the nineteenth century, as understood by their contemporaries. Infrastructure is one major focus, covering the building of railways and canals, bridges and tunnels, land drainage, the laying of submarine cables, and the construction of docks and lighthouses. Other key topics include developments in industrial and manufacturing fields such as mining technology, the production of iron and steel, the use of steam power, and chemical processes such as photography and textile dyes.

## Reports of the Late John Smeaton

Celebrated for his construction of the Eddystone Lighthouse near Plymouth, John Smeaton (1724–92) established himself as Britain's foremost civil engineer in the eighteenth century. A founder member of the Society of Civil Engineers, he was instrumental in promoting the growth of the profession. After his death his papers were acquired by the president of the Royal Society, Sir Joseph Banks, Smeaton's friend and patron. Using these materials, a special committee decided to publish 'every paper of any consequence' written by Smeaton, as a 'fund of practical instruction' for current and future engineers. These were published in four illustrated volumes between 1812 and 1814. As a consulting engineer, Smeaton carried out surveys and reports of existing structures, as well as drawing up proposals for new designs. Volume 3 contains mainly reports relating to bridges and harbours, including work at such major ports as Aberdeen, Dover and Hull.

Cambridge University Press has long been a pioneer in the reissuing of out-of-print titles from its own backlist, producing digital reprints of books that are still sought after by scholars and students but could not be reprinted economically using traditional technology. The Cambridge Library Collection extends this activity to a wider range of books which are still of importance to researchers and professionals, either for the source material they contain, or as landmarks in the history of their academic discipline.

Drawing from the world-renowned collections in the Cambridge University Library and other partner libraries, and guided by the advice of experts in each subject area, Cambridge University Press is using state-of-the-art scanning machines in its own Printing House to capture the content of each book selected for inclusion. The files are processed to give a consistently clear, crisp image, and the books finished to the high quality standard for which the Press is recognised around the world. The latest print-on-demand technology ensures that the books will remain available indefinitely, and that orders for single or multiple copies can quickly be supplied.

The Cambridge Library Collection brings back to life books of enduring scholarly value (including out-of-copyright works originally issued by other publishers) across a wide range of disciplines in the humanities and social sciences and in science and technology.

# Reports of the Late
# John Smeaton

*Made on Various Occasions,*
*in the Course of his Employment as a Civil Engineer*

VOLUME 3

JOHN SMEATON

CAMBRIDGE
UNIVERSITY PRESS

# CAMBRIDGE
## UNIVERSITY PRESS

University Printing House, Cambridge, CB2 8BS, United Kingdom

Cambridge University Press is part of the University of Cambridge.
It furthers the University's mission by disseminating knowledge in the pursuit of
education, learning and research at the highest international levels of excellence.

www.cambridge.org
Information on this title: www.cambridge.org/9781108069793

This edition first published 1812
This digitally printed version 2014

ISBN 978-1-108-06979-3 Paperback

# REPORTS

OF THE LATE

# JOHN SMEATON, F.R.S.

MADE ON

*VARIOUS OCCASIONS,*

IN THE COURSE OF HIS EMPLOYMENT

AS

A CIVIL ENGINEER.

*IN THREE VOLUMES,*

VOL. III.

LONDON:

PRINTED FOR LONGMAN, HURST, REES, ORME, AND BROWN,
PATERNOSTER-ROW.

1812.

# CONTENTS.

———

# C O N T E N T S.

TREW-

# CONTENTS.

ALTGRAN

Page

EARL

# C O N T E N T S.

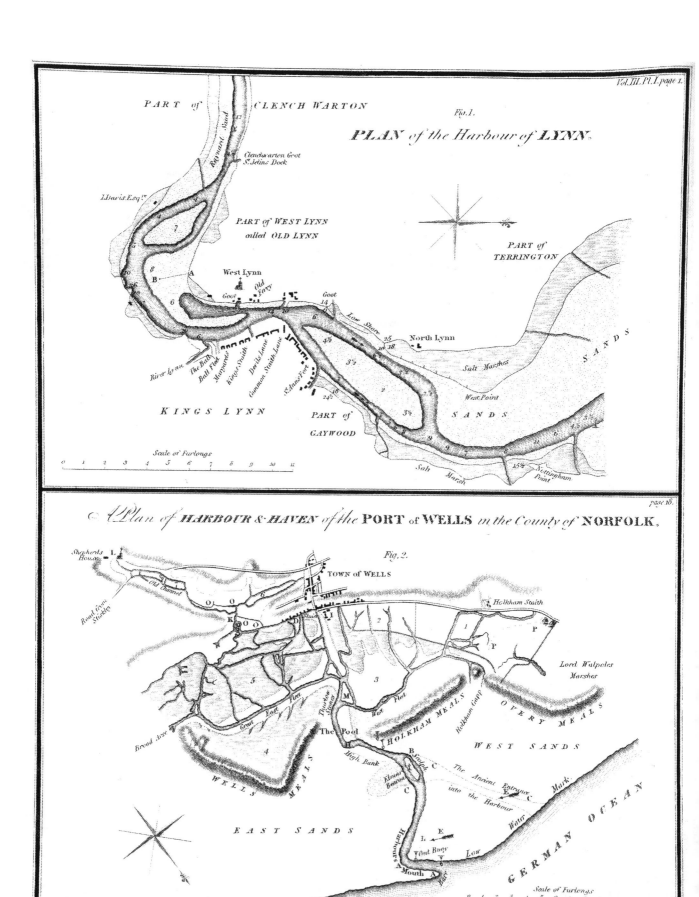

PART of CLENCH WARTON

Fig.1.

PLAN of the Harbour of LYNN.

Clench Warton Goot
St John's Dock

J. Davis Esqr

PART of WEST LYNN
called OLD LYNN

PART of
TERRINGTON

West Lynn
Old Ferry

Goot
Goot 14

Low Shore

North Lynn

SANDS

River Lean
The Bath
Bull Flat
Marsh
King's Staith
Devils Lane
Common Staith Lane
St Ann's Fort

Salt Marshes

West Point

KINGS LYNN

PART of
GAYWOOD

SANDS

Salt Marsh

Nottingham Point

Scale of Furlongs
0 1 2 3 4 5 6 7 8 9 10 11

page 18.

A Plan of HARBOUR & HAVEN of the PORT of WELLS in the County of NORFOLK.

Shepherds House

Fig.2.

TOWN of WELLS

Old Channel

Holkham Staith

Road from Stiffkey

Sluice

Lord Walpoles Marshes

Broad Arse

East Flea
Great
Thornton River

West Flea

HOLKHAM MEALS

Holkham Gapp

OVERY MEALS

WEST SANDS

The Pool
High Bank
Float Beacon

WELLS MEALS

Scolds

The Ancient Entrance
into the Harbour

Water Mark

EAST SANDS

Harbour
Float Buoy

Low

Mouth

GERMAN OCEAN

Scale of Furlongs
0 1 2 3 4 5 6 7 8 9 10 11 12 13

J.Farey Junr delin.

Published as the Act directs 1812, by Longman, Hurst, Rees, Orme and Brown, Paternoster Row, London.

Engraved by W.Lowry.

# REPORTS, &c.

## KING'S LYNN HARBOUR.

The REPORT of JOHN SMEATON, Engineer, relative to the Harbour of *Lynn* in *Norfolk*.

THE Matters to be reported upon as given me in charge by the Magiftrates of *Lynn*, were principally the following :

Firſt, The preſervation of veſſels lying in the harbour of *Lynn*, from the annoyance of winds and waves; not only from the ſea, but from the broad river above the town, when the wind happens to be in that quarter, and alſo from the raging tides that often accompany them, and drive the Veſſels from their moorings :

Secondly, The preſervation of the banks of the river, ànd more eſpecially thoſe near the town, from the aċtion of the winds and ſeas, accompanied alſo with too rapid a tide.

The better to enable me to acquit myſelf upon theſe ſubjeċts, I carefully examined the channel of the river *Ouſe*, from the roads near the *Ferror Beacon* to *German's Bridge*; and alſo the banks thereupon, and more particularly thoſe within two miles of the town

Note.—In the plan of the harbour, Fig. 1. Plate 1. the figures in the channel ſhew the depths in feet and half feet at low water; the other figures ſhew the height of the ponds, lower ſhores, ſalt marſhes, and banks above low water mark.

of *Lynn*. I have alfo carefully viewed and examined the circumftances attending the faid river, from *German's Bridge* to *Denver Sluice*; fo far as they appeared to me to have any relation to the bufinefs before-mentioned.

In the courfe of thefe enquiries, I had the pleafure to find, that the channel of the river, and particularly that part to feaward, not only from the accounts of the pilots who attended me, but from my own remarks, compared with former accounts and reports, was, at the time of thofe obfervations, *viz.* in the month of *July* 1766, in as good condition as it had been known for many years; and in fact without any material caufe of complaint. That a bar had formed itfelf in the upper mouth of the weft channel, and that the current at low water was wholly confined to the eaft channel, which had in confequence proportionally improved.

This is indeed a very material change in the ftate of the channel and harbour for the better; the contrary condition having been complained of for years paft, *viz.* the choaking of the harbour and channel to feaward with fands, and the great tendency of the current to take the weft channel inftead of the eaft; which latter (as it fcems) was that, which, according to the unanimous opinion of all thofe who have reported upon it, was the defirable channel to be maintained, and if poffible improved; the doing of which has been the fubject of much projection and altercation: but as nothing appears to have been done in confequence, but that Nature has very kindly brought matters into a more defirable ftate, it feems of the utmoft importance to propofe nothing for execution, that fhall counteract her intentions; nor while we are relieving one evil induce a greater.

According to all accounts, the point below the crutch on the eaft fide of the river, called at prefent *Nottingham Point*, projected much further to the weftward: and another point of land on the weft fide of the river, below *North Lynn*, but above the faid *Nottingham Point* according to the courfe of the river, projected further eaftward; fo that, to a perfon ftanding upon the quay called the *Common Staith*, at *Lynn*, the fea was land locked by thefe two points; but that now, by the gradual wear of one or both of them, an obferver ftanding at the fame place, will, at high water, fee the fea confiderably open between the faid points; and the wind being in that or nearly that direction, that is, betwixt N. N. W. and N. N. E. and a ftrong fpring tide of flood, the fhips are apt to ride very unquietly that lie within the point formed by the *Common Staith* and the *Ball*, which may now be reckoned the port of *Lynn*, and where the fhips chiefly lie to be loaded and unloaded from the merchants' yards.

This

This evil I find has been complained of for many years; and in the year 1741, a scheme was offered by Mr. *John Rosewell*; which was to build jetties from these two points so as to produce the same land lock as formerly : and this he imagined would not only redress this immediate grievance, of too much swell running into the harbour, but also others at that time also greatly complained of; that is, obstruction by sand, too shallow a channel, a high bar in the east channel, and too great a diversion of the current into the west channel.

That those jetties might be of some use in checking the swell of the sea, cannot be denied; but the ill effects they were likely to have by encreasing the tendency of the out channel to the westward, is set forth with great justness and spirit, in a small pamphlet, published in the year 1742, entitled, *Some thoughts on Mr.* Rosewell*'s and other Schemes, now proposed for amending* Lynn *Channel and Harbour\**; which, as I suppose it in every ones hands to save repetition, I recommend to a re-perusal, from the beginning to and including the 9th page. And indeed Mr. *Rosewell*'s reasoning, if such it can be called, concerning the improvement of the east channel by means of those jetties, has not been more solidly refuted by the author above referred to, than it has since been by Nature herself: for it seems very manifest, that in proportion as the points, particularly *Nottingham Point*, have been worn away, the channel has tended more eastward; and has improved in proportion as the west channel has decayed, which must be expected from an union of the two forces. As therefore the maintenance of the channel out to sea, in the good condition in which I found it, I look upon as a *primary* consideration; I entirely agree with the author beforementioned, that it would be a dangerous expedient to reinstate the two points abovementioned, by the building of jetties: on the contrary, I am of opinion, that where a channel must be maintained through a vast mass of sand, capable of shifting by winds, seas, currents, and every temporary impression of power thereon, that the more directly the waters make their passage out to sea the better.

Nor am I very clear, that the reinstating of the points would entirely cure the evil at present complained of; they being at too great a distance from the interior harbour, and must necessarily be placed at so considerable a distance from each other, that unless the entry was entirely spoilt, I apprehend the seas would still find the way within the heads : and in so long a fetch to the place where the ships lie, would gather and still disturb them at their moorings.

Nor can they be of much use in checking the too great indraught of the tides, at present complained of; for the passage between them must be very narrow to produce a

---

* This pamphlet is anonymous, but I since understand was wrote by Mr. *Elstobb*. See Appendix to this Report.

sensible

fenfible effect, and in proportion as it was contracted, it would wear deeper, and thereby ftill admit nearly the fame quantity of water to pafs. Yet, if they could be fo made as to check the raging tides, they would check the moderate ones too; and I believe all agree in opinion, that in the fituation of *Lynn* harbour, the greater the efflux, which in great meafure depends upon the influx, the better channel may be expected to be maintained: too great tides may be a partial and temporary evil, but it appears to me in the prefent cafe a fault on the right fide; and it feems that in this, as well as many affairs of human life, the judgment confifts in chufing *the leaft of two evils*.

I cannot however agree in recommending the expedient mentioned in the latter part of the pamphlet above quoted; that is, damming up the old river at the weft point, and cutting a new one through the land, down to the road; for this, if not made equal with the mean capacity of the old river, would check the influx of the tides; and if fo made, would be a moft monftrous expenfe, and at laft the event would be uncertain; for by producing an alteration in the fet of the currents, without the mouth thereof, it perhaps is out of the power of human forefight to fay with certainty, that (after fo great an expenfe) fome bar or impediment might not be thrown up, producing an equal obftruction to thofe then complained of; but thanks to Nature for relieving us from the neceffity of fo chargeable an expedient, and giving us the opportunity of *knowing when we are well.*

The fimple point of view, that in my opinion is neceffary to be attended to for the prefervation of the channel, is to do nothing *materially* to affect the indraught of the tides, or to diminifh the quantity of frefh waters coming down the river from above: as to the reft, it muft be left to Nature, and though in fuch a multitude of acting forces and difturbing caufes, the goodnefs of the channel muft by turns become *better and worfe*; yet the grand principles of prefervation being maintained in all the vigour poffible, after a wrong turn happens a right one will fucceed, as experience has fhewn to be the cafe; at leaft leave Nature to herfelf while tending right, it is time enough to help her when we are fure fhe is going wrong.

It is obfervable, that the point formed by the *Common Staith*, and running out betwixt that and *Common Staith Lane*, may be confidered as the jetty, whereby the fhips lying above in the interior harbour, are ultimately defended from the fwell coming in from the fea; and it is only the fwell that gets round this point from the fea that affects the fhipping; for in fact, the fhips that lie in the channel of the river, between the *Purfleet* and the *Ball*, are land locked by this point. This jetty has very much the appearance of having been artificially carried out as well as defended by ftone wharfing for this very purpofe.

purpose. The river is however still a furlong, that is, 220 yards, wide at this point, according to the plan; the channel lies over to the opposite shore, where it is ten or twelve feet deep at low water, whereas it gradually shallows towards the *Lynn* side, leaving the ground dry at low water for above a chain in breadth.

If I were obliged to recommend something in the *jetty way*, it would be to run out this wharf about a chain further, that is, twenty-two yards, being one-tenth of the whole width of the river. I am very clear that such a projection, immediately before the shipping, would afford more protection than six times as much work done at the two points below; which are at the mean distance of a mile and half; and this without being attended with any ill effect upon the channel below. The only objection that I see to this scheme is, that by narrowing the channel in this the narrowest place, it would in some measure check the influx of the tides; and throw the water more powerfully upon the opposite shore, just below the ferry, the defences of which are already too weak: but if this bank were effectually defended by a sufficient body of rubble stone, applied in such a way as I shall mention, when I come to speak upon the defence of the banks, the effect would be to scour out a channel so much deeper in this place, as, by creating an equal section, would furnish the same quantity of tide water as at present, and probably might deepen the channel on the *Lynn* side; at least something might be done to intice it to act in that manner.

It is observable, that at the end of *Common Staith Lane* there is a considerable prominence of rubbish that has gradually been formed, and that something of the same kind· has been thrown out partly from the lanes, and partly from the houses, quite away to the. *Purfleet*, and higher: I look upon it, that it is owing to this body of rubbish, that the channel of the river edges over so soon to the west side; if this body of rubbish were dug away as near to the buildings as can be done with safety, and taken down to or below low-water mark as far as the *Common Staith*; that the ebb tides, accompanied with the land floods, would improve the channel, and enable the ships to lie not only lower down, but closer in, and thereby more effectually shelter them from the seas, which would be broken off by the *Common Staith* point; and this is a step that I would recommend, whatever becomes of the proposition of an extension of that point.

I come now to the defence of the shipping when the wind is at S. W. or westerly; that is, when it blows right down the broad river which extends above the town; and with this view it obviously occurs, that if a jetty or break-water were carried out upon the sands from the point above *Old Lynn*, this would in some degree shelter the shipping that lie from the *Ball* downwards: but yet, unless the jetty be extended from A to B in the plan,

that

that is, the length of a furlong, or 220 yards, the veſſels lying at or near the *Ball* will ſtill be expoſed almoſt as much as before: on the other hand, this work will throw the current more forcibly into the bite or great hollow oppoſite this point, where the water ſets already too hard, and is making daily depredations. I would therefore beg leave to ſuggeſt, that as no ſhip lying from the *Ball* downward, can be ſubject to a direct fetch of more than a mile in length; and ſince in this length no great ſwell can be raiſed, but only a ſhort chopping ſea or windwaſh, which would ſignify little, independent of the force of the wind and a ſtrong tide of ebb, both which muſt remain; I ſay, I would put a query, whether it would not be adviſeable at once to endeavour to encreaſe the ſecurity of the veſſels in the harbour, by improving their moorings; which I apprehend might be done by putting down a row of ſtrong *dolphins* on the weſt ſide of the channel in the middle of the river; and continuing the line lower down, and increaſing in number thoſe that are on the eaſt ſide. This meaſure, independent of every thing elſe, would be a great ſecurity againſt diſturbance, both from land and ſea. However, if the magiſtrates are inclined to begin a jetty or break-water from the S. W. point in or near the direction ſpecified in the plan, in order to try the effects of it, I would not be underſtood to diſcourage the attempt; and if not carried beyond the limits there ſpecified, I apprehend any ill effects that it might have upon the oppoſite bank might be guarded againſt. It may be begun with rice and rubble ſtones, and afterwards by ſinking old veſſels; or continued with the former materials, as ſhall appear in the execution beſt to anſwer the end.

I come now to the ſubject of the banks: and on this head, as I have already declared that my opinion is in general, to diſcourage all attempts to prevent the free influx and efflux of the tides and land waters, in order to preſerve the channel out to ſea in the moſt effectual manner, upon which both the navigation and drainage dependent on the *Ouſe* entirely hinge; nothing remains to be done with the banks, but to make them ſtout enough and high enough to ſtand againſt all extremes: my buſineſs will therefore be, to ſhew how this is to be effected. To give directions about the particular parts would be endleſs, and will be the proper buſineſs of the ſurveyor, who ſhall be entruſted with the execution thereof; I ſhall therefore content myſelf with ſhewing the general principles, which will eaſily be applied by the judicious artiſt to the particular caſes.

In the firſt place, I entirely diſapprove of all jetties built into the ſtream, as a defence for ſaving the banks or foreſhores from the action of the water, as I am convinced from many obſervations that they have a direct contrary tendency; for they ſeldom fail of producing a deep pit, either oppoſite to, or on the downſtream ſide of the jetty, which

tends

tends to undermine the banks, and even the jetty itself ; fo that thereby the *rent is made worfe*.

Whatever works are attempted for the prefervation of the foot or banks of forefhores, when too hard a *fet* of the water tends to undermine them, ought to be difpofed parallel, or according to the direction of the ftream ; fo that the water, inftead of being ftopped or thrown off, fhall glide gently by, with the leaft interruption poffible ; for thereby the water gets away with the leaft poffible action upon the banks, and confequently wears them or their defences the leaft. For this reafon all angles and fudden turns are to be avoided as much as may be, and where a turn muft be made, as very frequently happens, let it be made with as eafy a fweep as poffible, keeping it as near as may be the fame with the natural, that is, the general *bend* of the river, cutting or rounding off all fmall fudden turns, angles or extuberances, which may happen in the general fweep ; which directions will hold, as well for forming the bafe and middle lines of the artificial banks themfelves, as for preferving the foot of the natural forefhores, whereon the other is founded. To act otherwife is to oppofe a natural current, and to oppofe a natural current is to give it fomething whereon it may inceffantly act, towards the deftruction of the oppofed work ; which in the cafe of banks and their defences, is to do fomething contrary to the very intent of doing it.

This doctrine cannot be better illuftrated than by the two great jetties, in the turn above the *Ball-fleet*; oppofite and juft below which are five and fix fathom water ; whereas the natural bottom or depth is not more than as many feet. By this means their foundations are fapped, as well as the adjoining banks, which in confequence muft be fubject to very expenfive repairs, as the bite of the river will grow deeper on every alteration.

This is alfo the cafe with a number of jetties on the weft fide of the river below the ferry : to fay the truth, I fcarcely ever faw an inftance where this fort of works, when applied as a defence to the banks, did not do mifchief ; it would be very well that they were all removed, or if this is thought too much expenfe, at leaft never more to be repaired in their prefent form.

Where the fides of a river or forefhores grind away by too great ftrefs of water, the moft infallible, fecure, and lafting method of doing it is, after the fudden ex-tuberances and irregularities of the curve are taken off as low as the water will admit of by hand, to line up the foot with rubble ftones, thrown in promifcuoufly, fo as to

form

form their own natural flope againft the fhore, till they appear above water, at low water. This being done for the whole length of the galled place, with the regularity before defcribed; and being further fupplied with a frefh quantity after a fettlement has happened, in confequence of fucceeding great tides and floods; will prove a lafting fecurity againft further depredations at the foot; and the foot being well fecured, Nature will, in moft cafes, where there is room to do it, form of herfelf fuch a natural flope to neap tide high-water-mark, as will need no artificial defence. The chalk and clodlime rubble that is brought down the river, feems a very good material to be employed in this way, but the rubble of any ftone that will bear wet and dry will do, which can be procured cheapeft; the larger and more irregular, both in point of fhape and fize, the better.

The fame defence, where ftone is wanting, may be procured by the application of ftakes and rice; but difpofed, as before directed, even and parallel to the ftream*, without jettings or extuberances; but nothing can be done with thefe materials comparable to rubble ftones, on account both of want of weight and duration.

It may poffibly be objected, that a fufficient quantity of rubble ftones, to fecure the foot of the banks and forefhores in this manner, will be expenfive; and in fome cafes poffibly it may; but it muft be confidered, that here will be no expenfive tide-works, with a number of men, and that there will be very little additional charge beyond that of bringing down the ftones; but in preference to all methods that are in themfelves ineffectual, the cheapeft way is to do nothing.

The bafe of the bank and forefhores being fecured as above defcribed, it has already been faid, that where there is room enough, Nature will herfelf form a flope upon the forefhores up to the high-water-mark of neap tides, which will be of itfelf a fufficient defence againft waves and currents; and where there is no want of room to form a fufficient flope above the neap tide high-water mark to the top of the bank, there is no better way than to flope and turf it over, in the manner that has already been put in practice, but above and below the town of *Lynn*; and alfo on the other fide of the water. I can give no better directions than thefe fpecimens will afford, which feem done with great judgment; only that, inftead of ufing any wood at the foot of the turf about neap tide high-water-mark, I would advife to rely wholly upon a *lay* of rubble ftone; and if fome of it is broken fmall fo as to fill up the interftices of the larger pieces, this will form a more complete union between the rubble and the turf, than if compofed of large ftones only.

* The ends outward, and a little inclining down ftream.

Where

Where the artificial banks ſtand ſteep upon the foreſhores, I obſerve that the practice about *Lynn* has been to defend them by a boarded wharfing. This at firſt ſeems to promiſe much ſecurity; but experience has proved that this is only *outſide :* and if we conſider the action of the waves upon it, we ſhall find this method not at all calculated to ſuſtain their ſhocks for any length of time, the very union of the parts being the cauſe of their deſtruction.

If the ſea daſhes againſt the end of a faggot or a ſtone, theſe having no ſolid connection with their neighbours, the impreſſion goes no further; but the tremors raiſed in one part of a boarded wharfing is communicated to the whole, or to a large area; which, accompanied with the great ſhocks that muſt unſue when a conſiderable part of the ſurface is ſtruck together, theſe agitations by little and little ſhake and looſen the earth behind, which by the riſe and fall of the tides is by degrees waſhed out at the foot and through the crevices, which brings the whole to ruin. This, accompanied with the ſpeedy decay of all wood work expoſed to wet and dry, makes it ſeem to me quite eligible totally to diſcard this method in all future repairs. In lieu whereof I would recommend the following: wherever the foreſhores are not broad enough from low water to neap tide high-watermark, to ſtand by a natural ſlope, I would adviſe them to be reduced to a ſlope of two to one; that is, to batter two feet to one foot perpendicular; and this being covered a foot thick with rubble ſtone of all ſizes bedded together, and footed upon the rubble, ſuppoſed to be thrown in to ſupport the ground under low-water-mark, will, as I have often experienced, make a very laſting defence, and is capable of great reſiſtance in all ordinary caſes; but if expoſed to the waves of the open ſea, the cover muſt be increaſed in weight and thickneſs. This method will anſwer the end, if the batter is three to five, or even one to one; but the latter I would not recommend without extreme neceſſity.

With reſpect to the upper part of the bank, from the high-water-mark of neap tides, to the high-water-mark of the equinoctial ſpring tides; I would, in all caſes where it is poſſible, ſhift back the banks, till they will admit of the ſlope and turf method above referred to; but if this be not poſſible, to continue the ſame batter as the foreſhore, and to face the artificial banks with ſtone in the ſame manner as already deſcribed.

I do not know of any ſecurity againſt *inundations,* in a country that is defended by banks, otherwiſe than by making thoſe banks, not only ſufficiently ſtrong, but high enough to ſuſtain the greateſt extreams, without being overflown: and it is from *experience alone,* that theſe extreams are to be learnt.

The greateſt extream of a flood or tide that has been known at *Lynn*, appears to have been one on the ſecond of *December* 1763, when the water flowed one inch deep in the compting houſe of Mr. *Elſden* of *Lynn*, and left a very viſible mark, which he ſhewed me. This mark ſhould at high-water of ſome ſpring-tide, be transferred to the Cuſtom-houſe, or ſome public place near the river, to which recourſe may conſtantly be had. Though this is the largeſt tide we know of, yet there are accounts of ſeveral tides within a few inches as high as this ; I do not therefore look upon the banks to be ſafe, unleſs they are proof againſt ſuch another tide as this ; and to this, as a ſtandard, I would conſtantly refer : I cannot, however, look upon them as *proof*, unleſs they are made up at leaſt one foot above the level of this mark, and maintained, after ſettlements, at the height of half a foot above the ſame. It is not, however, neceſſary, to preſerve the ſame ſlope above the high-water-mark of equinoctial ſpring tides to the extreameſt height, as below that mark ; becauſe they will but ſeldom come to a ſtreſs ; yet when they do, they muſt be *ſufficient*, or they anſwer not the end.

In order to this, they ought to be three feet at leaſt broad at the extream height ; and at leaſt three times as much broader at the level of the equinoctial ſpring-tide mark, as the extream height exceeds that height. The artificial banks below that mark ſhould be at leaſt four times as much broader, in their baſe or ſeat, upon the natural level of the ground, as the perpendicular height of the ſaid extream tide-mark is above the natural level of the ground, whereon each part of the ſaid bank reſpectively ſtands. Theſe proportions will be very ſufficient in the neighbourhood of *Lynn*, where the earth is good ; but where the earth is looſe, ſandy or moory, the baſes or ſeats and tops ſhould be reſpectively broader.

Theſe are the beſt methods that I know of putting the town and country in a ſtate of ſecurity ; every other attempt to leſſen the tides themſelves will not only be vain and fruitleſs ; but even, if effected, hurtful to the general drainage and navigation of the country.

If the expenſe be objected to of making up the banks in this manner, I anſwer ; that if we take the compaſs of a few years, it will be a great ſaving : but ſince this is the only permanent and ſecure method that I know of ; if it is not worth while to put the properties into a ſtate of ſecurity, then the proprietors muſt be contented to poſſeſs them in a ſtate of inſecurity :

After having declared it as my opinion, that every attempt to check the free influx and efflux of the tides (that is, while it remains an open river) is likely to be hurtful to the

ſcour

ſcour of the ſea channel ; it ſeems natural to be aſked, if I do not look upon *Denver Sluice* in this light ? I anſwer, that *Denver Sluice* is much too far up from the out-fall to work any conſiderable effect either way ; but that the *natural* effect of it, as well as the hundred foot drain, ſo far as they operate on the harbour of *Lynn*, and ſea channel below, muſt be rather beneficial ; as may be demonſtrated from a circumſtance mentioned in *Badſlade's* hiſtory, page 46, where, ſpeaking of the river *Ouſe*, he ſays, " The low-water-mark up " the river is much lower in neap than in ſpring-tides ; whereas down to ſeaward, the " low-water-mark of a ſpring-tide is lower than that of a neap." This being the caſe, (which is not a peculiarity of this river,) it is evident that the ſpring-tide of flood, in the remote parts, could not get back again the ſame tide ; but, inſtead of returning back, ſpent itſelf up the river, and in the fens, before that ſluice was built ; it is therefore, at that length, more uſeful to check the tide of flood, in order to give it a better recoil, than to ſuffer it to ſpend itſelf in the fens at the ſpring-tides, and languidly to return in the neaps ; without power in itſelf to operate, or to co-operate with ſuch a tide as would give it power. The cauſe, therefore, of the univerſal filting of the channel of the *Ouſe*, after the firſt erection of *Denver Sluice*, and cutting the hundred foot river, ſo univerſally and loudly complained of, muſt be ſought from other ſources, than a *natural* tendency of theſe works to produce this effect : but as this will lead me into a freſh and large field of matter, not immediately conducing to the buſineſs in hand, I ſhall forbear the purſuit of this ſubject any further at preſent.

I cannot conclude my Report without obſerving, that it ſtrikes me, that the ſureſt, and I may add the cheapeſt, way of putting the ſhips into a ſtate of *perfect* ſecurity, would be by building wet docks in the manner of *Liverpool* ; for which, the ſituation of *Lynn* ſeems to afford a noble opportunity, particularly in the flat ground below the block-houſe.

J. SMEATON.

*Auſthorpe,*
14th *September* 1767.

APPENDIX.

# APPENDIX.

Containing an Extract from a Pamphlet printed in the Year 1742, said to be wrote by Mr. *Elstobb*; intituled, Some Thoughts on Mr. *Rosewell's* and other Schemes, now proposed for amending *Lynn* Channel and Harbour; in a Letter to the Merchants, Owners, and Masters of Ships, belonging to the said Place.

GENTLEMEN,

THE present state of your channel is now not only become the common topic of conversation, but also the laudable care of the magistracy, whose ready disposition to do service to the town, is evidently manifested by the early steps they have taken in this affair; in consequence of which, Mr *Rosewell*, by order of the honourable the commissioners of the navy, has lately inspected the harbour and channel to seaward, down as low as the road, and has proposed a method, to remedy the evils, which he observed at present to attend them, and to prevent their growing bad for the future.

But as I think the remedy he proposes, will not remove the evils he complains of, but will rather increase them; so I cannot forbear giving you my sentiments thereupon, tho' without any design of depreciating that gentleman, who I doubt not is sufficiently judicious in things of this kind. But as he came an entire stranger to the place; and as I am informed, took but a transient view of it, and formed most of his sentiments, from the information of people, prejudiced in favour of particular notions and opinions concerning the thing, it is not to be wondered at, that he is so far wide of the matter.

His opinion it seems is, that *the badness of the channel to seaward, is owing to a bar of sand, lately grown up cross the east channel, about three miles below the crutch-point.* And he says, he is informed, that *the said crutch-point on the east, and the point on the west side of the harbour's mouth, formerly landlocked each other, by which means the harbour was preserved in a great measure from the fury of the north-west winds, and flood tides, was rendered safe, and the channel maintained deep and good ; but by the continual force of the said winds and tides against the west point, and by the violent action and force of the south-west winds and ebb-tides, against the crutch-point, on the east side, both these points of land are now much shortened, and the harbour laid more open; so that the fury of the north-west winds tears down the marches, endangers the banks, and threatens the ruin of some of the lands on the east side of the river, between the block-house and the crutch ; the consequence of which may be* (he says) *the entire loss of the harbour. Besides, since the points are thus become open,* (I think he says), *it occasions the flood to form an eddy, by which means the silt subsides, and has formed the bar which he so much complains of.*

The

The remedy therefore which he propofes is, to *erect two jetties, or wooden piers, one on the weſt-point, to ſtretch over to the eaſtward; and the other on the crutch-point, to ſtretch itſelf to the weſtward; and ſo to landlock the harbour, as it formerly has been.*

By this means (ſays he) *the harbour will be preſerved from the violence of the north-weſt winds and flood tides, will be rendered ſafe and quiet, the eaſt-channel recovered and the bar be ſcoured away.*

This, gentlemen, is his opinion, and the remedy he propofes. And now, with all due ſubmiſſion, I ſhall beg leave to give you my ſentiments thereon.

'Tis to be obferved, that in the prefent ſtate of the harbour and channel, there are but two evils which have fallen under his notice.

The firſt is a bar of ſand on the eaſt channel, three miles below thefe points, which bar he ſuppofes to be lately grown up, and formed by means of an eddy in the flood tides, occafioned by the ſhortening or opening of the ſaid points. But how the alteration of the ſtate of thefe points could any way form ſuch an eddy as to occafion a bank of ſand in the channel, at three miles diſtance, is to me quite unconceivable. Befides, the bar is no new formed thing as he ſuppofes, but is of old ſtanding, as is well known to the gentlemen navigators; tho' 'tis not always in the ſame ſtate, but often fluctuating and changing: being ſometimes higher, ſometimes lower, according as feafons have proved wet or dry, and the quantities of the ebbs and back-waters have been greater or lefs. Thus, in ſome very wet feafons, it has been almoſt worn away, and the channel for a while maintained good; and in dry feafons again, it grows up, and the channel becomes bad; which has lately been the ſtate of it, tho' now by the increafe of the back-waters 'tis already much mended. However, let the duration of it, and the formation of it be as they will, the removal of it is the thing.

And this I believe every one will agree muſt be effected, by the ebbs and back-waters, which ſhould be convey'd as entirely, unitedly, and directly to it as poſſible, ſo that they may exert their whole undiminiſhed force and power thereon, without meeting with any ſtops or impediments in their way. For the force of water in ſcouring away a ſand, or deepening a channel, is always proportional to the quantity acting in a given time, and the vigour or force with which every equal part of that quantity does act. Thus, for inſtance, if a tun of water was to pafs over a ſand in a minute's time, and every gallon of that water exercifed a force on that ſand, which may be reprefented by 1 or unity, then the whole force exercifed in a minute's time upon that ſand, may be reprefented or expreſſed

by

by the number 252. But if the quantity of the water acting in the same time be doubled, and the force with which every gallon acts, be doubled also, as it will be if the velocity of the current be doubled, then the force which the water exercises upon the sand in a minute's time will be four fold what it was before, and in this case may be expressed by the number 1008.

What then can we think of this gentleman's jetties, which, instead of conveying the back-waters and ebbs directly to the bar without obstruction, must necessarily divert them from it, and turn them over to the westward, giving them such a course as will direct them much more into the west channel, and cause them to strike more directly against the brest sand than at present, and so both lessen the quantity acting, and also the force with which every proportional part does act; so that if the bar increases and grows upon us now, we may then expect the decay of the east channel will be much accelerated; for sure 'tis not a little absurd to imagine, that turning the course of the ebb-water more to the westward, can promote the removal of a sand which lies in the east channel, or any way deepen or mend that. Too much of the ebb-water does already go down the west-channel, which lies too directly in its present course; but if its direction be changed still more to the westward, as it certainly will be by these jetties, then the principal part of the ebb-water will be turned into that channel, and other slade ways, which can afford it a passage to the westward; and the east channel will be almost deserted, and left almost dry every ebb, and consequently in a little time would choak up, and be entirely lost.

So far then is the method this gentleman proposes from being a certain remedy to the east channel, that in all probability it will be the utter destruction of it. And thus much for the first evil he complains of.

The other evil is, the washing away of the marshes, and the danger of drowning the lands on the east side.

This (I shall beg leave to observe) is a thing which more immediately concerns the land-owners than this corporation. The security of their estates is certainly as much their own proper concern, as the security of those on the west side is the proper concern of the owners of them. But (says this gentleman) *the drowning of these lands will be the ruin of your harbour*.

This I must confess I do by no means apprehend. Suppose that two or three hundred acres of these lands were every tide laid under water, not one gallon the less water would flow up the river than does at this present, and all the waters, which would then cover
these

thefe lands in the time of flood, would be an additional quantity to return in the time of ebb, which would certainly very much help to fcour away the bar, and maintain a good channel. But fuppofe that the fecurity of the harbour does depend upon the prefervation of thefe lands, are thefe jetties like to be effectual for that purpofe? I conceive not. 'Tis true, they will leffen the beat of the tides during the time of high-water, which by the opening of thefe points, when the north-weft wind blows, is now very great, and wafhes away the furface, and turf of the land very faft; this I fay would be in a great meafure prevented. But then the flood-tide would be carried continually and directly into the crutch-bite, and fo would the ebb alfo; and by the continual and violent action of both, the lower fhores would be ground away, and the bite extended more and more to the eaft-ward, and the lands and banks will be as effectually ruined by the undermining of the fhores, as by the wafh and beat of the waters upon the furface; tho' perhaps not alto-gether in fo fhort a time.

Now, as I think this will certainly be the confequence of this fcheme, fo I thought it an incumbent duty to apprize you of it, that you, whofe interefts depend upon the prefervation of the navigation of this port, may make proper reprefentations againft it.

I know fome gentlemen are mightily alarmed at the wearing away and opening of the two foremention'd points, as if the ruin of the harbour muft immediately enfue, and as if the goodnefs of the channel entirely depended upon the projecting or extending of thefe points. But it may be remembered, that fome years ago, while *Denver* fluice was ftanding, and the reception of the flood leffen'd, and the back waters prevented from coming down, the quantity of the ebbs was then fo fmall, that it was all carried off by the weft-channel, and the eaft-channel was fo deferted, that at low water it was left almoft dry, which made it foon grow fo fhallow, that at the time of high water, there was not water fufficient for any loaden fhips, all which were forced to go round at the back of the fands, and come through the weft channel up to town; and at that time thefe points were both far extended, and landlocked the harbour. 'Tis there-fore very evident, that the goodnefs of the channel does not depend upon the extenfion of thefe points, nor do I apprehend that there is fuch imminent danger from the opening of them, as fome are inclined to think.

The channel is now in as good a ftate as it commonly has been in for fome years. And if we fhould in a little time have a continued wet feafon, which may happen, it will quickly be much mended. And I think it is much the better way, to truft to nature for a remedy, than to execute at a great expenfe fuch fchemes as in all probability will be very per-nicious and hurtful to the channel.

Some

Some gentlemen it feems do plainly difcover the impropriety of Mr. *Rofewell's* fcheme, but yet think fomething ought to be done for the prefervation of the lands, and the amendment of the channel, and therefore are for erecting a jetty crofs the crutch-bite, all along the eaft fide, towards the block-houfe ; and another jetty on the weft-point, to extend this point more over to the eaftward.

The firft, certainly, fo long as it ftands, if made fufficiently, high, will be a great fecurity to the marfhes, and lands on the eaft fide, and would prevent both flood and ebb from encroaching on that fide, and likewife affift them in fcouring away that pernicious point of fand, which extends itfelf from the weft point, fo far over to the eaftward, into the crutch-bite, for though the turf and vegetable foil at the weft point are much wafhed away and fhortened, the fand ftretching from that point is by that means much lengthened, and determines both the flood and ebb more over to the eaftward than before ; and fo far will it be from doing fervice to the channel, to extend that point, that if the vegetable foil was only fecured in its prefent limitations, and the point of fand intirely fcoured away, it would certainly be much better for the channel, for then the ebb would not be fo inclinable to go down the weft channel, but would proceed more directly to the bar, and act upon it with much greater force than now, and confequently keep it lower, and maintain the eaft channel better ; and then perhaps the weft channel might gradually grow up. And happy would it be for *Lynn*, if that channel was intirely ftopped up, for it is the quantity of ebb that goes down that channel, that fo weakens the current in the eaft channel, that inftead of fcouring away the points of fand which project into it, it is ftopped and retarded by them, and formed into vortices and eddies, which throw more and more filt upon them, and in fome places (as at the bar) extend them quite crofs the channel ; for it is by this means, and by the flood-tide coming up the weft channel, and forming an eddy at its entrance into the eaft channel, that the bar fo much complained of is formed : for it is obfervable that this bar lies not far below the mouth of the weft channel, juft about the place where one might expect fuch eddies would occafion it.

A jetty therefore at the weft point, feems to me quite wrong, and I think will be very hurtful. If therefore a jetty muft be erected, I would advife it, not upon the weft point, but about a furlong to the weftward of it, to extend itfelf to the northward, along the eaft part of the breft fand. This would allow the ebb to fcour away the fand, that ftretches from the weft point and turn it more directly down the eaft channel, towards the bar ; and would at the fame time very much fecure the eaft fide from the violence of the wefterly and north weft winds, and be a great fhelter to the harbour too from thefe winds.

And

And if the weſt channel could be ſtopped up, by the ſinking of old hulks loaded with ſtones, or any other means, ſo that the ebbs might be prevented from taking their courſe that way, we might then have great hopes of maintaining the eaſt channel, in a much better ſtate than it has ever yet been in.

But as ſome gentlemen are apt to imagine, that a jetty on the weſt point, by turning the ebb-water more over to the eaſtward, will direct it more into the eaſt channel, it muſt be acknowledged that it would have this effect, if this channel were ſituate to the eaſtward of that point. But the eaſt channel itſelf is ſituate far to the weſtward of that point, and conſequently the more the ebb is carried to the eaſtward by the jetty on its ſouth ſide, the more weſterly muſt be its direction on its north ſide, to arrive at that channel; and when once it has received this weſterly direction, the weſt channel and other ſlade ways lie ſo fair to receive it, that in all probability very little of it will go down the eaſt channel.

The way then to direct it down the eaſt channel, is not to turn it more to the eaſt at the weſt point, it going far much to the eaſtward at that place already; but the way muſt be to let that point of ſand wear away, that it may take its courſe more in a direct line for that channel. Some may object, that the wearing away of that point will lay the harbour more open, and make it more turbulent and unſafe than at preſent. But in anſwer to this, a jetty from the weſt marſh, to extend northward, along the eaſt part of the breſt ſand, will in a great meaſure prevent that. However, it is to be obſerved, that the ſecurity of the harbour, and the goodneſs of the channel, are two different things, and what procures one, may be pernicious to the other. The goodneſs of the channel will be promoted by the quantity of water that comes into it, and flows through it, but the greater the quantity of water, the greater the agitation, and the more turbulent. On the contrary, the leſs the quan‐ tity of water, the leſs the agitation, and the quieter the harbour, but then the worſe the channel. When ſuch works are undertaken, therefore, the end and purpoſe ſhould be well conſidered, and things inconſiſtent in themſelves ſhould not be purſued at the ſame time.

However, it is not to be expected, that theſe, or any other means, can render the channel *conſtantly* and *unchangeably* good, ſo long as its courſe lies betwixt and amongſt ſuch looſe and moveable ſands. A rage of winds and tides, or an extraordinary time, either of drought or rain, will occaſion great alterations in it, ſometimes for the better and ſometimes for the worſe, becauſe even when the back waters are the ſtrongeſt, and ſo the moſt likely to preſerve a good channel, they will ſometimes ſcour away the ſands in one place, and let them drop in another place, where they may be more prejudicial to the channel than before; and thus will the ſtate of the channel be always liable to changes and alterations.

# WELLS HARBOUR.

(See the Plan, Plate I. of this volume, Fig. 2.)

The REPORT of JOHN SMEATON, Engineer, upon the State and Condition of *Wells* Harbour in the County of *Norfolk*, and how far the fame may be affected by the Imbankment of the Slade Marfhes.

HAVING carefully infpected the prefent Condition of the Harbour of Wells in the county of Norfolk, and the Slade Marfhes, and other premifes, the object of a fuit wherein Sir Martin Brown Folkes Baronet, and Robert Hales Efq. were plaintiffs, againft George Chad Efq. and other defendants, and having alfo carefully infpected the feveral Plans and Papers that have been produced in evidence in the faid caufe, the following facts drawn from my own view of the premifes, and in part from the evidence produced, appear to me to be very well afcertained and agreed upon.

1ft. That the harbour of Wells has of late years grown into a worfe ftate than it formerly was in, and particularly fince the imbankments that took place in and about the year 1719; and,

2d. That for remedy of the complaints that then fubfifted in the harbour, in the year 1738, a work was conftructed that had the name of Freeftone's Sluice, fo called from the projector or builder thereof, whofe name was *Freeftone*.

3d. That this work had a beneficial effect in fcouring away the mud and fand that annoyed the Harbour and channel from the mouth of the faid fluice, down as far as the Pool.

4th. That at that time, that part of the channel called the *Pool* was fo deep, that at low water two or three tiers of veffels could lie afloat, and fwing round.

5th. That in the year 1758, the laft imbankment was erected, which is the matter now complained of, and is the object of the prefent fuit.

6th. That

6th. That the extent of ground inclofed, and defended by the faid imbankment, is nearly 17 acres in the Eaft Marfhes, and 47 in the Slade Marfhes, together making about 64 acres, befides and interfperfed in which are creeks formerly and now containing water to the amount of about three acres, making altogether a furface of about 67 acres, or thereabouts.

7th. That befides the above, there is not only a much greater quantity of imbanked lands, which was chiefly taken in about the year 1719, but ftill a much greater quantity than all the imbanked land put together, ftill remaining unimbanked, over all which the fpring tides *ufually* flowed before the imbankment, but the neap tides *rarely*, and which now is the cafe with the greateft part of the unimbanked marfhes.

8th. That no frefh water river makes its way to fea through the channel of the harbour of Wells, nor indeed any confiderable quantity of frefh water of any denomination; and that not only the channel that forms the harbour, but the feveral branches and creeks into which it is divided (as is particularly diftinguifhed in the map made by Beiderman under the direction of Mr. Mylne) are all fupplied with fea water on tide of flood, which reflowing back to the fea on tide of ebb, thereby forms a *back water* and produces a *fcour* that tends to keep the channel of the harbour open.

9th. That in the year 1765, Freeftone's Sluice having been originally conftructed in a flight manner with fafcines, ftakes, piles, &c. the mouth thereof was fo much widened, that its effect having been for fome time paft greatly impaired, another was built upon a new fite, which when erected, reproduced the fame effect, as to clearing the harbour and channel down to the Pool.

10th. That the fluice of 1765 which had been conftructed with greater ftrength and care than that of Freeftone, though upon the fame model, in the year 1777 was found to have been nearly deftroyed by worms; on which occafion Mr. Wooler was called in to advife the commiffioners, who reported his advice to build a new fluice in a new fituation, upon a new conftruction, and with fuch materials as the worm could not touch.

11th. That on account of the expenfe eftimated by Mr. Wooler at £2,000, this work was not undertaken, but in lieu thereof, proceedings were gone into that terminated in the prefent litigation; on fuppofition on the part of the commiffioners, the defendants, that the taking down the Eaft Marfh Bank, whereby the tides would be readmitted upon the furface of the 67 acres before mentioned, would reduce the ftate and condition of the harbour in every refpect, or in all the moft material refpects, to what it was before the erection of that Bank in 1758.

12th. That however, the Sluice fince Mr. Wooler's report has been repaired, and it appeared at the time of my view to be of the fame conftruction that it had formerly been,

according

according to its defcription, and tending to produce the fame effect, which it is reported to have formerly done.

13th. That during the whole interval fince Freeftone's Sluice was firft conftructed in the year 1749, to the prefent time, the operation of the fluice has conftantly been (when in order) nearly the fame, viz. that of clearing and keeping in the fame good condition nearly all that part of the harbour and channel that lies between the mouth of the fluice, and the upper or fouth end of the Pool.

14th. That during a period of years fince the year 1749, the Pool has been filling up, or growing fhallower, infomuch, that at low water, at the time of my view, there was not above fix feet water.

15th. It further appears from evidence, that, for a long period of years preceding all the imbankments herein mentioned, the outfall or mouth of the channel of the harbour has lain very confiderably more to the weftward than at prefent, and that it has from time to time changed more to the eaft.

16th. That formerly, that is, within the compafs of 20 years, the direction of the channel out to fea was N. W. and that at that time the flow of the tide being in the direction of the channel, during all the time of tide of flood, it was very eafy with a proper wind to bring veffels from the fea through this channel into the pool and harbour.

17th. But that fince the channel has come more to the eaft, and the direction gradually veered about, fo as to have come now to a N. E. or N. E. by N. direction, as it was at the time of my view, it is a fact, that whenever the tide has rifen above the furface of the Broad Sands, the tide ftill keeping its former direction from N. W. to S. E. drives the veffels acrofs the channel upon the eaftern bank or fand, and therefore without a ftrong leading wind veffels cannot enter without danger, and in fact fince thefe changes have taken place, many fhips have been loft, notwithftanding the fkill of the pilots, to the no fmall difadvantage of and difparagement to the port.

18th. That the diftance from the quay to the outfall of the channel is at prefent betwixt three and four miles, and that the channel both now lies and ever has lain, through large broad open fands, from the northward or out end of the pool to its outfall into the fea, which fand being perfectly clean, and free from all particles that might create tenacity, when dried by the fun is capable of being blown by the wind, and moved by the common agitation of the fea, and very confiderable changes are wrought by the action of the fea in great ftorms and tempefts; and that this is the defcription of the fands in general that lie many miles extended on this coaft.

19th. That

19th. That the breadth of this fand, or diftance from the pool to low water mark, has confiderably increafed within the memory of man.

20th. That the channel from the *Shelf* without the north part of the pool to the fea has now as much, and about the fame draught of water or flow of tide at high water, as it has been known to have in man's memory.

21ft. That the port of Wells is of great confequence not only from its fituation with refpect to the furrounding country, but alfo from the traffick fettled in the place, and therefore that nothing ought to be done that can be of real detriment thereto.

I have thrown together the above leading facts not only for my own eafe in compofition, by avoiding more tedious defcriptions, but that from this general view of the ftate of facts, I may be the more readily followed in my reafonings thereupon.

From the above there naturally arife the following queftions :

1ft. What is the natural caufe of the decay of the harbour of Wells?

2d. Whether the imbankment in queftion made in the year 1758 have materially contributed thereto?

3d. Whether the removal of that imbankment will contribute to a material amendment thereof?

Queftion the firft.—What is the natural caufe of the decay of the harbour of Wells from what it has been in former times?

To have a clear and comprehenfive view of the caufe of *decay*, it will be neceffary to fhew the natural caufes by which the port of Wells has been *formed*.

We are all apprifed that notwithftanding the annual downfalls of rain and fnow upon the land, which run into the fea in every part of the habitable globe, the quantity of water contained in the fea is never the *greater* ; and this may reafonably be expected when we confider that the exhalations which form thofe rains and fnows, are chiefly raifed from the fea by the power of the fun and winds.

The floods and torrents that make their way from the furface of the land, in confequence of the rains and fnows, take along with them great quantities of clayey, earthy. and fandy matter (intermixed with many other kinds of matter, which it is needlefs here to enumerate)

enumerate) down into the channels of the rivers, and are by the violence of the torrents carried to the sea, and there in appearance difperfed.—Thefe kinds of matter, not being capable of being raifed again, and returned back to the land, as the water is, it neceffarily follows that the quantity of fandy and earthy matter that is depofited in the fea, is in a continual though apparently very flow ftate of increafe, and by the conftant flux of the tides, and the agitation of the winds, thofe materials fo depofited are capable of being removed from the parts adjacent to their entry into the fea, not only to the neighbouring, but even to very diftant parts; and though the remarkable quantity of fandy foil in the county of Norfolk may in part account for the vaft quantities of fands upon the coaft thereof, extending as it were in one continued chain from the coaft of Lincoln-fhire, very much infefting the great bay between the two counties called the *wafh*, and ftill extending coaftways far beyond the port of Wells, even to Cromer, as I have been informed, yet it is by no means neceffary to fuppofe that they have *all*, or in the greateft meafure, proceeded from the land of the two counties, to the coafts of which they now lie contiguous; they may as well have proceeded from parts far more diftant, and that brought hither by the inceffant action of the winds and tides, they find a place better adapted to their reception and repofe than thofe from whence they came; and indeed it may alfo be as well fuppofed that they may in *part* have come from thofe coafts which by the particular fet of the wind and tides have been *wafting* for ages paft, as to fuppofe that they all proceed from the high lands of any place or kingdom whatfoever.

It is however of no confequence to our argument to point out *whence* they came; it is fufficient that they have come, and that being here depofited, and finding it a place where they are lefs liable to be carried away than they were to be brought hither, the quantity upon the whole muft increafe, and fince there appears no power of nature by which they can be returned to the high grounds, or coafts from whence they were brought, it muft follow that they muft continue to increafe, till fome contrary power of Nature fhall take place that we are not acquainted with, or the place of their reception become entirely filled.

It follows then that the attachment of fandy and earthy matter to this coaft is *in a progreffive ftate of increafe*, which it is as much out of any human power to prevent, as it is to prevent the fandy and earthy matter from being wafhed down from the high grounds into the fea.

We need not therefore extend our views further to be enabled to fee clearly, that in fome former age, that is, in fome ftate of the progreffion of this work of Nature, there was a time, when nothing more than a naked fand lay againft the bare coaft of the

elevated

elevated ground upon which the town of Wells now ſtands, which we may ſuppoſe ſimilar to that which now lies between *Wells Miels* or *Marram Hills* and the low water at ſea, which being upon one regular decline, the tide water can flow and reflow over the ſurface without channel or creek; and in this ſtate, as there would be no harbour, there would not on that account be need of a town: but let the breadth of the ſand gradually increaſe to double, and then the declivity would become too ſmall for the tidal water left by the flood to make its retreat ſo as to keep pace in its return with the ebb at ſea; and therefore a body of water being thus left behind, and having a ſenſible declivity towards the ſea, would naturally make its way into the loweſt ſlades, and there cut a *gully*, which (if not formed gradually as may be ſuppoſed) would ſoon be enlarged by the influx and efflux of the tide; and thereby a ſcour would be produced through this paſſage enough to keep it open, in a degree ſufficient to let the water in and out, till it became ſo quiet as in a manner to ceaſe the operation of ſcouring; but the breadth of the ſands gradually increaſing, a greater body and ſurface of water will want a paſſage, and the power of ſcouring will increaſe with it; the gully will therefore by the ſame ſlow degrees enlarge to a *creek*, or *fleet* as it is here called, and leſſer ones will be formed to conduct the water more readily into the larger.

The natural progreſs attending this operation is, that the parts of the ſand furtheſt diſtant from low water, being leſs liable to agitation from winds and tides, the ſand intermixed with the finer particles of earthy and clayey matter brought in by the tides, are the moſt readily and quietly lodged there; and particularly the earthy and clayey particles will find a reſting place, which binding the ſand together, the whole will continue to riſe, till at a certain height, and in conſequence of a certain length of time of abſence of the ſalt water each tide, and expoſure to the ſun and air, the ſurface becomes fitted for vegetation, and by degrees will grow a certain ſpecies of graſs, and become a ſalt marſh; the graſs again entangling and locking up the earthy matter, will cauſe it to continue to riſe, and in an increaſed ratio, as the water upon the plain graſs ſurfaces becomes more ſhallow, and in conſequence more ſtill, and free from the agitation of external diſturbing cauſes.

During this period, as the marſhes have increaſed in height, they have alſo increaſed in breadth, and in conſequence a greater body of water will be left upon them; the gullies and creeks therefore, as they multiplied in number with the increaſe of breadth, the larger ones would increaſe in ſize and depth, and if all were ultimately collected into one, as has been the caſe with the channel of Wells Harbour, the ſcour would be ſufficient to maintain a channel through which veſſels might be brought from the ſea, and

thus

thus an ufeful Harbour would be formed, which would increafe in depth and utility by the continuance of the forming powers, but yet, only to a *certain degree*.

I have faid that, as the marfhes increafed in breadth and height, more water wanting a paffage to the fea would be left upon them; and this, fo far as regards breadth, is felf-evident; but as the increafe of height diminifhes the depth of water upon them at high water, that the quantity left behind upon the whole wiil be greater, needs fome explanation. In refpeʧ to this, it is obvious to every one who views the fubjeʧ, that while the depth of water upon the marfhes is confiderable, the water makes its way to fea by fettling gradually, and paffes off in the nearest direʧion over the marfh furfaces, without having any need for the gullies and creeks as drains; it is therefore perhaps only the laft half foot that may need the gullies; which, however, being limited to fome certain thicknefs (be it half a foot more or lefs), and this thicknefs much lefs than the depth at high water, the quantity fo left will be in proportion to the quantity of furface; and the number of gullies, being alfo in proportion to furface, the aggregate of the whole and the fcour thereof in the laft channel will alfo be proportionably increafed; it is likewife remarked, that in faʧ, the fcour is not very material till the water upon the marfhes has ebbed near their furface, that is, until it is juft retreating into the gullies, when the principal fcour begins.

Under thefe circumftances, the fcour would increafe and confequently the goodnefs of the harbour of Wells would naturally improve, while the neap tides covered the furface of the marfhes; but as the fame progreffion would in time caufe the furface of the falt marfhes to rife above the ordinary neap tides, the fcour would then begin to diminifh, becaufe, being not only deprived of the efflux of the water from the grafs furfaces into the gullies a number of times in a fortnight, the fcour of thefe gullies becoming immediately lefs, they would themfelves begin to choak up, and contain lefs *following* water; and therefore, in both refpeʧs, the fcour being diminifhed in each particular gully, the fcour of the whole muft diminifh.

It is moft probable that this harbour, from the flow progrefs of the changes above fpecified, may have continued in a very flourifhing ftate for a long term of years, reaching backward beyond all record; though it is likely that for a part of that time it may have been in reality in a ftate of decay, and which we will now fhew will naturally refult from the fame progreffion of caufes that carried it to its *maturity*; fo that after having clearly feen from what natural caufes the Harbour has been produced out of the fea without a frefh water river to give it birth, we fhall then fee the natural caufes to which its *decay* muft be attributed.

From

From what has been laid down, it will appear most manifest that the rising or elevation of the surface of the salt marshes, by a fresh accession of sea mud which they will acquire more or less every time they are overflown, will not stop at the *neap tides*, but will gradually rise higher and higher towards the high water of spring tides; and if after that they were only to be overflown in the great springs or raging tides, yet as every one of these tides will deposit something, they will ultimately be shut up at the height of the extreme high water, though no *imbankment* whatever was to take place; and this will also happen in succession to the gullies, creeks, &c.

For as the surface of these marshes rises higher and higher from the neap towards the spring tide mark, they will be less and less often overflown, and the gullies made by the reflux of the tidal water from their surface will become less and less capacious, and in consequence of a want of reflow, the creeks will suffer the same fate, and lastly, the *fleets* and *main channel*: But as the tide water flowing in through the channel, fleets, creeks, and gullies, to the several extremities of its branches must flow back the same way, it is the extremities that will be first landed up, because every part betwixt such extremity and the sea will have water beyond it to flow over or throught it, and thereby producing some degree of a scour, will keep open a passage either greater or less, while any water can get beyond it; and hence we must expect, that the parts of the channel most distant from the sea will be those that in a state of nature will soonest lose their depth and capacity, till progressively, from the extremities towards the sea, the gullies, creeks, fleets, and main channel will become solid land : and so far it appears from testimony, that long before the imbankation in the year 1758, nature had got on so far in its progress in the decay of Wells Harbour, that it was much complained of; and the upper part between the quay and the pool was got so bad, that the goods were in general lightered up to the quay, and that for remedy thereof, Freestone's Sluice was built in the year 1749, which produced a good effect in clearing that part of the Harbour. As then it plainly appears from the preceding discourse, that the progression of nature has no tendency to cure the evils complained of, but still to increase them, they are in consequence incapable of any remedy, except what can be applied by the ingenuity and labour of man.

We now come to the 2d question, whether the imbankment made in 1758 has materially contributed to the decay of the harbour of Wells ?

From what has preceded it has been clearly shewn, that as the keeping open and maintaining the channel entirely depends upon the reflow of the tide water, or *back water* as it is called, whatever cuts off and diminishes this, must be a detriment to the scour, and consequently to the maintenance of the channel ; I therefore do not hesitate in saying that all the imbankments both eaſt and weſt of the town of Wells, the water received upon the ſurfaces of which, and into their gullies and creeks, uſed to make its way back to the ſea by the channel of Wells Harbour, muſt have co-operated with the progreſſion of nature, and thereby tended to bring on more ſpeedily a general want of depth, which, as it has been already remarked, will be firſt perceived at thoſe parts of the tidal flow the moſt remote from the ſea ; nay, we may go ſo far as to ſay, that if a bucket of water were taken up in the upper part of Wells Harbour, and not ſuffered to reflow, it might prevent the diſplacing of ſome particle of matter that had been lodged therein by ſome preceding tide. In this ſentiment I ſuppoſe every able engineer and ſkilful perſon that has viewed the premiſes will join as a *general opinion*, but though very true as a general ſentiment, yet the whole merit of the queſtion depends upon the *quantum*, the *how much* damage could reſult from thoſe artificial means ; and as in this it appears that the opinion of different men have widely differed, I will proceed to ſtate my own with all the preciſion and clearneſs I am able.

Now, if the breadth of the ſands continue upon the increaſe, we muſt conclude that the reflux of the tidal water from the whole ſurface covered continues, upon the whole, if not greater, as great as formerly, and therefore will maintain as good a channel to ſea at the outfall at low water ; nor does it appear from any teſtimony that, from the ſcalp to the outfall of the channel into the ſea, there is a leſs depth at high water than formerly exiſted ; the complaints and appearance of a diminution of depth of the harbour are from the lower or north end of the Pool upwards towards the town. It appears therefore, that the upper part of the channel that lies contiguous to the main land is *landing up*, and that in fact the harbour is moving *towards the ſea*.

It does not appear from teſtimony that the filling up of the Pool, or its growing ſhallower, ſo that ſhips cannot now lay afloat or ſwing round in tiers as they uſed to do at low water, has been a matter very much complained of till after the imbankment of 1758 ; but this does not diſprove that the fundamental cauſe thereof may have exiſted long before the ill effects have become ſenſible to ſeamen, and thoſe uſing the harbour ; and indeed it clearly appears to me, that the filling up of the Pool ſooner or later, is only a link in the general chain of cauſes that muſt have operated ſo as to produce this effect, whether any imbankment had been made or not.

To ſee this clearly we have nothing to do but to advert to the ſituation of the Pool, for it begins juſt below or about where the *Weſt Fleet* falls into the main channel. This Fleet received the drainage and back waters from the *Holkham* marſhes, which appear to have been anciently of much greater extent than at preſent, and conſequently the back water or

reflow

reflow muſt have been very conſiderable from this great extent of ſurface. as well as that which ſtill lies unembanked, and which being joined by the great reflow of back water from the Eaſt Fleet, the Little Fleet, and the Haven Creek, with all their extenſive dependencies, muſt have formed a great and rapid ſcour of back water, eſpecially when it is adverted to, that it is confined within a narrow compaſs between the Holkham and the Wells Marram Hills or Miels ; we muſt therefore expect no leſs than a very deep channel in this extent, well deſerving the name of *The Pool.* And that this confinement of the whole water both of flood and ebb, between the hills juſt mentioned, gave an additional ſcouring power to the tidal water, appears further from this, that the Pool never appeared to be of further extent northwards till this opening between the hills being cleared, the tidal water having an opportunity of eſcaping on tide of ebb over the broad ſands lying without thoſe points, its force became diſſipated, and the great and ſudden ſcour ceaſed, that cauſed this deep water even in that age when the whole ſcour was in its greateſt degree of ſtrength and perfection, and the depth and goodneſs of the harbour conſequent upon it. But whenever (as has been ſhewn) the ſcour beyond it became diminiſhed by the riſing of the marſhes above the high water of *neap tides,* the depth of the pool would begin to diminiſh, for the ſcours would then be unable to ſcoop out the ſands from ſo great a depth as before, that would be continually brought in by the N. W. winds at *Holkham Gap,* from the broad ſands lying to the north thereof, and then by the Weſt Fleet carried down into the Pool.

Nor is the Holkham Gap the only ſource from whence the ſands brought down by the Weſt Fleet might be collected ; for I look upon it that the inſide, that is the ſouth ſide, of the Holkham Miel Hills, is continually melted down by winds and rains into the great area or bay drained by the Weſt Fleet, while thoſe hills are continually ſupplied with freſh ſand blown up by the force of the N. and N. W. winds upon and over them, from the broad ſands that are yearly extending more and more from thence to the north-wards into the ſea ; the higheſt parts of which broad ſands being frequently left at neap tides long enough to be dried by the ſun, are by the wind capable of being blown up in great quantities, ſo as to raiſe and continue thoſe hills far above the high water of any tides whatever, though compoſed of nothing more than a blown ſand from the ſea, ſome-what united by the bent graſs that grows up through the ſame, of which there are very many examples in various parts of the kingdom.

In that age when the ſcour at the pool was in its full perfection, it is probable that it was very much deeper than it has ever been reported by any teſtimony or record now extant, and ſo long as it continued deep enough for the purpoſes of ſhipping, there would be no cauſe for complaint ; and till it became too ſhallow for ſhips to ſwing at low water,

its

its diminution of depth would be little regarded. There is therefore no doubt but that the Pool was growing fhallower long before the imbankment of Holkham marfhes, as a confiderable length of time muft have elapfed between the period when Holkham imbanked marfhes were juft rifing above the high water of neap tides, and their acquiring the height at which they were imbanked; and which height I judge was not materially different from the prefent.

When the ground of Holkham marfhes was become high enough for embankment, the natural fcour arifing therefrom was diminifhed from what it had been, and though they had not been imbanked, would have been ftill much lefs at this day. Yet fo great a furface as 560 acres, the drainage of which plainly appears by the map to have made its way out to fea by the Weft Fleet, being cut off all at once in the year 1719, and 108 acres more in the year 1721, the whole of this together amounting to 668 acres, may reafonably be prefumed to have had fome fenfible operation in accelerating the effects that nature in her progreffion would afterwards have brought on, though no artificial imbankment had been made; but it by no means follows, that the effects of thofe imbankments muft be *immediately* perceived at the Pool. This by flow and imperceptible degrees muft be fuppofed going on as before, and by lefs flow but yet imperceptible degrees might be going on after; yet this muft be attended to, that in proportion as the fcour of the weft fleet was diminifhed by the lofs of the imbanked marfhes, the power of the fame fcour would be diminifhed to carry down the fand continually into the pool, as it before ufed to do, fo that till the great bay that ftill continues to be drained by the weft fleet, as it were gorged with fand from Holkham Gap and Holkham Miels, it would not be brought down in fo great quantities each tide as before the Holkham imbankment; and therefore as it appears that the whole progrefs of this bufinefs is very flow, it muft be expected to have been a feries of years between the *caufe* from the Holkham imbankment and the *effect* becoming more perceivable in the fanding up of the pool; and in cafe this neceffary period of years extended from the year 1720 to the year 1758, that is near 40 years, at which time the laft imbankment was made, then thofe would become *cotemporaneous* events, and it would be no wonder that the united effects of the Holkham imbankment, and that of the Church marfhes in the year 1719; and alfo of the Weft marfhes of Wells in the year 1720, making in the whole 572 acres, together with the additional lofs of fcour from the furface of the falt marfhes which yet remain unimbanked to the amount of between 15 and 16 hundred acres, the drainage of all which have their outfall through the Pool; I fay it would be no wonder if under thefe circumftances the united effects tending to land up the pool fhould be charged to the account of the imbankment made in the year 1758, if it were poffible to fuppofe that the imbankment made that year of 64 acres could have a fenfible effect, for the object of this imbankment amounted to no more than $\frac{1}{34}$ part of

the

the whole quantity concerned in producing it ; when it remains *problematical*, whether the effect of the 572 acres taken in before, which is near nine times as much, would *alone* have been fenfible, if the effects thereof had not been mixed with what might have arifen from nature's fimple progreffion, in cafe its operation had been no ways difturbed by that of art.

I find myfelf therefore forced by fair induction to infer, that though ftrictly fpeaking the effect of every thing that has an effect muft be *fomething*, yet that the imbankment of the flade marfhes in the year 1758, could not in any fenfible degree capable of meafure or eftimation contribute to the landing up of the pool.

The effect thereof upon the outward part of the channel muft be ftill more remote ; for it does not appear from any principles of art, obfervation, or practice, that when a back water, even affifted by a frefh water river ever fo large, makes its way through broad moveable fands, it has any tendency in itfelf to make its way out to fea in this direction more than in that. The natural tendency of water is to make its way in whatever direction it finds the *greateft declivity* ; and if that happen to be in the fhorteft direction, it has no natural tendency to gain a longer courfe, as that would leffen the declivity : If therefore water is found going in a courfe that is not the fhorteft, we may conclude (and on examining we always find) that this longer courfe is owing to the intervention of fome object fo placed, that the water can have in that particular part a more fpeedy defcent in a direction different from that which would form the fhorteft line of the whole defcent ; and from caufes of this kind a ftream may have a courfe in every poffible degree meandering that we frequently obferve in nature.

If therefore the courfe of the out-fall channel to fea at low water was in a north-weft direction through the broad fands, as it feems very well attefted to have been in former times; and if at thofe times the courfe of the channel correfponded with the general fet of the tides upon the coaft, fo that veffels going in or out through the out-fall channel were not driven out of their proper courfe by a fet of the tide *crofs* the channel, as is faid to be the cafe at prefent; it may be imputed to fome of thofe lucky caufes that operated in favour of thofe times, as no ftrength of back water alone could have a natural tendency to produce this effect : And if the accidental operation of contrary caufes, as fuppofe winds and tides, either confidered alone, or as acting in correfpondence with the regular progreffion of nature already defcribed, has brought the direction of the out channel into a fituation lefs favourable to navigation than it ufed to be formerly, at the fame time that it has produced a fhorter courfe for it to fea, than it would have if returned to its former direction; I am therefore clearly of opinion, that no increafe of back water, even if aided by a frefh water river, if artificially brought

down

down this channel, would ever caufe it to return to its former north weftern direction, fo as to difembogue itfelf, at a place confiderably more weft than the prefent outfall.

I do not however mean to fay that it *cannot return*, but that if it does, it muft be in virtue of caufes operating in a contrary way to thofe which have brought the mouth of the channel from the weftward towards the eaft, and not in virtue of any change that can be expected to be wrought fimply by an increafe of back water; for that would be to make a more rapid ftream have a tendency to go a longer journey to fea in preference to a fhorter.

It is alledged indeed, that, from a general increafe of the breadth of fands upon the coaft, it is now further from the north end of the pool to fea than it ufed formerly to be, but this muft be underftood of the ftraight line from the faid point to low water, without regard to the direction of the channel. For though the prefent channel fhould be as long from the pool to the fea, or longer than formerly; yet from the general increafe of the breadth of the fands, were the channel to go much further weftward than at prefent, it would have a longer courfe to fea at low water.

What may have been the particular caufes that have brought the mouth of the channel more to the eaft than formerly, does not appear material to the prefent queftion; but to give fatisfaction as far as I am able, I will hazard the following conjecture: It is an obfervation univerfally agreed upon, that the N. W. winds make the higheft tides in the whole German Ocean; the N. W. winds, then, accompanied by *higher* tides will produce more agitation, and confequently carry the fands from the N. W. eaftward, in a greater degree than that in which equal winds from the fouth-eaft accompanied by leffer tides will bring them back; and according to the local direction of the coaft here, they directly tend to accumulate them upon the fhore toward the N. E. If this is a true folution, as I apprehended it to be, there feems not the leaft likelihood that the direction of the channel fhould ever be permanently removed to the weft, but rather that it fhould be carried further to the eaftward than it now is; and what gives ftrength to the above conjecture is, an obfervation I had the opportunity of making upon my view, viz. that the outfall channel of the harbour of Blakeney (a few miles eaftward of Wells, and fituated in a fimilar manner, in regard to the courfe of its outfall channel through the broad fands) has fhifted alfo more to the eaftward than it was, by above half a mile in the laft feven years, as appeared from the marks I was fhewn for its entry at that period, compared with its prefent place; notwithftanding that Blakeney Channel has two large frefh water rivulets that make their way together through this channel to its no fmall advantage towards keeping it open. The place alfo of the outfall at low water of the eaft difcharge from the falt water creeks called Wareham Deeps, I was fhewn to have removed its place further eaftward, than it was at Michaelmas laft, by

feveral

feveral hundred yards, and that chiefly within the compafs of the preceding three weeks to my view, during which very ftrong north wefterly winds had prevailed. Wareham Deeps lie betwixt Blakeney Outfall Channel, and that of Wells.

There is one thing more refpecting the outfall channel of Wells Harbour that it may be proper to touch upon, and that is, the bar that is defcribed in Beiderman's map to lie acrofs the mouth of the channel fo as to be prejudicial to the entry of veffels; but after what has been faid, be the impediment arifing from hence greater or lefs, it cannot be imputed to the imbankment of marfhes; but to the fport of winds and feas at the place, which will further appear from this circumftance; that when I was examining this place at dead low water of a fpring tide, viz. the 15th of March laft, I did not obferve there was any appearance of a bar, the courfe of the channel being right out to fea, at N. E. by N. by the compafs, yet in another feafon it may probably return.

Hence, from the above premifes, I muft entirely acquit the whole of the imbankments, from having been in any degree contributory to the difagreeable effects arifing from the change of the outfall channel of Wells Harbour.

What I have further to fay, will come moft naturally under the third and laft queftion propofed.

Queftion the 3d, Whether the removal of the imbankment of the Slade Marfhes will contribute to any material amendment of Wells Harbour?

It appears clearly from what has preceded, that the progreffional operation of nature, which originally formed the harbour of Wells and brought it to maturity, has alfo occafioned it to grow more and more into a ftate of decay; and will finally clofe it up, and convert into firm ground, fit for arable purpofes, and thofe of pafturage, the very fpot where fhips have rode at anchor; and that this being the progreffion of nature cannot be countervailed in any degree, but by the induftry, art, and hand of man.

It is a fact well eftablifhed by evidence, that, previous to the erection of Freeftone's Sluice in 1749, the upper parts of the harbour, as far down as the pool, had got landed up to that degree, that the fhipping, which chiefly laid in the pool, were obliged to have a great part of their ladings brought and carried to and from the town by lighters. That upon the erection of Freeftone's Sluice (which as far as appears was the very firft attempt in this harbour to counteract the operations of nature by art this

ude

rude and fimple piece of art fucceeded fo far as in a very great meafure if not effectually to relieve the diftrefs that they then laboured under ; viz. that of the quays being in great meafure inacceffible to fhipping ; a conftruction fo rude, that though bearing the *name* of a *fluice*, it would feem as if it had been one of the firft attempts to obtain relief by art, before fluices had been invented ; and hence we may infer, that if this did fo much, what might have been effected by a *real fluice*, built upon a proper and regular conftruction. This fluice however was attended with a beneficial effect fo long as it lafted ; and when it went to decay, the part of the harbour affected by it reverted to its former ftate.

In the year 1765 they again fet about to relieve themfelves, not by building a fluice upon a better plan, but by erecting a new one upon the fame plan, upon frefh ground, with better, and as they expected more durable materials : And this fluice reproduced afrefh the effect of the former, which has indeed continued to this day. But in the year 1777 it was difcovered to be defective and liable to fail on account of the timber wherewith it was built being eaten with a fea worm, unknown in thefe parts before; that had attacked it. On this occafion Mr. Wooler, an ingenious engineer, very competent to the bufinefs, was called in, who very judicioufly advifed, not only to build a new fluice upon a new foundation, and of more durable materials, but of a different conftruction, fo as to give a better effect to the iffuing waters, as the beft means for preventing the harbour going to decay. This falutary advice however appears to have been rejected by the commiffioners of the harbour, on account of the expenfe ; in lieu thereof, they adopted another expedient, which, as they thought, was liable to be attended with lefs expenfe, and quite as effectual. Having obferved, on popular grounds, that many things had gone wrong with the harbour fince the laft imbankment in 1758, they fuppofed that they were the effects of that meafure ; and inferred that by removing the caufe, the effects would ceafe, and every thing come right again. It fhall therefore now be my bufinefs fairly and fully to examine what foundation there was, or may be, for fuch a fuppofition.

As the commiffioners of the harbour I prefume did not pretend to profeffional fkill in *civil engineering*, they could be no otherways blameable for misjudging in a matter dependent on that art, than that had they fully examined the queftion upon the like popular grounds, which are alike intelligible to all men, they would have feen there was no juft foundation for their expectations. For in the year 1749 the waters of the Slade Marfhes had not been interrupted in their operation, and yet this harbour was become choaked, and had got into fo bad a ftate, as to call aloud for immediate relief; what reafon then had they to expect it in the year 1778, when, according to the natural progreffion, every thing had got worfe?

In

In 1758 Freeftone's Sluice had gone to decay, and the harbour was fpeedily reverting to the ftate in which it had been before the year 1749. Now had the operation of the Slade Marfh waters been obferved to have been of any material confequence, the imbankation of 1758 would have been oppofed and objected to at the time, which yet does not appear; and had any bad effects appeared to have arifen from the imbankation in the year 1768, that is ten years after the imbankment, and three years after the rebuilding of the fluice, when an application to Parliament was neceffary to get frefh powers to defray the expenfes incurred by that erection, &c., it would have been natural to endeavour to get fome equivalent for the damage, or at leaft, as fluice-building muft then appear to be the beft expedient, to have got powers over the grounds in the unimbanked marfhes to enable them to erect fuch proper additional works, as might from time to time be neceffary; and which might have been expected upon eafy terms, and without depending on leave being always given in cafe any damage to the harbour had been felt or apprehended from the imbankment, becaufe the more amicable terms fubfifted at that time between Sir John Turner and the commiffioners, the lefs he could have oppofed or denied fo reafonable a requeft. But yet nothing of this appears, or indeed any other, till after the year 1777, when Mr. Wooler had reported that a proper fluice was " of the utmoft importance towards keeping the " channel of the harbour open, and clear of the fands that are conftantly brought in by " the tides;" and that fuch a work would probably coft the fum of £2,000. Thefe I fay are popular arguments, which as every one can equally fee the force of them, ought to have induced the commiffioners to feek relief, by purfuing means of reducing Mr. Wooler's advice to execution, rather than deliver the Slade Marfhes once more to the empire of the falt waters: but as a profeffional man, I conceive it will be expected from me to give a direct proof of the efficacy of this idea; of this I fhall therefore endeavour to acquit myfelf in the cleareft manner poffible.

Refpecting that part of the laft imbanked marfhes that lies weft of the ancient imbankments, comprehending 16 acres, it is evidently of no more account than any other 16 acres that lie immediately upon the haven creek, and whofe waters immediately ebb within the tide, without paffing through the fluice; that is, they would have no other effect in fcouring than as making a part of 1770 acres, the water from which makes its way to fea by the Pool; but with refpect to the forty-feven acres that lie eaftward of the ancient imbankment, and are called the Slade, or Slade Marfhes, they appear to me to have a different import.

The reafon why the waters paffing the fluice have a greater effect in fcouring than thofe which return to fea without paffing the fluice, is, becaufe by the contracted opening of the paffage of the fluice, the waters that lie in the creeks behind it are detained from ebbing fo quickly as they otherwife would have done; that is, their numerous mouths when always open, reduced the level of the water contained therein, to nearly the fame level as that of the water in the main channel of the harbour, being ftopped by dams made acrofs and united by crofs paffages into one, and the mouth of this being contracted by the work called the fluice, a body of water is held back in thefe creeks, as refervoirs, which not being able to efcape fo faft as the tide ebbs in the main channel, it follows, that a body of water by thefe means is vended upon, and after the half ebb, which difcharging itfelf into the harbour's creek, forms a fcour when the depth is fo much leffened as to operate with power in grinding the bottom, which otherwife would have been fo languid as not to have ftirred a grain of fand or mud, in which cafe its effect would be little or nothing. This artificial fcour thus procured, in fome degree imitates the effect of a frefh water river, which in thefe fituations is very greatly beneficial, not from any virtue there is in frefh water preferable to falt in thefe cafes (if any thing rather lefs on account of its lefs fpecific weight), but from its having a fall from the land, and proceeding therefrom continually it not only ftrengthens the ebb, but running to fea at low water when the fall being greateft, and the fandy bottom expofed to its action, it continues to work at a time when it can operate to the beft advantage; and when the ordinary current of a river is affifted by extraordinary land floods and frefhes from downfalls of rain and fnow, and this operating at low water, when, as juft remarked, the fall is the greateft, in fuch cafes it is capable of producing extraordinary effects, and of keeping a harbour continually open with a channel of a given magnitude, though loaded with fands in any poffible degree: for a frefh water river has this peculiar advantage, that at the fame time that it ftrengthens the fcouring power of the ebb, it operates moft forcibly at low water, when there is the leaft to obftruct its operation; it oppofes the tide of flood from the fea, and thereby prevents its bringing fo much fand and filt into the harbour as otherwife it would.

The defect therefore of this fluice of Wells is, that though it retains the waters fo as to be behind the general ebb, and thereby ftrengthens the latter part of it confiderably; yet being at low water all fpent, when the greateft good might otherwife be obtained, it lofes that good effect which would be had from a frefh water river, or from a proper fluice; that is, one that will retain the water wholly till a proper time of tide, and then being let go in one collected body, is capable in a fhort fpace of time of producing marvellous effects; and yet I fhould not expect fuch a fluice either to clear the Pool to the depth it had fifty years fince, or to carry the outfall channel to the Northweft.

By fluices of this kind, which are the only expedients art has found that is comparable to a river, the greateft part of the fea ports in Flanders and Holland are kept open, and

under

under circumſtances more unfavourable than the port of Wells, many of which are built not only with great expenſe, as to the uſeful part, but with much magnificence, as relying upon the durable utility of their conſtruction. Thoſe of the Wells traders who have occaſion to viſit the port of Oſtend, ſo much reſorted to at the preſent time, will ſee a remarkable example thereof in the grand Sluice of Oſtend re-erected in the year 1755.

The Slade then conſiſting of 47 acres (or ſomewhat leſs, on account of the high grounds incloſed therewith), it appears to me might derive ſome occaſional advantage in reſpect of ſcouring, from the very circumſtance of their ſituation lying behind the ancient imbank-ment; inſomuch that what is alledged by ſome of the witneſſes may at ſome particular time or times be true; videlicet, that when the tide was ſpent at the town, it came down from the Slade.

It ſeems well atteſted, and from the nature of the thing (as it appeared on my view) muſt be the caſe, that the ſurface of the Slade never uſed to be overflowed, but in extreme ſpring tideſ, here called *rages,* and the condition of it in growing ruſhes and other vegeta-bles peculiar to fenny freſh waters before its imbankment, ſhews this to be the general caſe; and this would naturally and neceſſarily ariſe from the very contracted channel and opening left between the N. E. bank of the ancient imbankment and the high land ſo narrow, that it does not appear, except in ſuch caſes, that there would be a ſufficiency of time at high water for the whole ſurface of the Slade, conſidered as a pond, to fill any thing near the utmoſt height of the ſea; but yet whatever water was at ſuch times brought upon it, would, upon the ſame principles as the ſluice, be left behind, and retained by the ſame narrow paſſages, ſo as to require poſſibly the whole of the ſucceeding low water entirely to vend it; but then it will follow, that by how much it was the longer in vend-ing, it would come down the more leiſurely, and toward the latter part of it the more drib-bling, in proportion as its channel to let it out grew more contracted by the ſurface of the water being lower therein: ſo thac although it might at thoſe times come down in good quantity in the firſt quarter's ebb, where meeting the remains of the waters from the 16 acres of Slade Marſh, and other waters from the unimbanked marſhes, that ſtill fall into the haven creek without paſſing the ſluice, and might make a ſenſible increaſe of the current; yet as this muſt greatly fall off after the top waters were gone, and give but little aid in the latter part of the ebb, when it was moſt wanted, muſt equally appear plain and clear. But yet whatever good effect might be aſcribed to the water from the Slade, or in reality it might have, when it could operate in the manner I have pointed out, yet as it appears from equal teſtimony that thoſe rages happen but ſeldom (four or five times in a year), they could be of no material benefit, becauſe their power of ſcouring and grinding the

bottom,

bottom, inafmuch as the fea is inceffantly bringing in a frefh acceffion of fand, filt, and mud, muft be in proportion to their *frequency*, fo that if it were to be fuppofed (what it does not appear to me reafonable to admit), that the effect of the Slade waters was when they happened even equal with that of the fluice, yet the fluice, if we put the neap tides out of the account, operating 365 good tides in a year, and thofe but five, the benefit could only be as 70 to 1 ; an effect fo fmall, that, when mixed with many others, could not be perceived; and though in the eye of reafon every thing that operates at all muft have an effect, yet it clearly appears to me that the effect to be expected either by the fhutting up or opening thofe marfhes in the manner they were before the imbankments in 1758*, could procure no fuch beneficial effect upon the harbour as to prevent the neceffity of fupporting and continuing the fluice, or even to be of any meafurable or eftimable degree or value; and that this explanation of the fmall utility to be derived or expected from the Slade marfh waters, is in reality the true one, is proved by the facts already ftated; viz. that before the year 1749 they had proved *totally ineffectual*, and had always proved fo ever fince when the fluice was out of order. I muft therefore conclude in the fentiment of Mr. Mylne, that whoever would find a caufe for the alteration of the courfe of the out-channel, for the filling up of the Pool, for the landing up of the harbour, channel, or creek, and in general the decaying ftate of the harbour of Wells, muft feek fome caufe far more extenfive than the imbankment of the Slade marfhes in the year 1758, and the remedy from human induftry and art in fomething more powerful and better adapted than any of the fluices there applied appears to have been.

*London,*
4th *May* 1782.

J. SMEATON.

---

* I fay in the manner they were before the year 1758, that is, when there was no bank at all ; becaufe if a partial breach was made in the bank, the waters iffuing through this breach would have an effect fimilar to the fluice, till worn by the entering and iffuing waters too wide to produce the effect, as was the cafe when Freeftone's Sluice was worn too wide, it ceafed to do its duty.

REFERENCES

## REFERENCES to the Plan of Wells Harbour.

### Plate 1. Fig. 2.

|   |   | A. | B. | P. |   |   |   |
|---|---|---|---|---|---|---|---|
| 1 | Holkham Marſh, imbanked about the year 1719, by the late Lord Leiceſter, including creeks - - - } | 560 | 0 | 0 |   |   |   |
| 2 | Wells Weſt Marſh, imbanked 1719, by Sir Charles Turner, including creeks - - - } | 108 | 2 | 12 |   |   |   |
|   |   |   |   |   | 668 | 2 | 12 |
| 3 | Weſt Salt Marſh - - - | 588 | 2 | 0 |   |   |   |
| 4 | Lodge Marſh - - - - | 266 | 2 | 35 |   |   |   |
| 5 | North, or Out Salt Marſh - - - | 717 | 2 | 25 |   |   |   |
|   | Channels and creeks - - - - | 146 | 1 | 12 |   |   |   |
|   |   |   |   |   | 1,719 | 0 | 32 |
| 6 | Eaſt, or Church Marſh, excluſive of the ancient creeks - | 106 | 3 | 2 |   |   |   |
|   | Creeks in ditto - - - - | 4 | 0 | 0 |   |   |   |
|   |   |   |   |   | 110 | 3 | 2 |
| 7 | Warham Slade, excluſive of the ancient channel and creeks - | 59 | 1 | 36 |   |   |   |
|   | The channel and creeks in ditto - - - | 7 | 0 | 32 |   |   |   |
|   |   |   |   |   | 66 | 2 | 28 |
|   |   |   | Acres | 2,565 | 0 | 34 |   |

A A The preſent entrance into the harbour.

 B The courſe of the Old Channel by the Scolph.

C C The weſt ſide of the antient entrance into the harbour.

 D Friſton's Jurties.

E E The arrows ſhewing the ſet of tide over the ſands on this coaſt, for the laſt three hours of flood and the firſt three hours of ebb.

 F The place of the ſecond buoy at the turn by Broom's wreck.

 G The preſent navigable channel at the Scolph, ſince the imbankment made by Sir John Turner.

 H The preſent Pool.

 I The Quay.

 K The bank made by Sir John Turner in 1758.

 L The place to which the water uſed to flow before the imbankment made at K.

M N The places where the waters from the weſt marſhes empty themſelves into the main channel

P P The line from which it is ſuppoſed the water has drained towards Wells Harbour.

## ABERDEEN HARBOUR.

The REPORT of JOHN SMEATON, Engineer, upon the Harbour of the City of Aberdeen.

(See a Plan, Fig. 1. Plate 2.)

THE principal complaint attending this harbour is the difficulty of entry, occasioned by a barr a little without the harbour's mouth, and a shifting bed of sand, gravel, and shingle on the north side of the entry, which, by the action of the seas, when the wind is in the north easterly quarter, drives into the main channel, choaking it up in different degrees, according to the violence of the sea, the state of the tides, and of the land speats, floods, or freshes in the river Dee, which here falls into the sea.

When I was there, which was in the month of August 1769, the entry was then said to be in a good state; and on sounding it upon the 7th of that month, which was the 6th day after the new moon, (and consequently the tides in a mean state between spring and neap), I found full four feet of water upon the bar at low water, and at high water the same day full fourteen feet; but it is said that the ordinary spring tides make but much about the same depth upon the bar at high water, and that at low water the barr is left with only the run of the river over it. The neap tides it is said usually make ten feet water upon the bar, but this is to be understood (I suppose) at such times as the entry is in a good state. On sounding at low water, I found the body of the bar to be composed of loose stones of different sizes, and the whole intermixed and compacted together with gravel, over which was a layer of sand from six inches to a foot in thickness, which after great land freshes is said to be quite swept away, and the stones and gravel left bare, which is its best state. The bar is but of short extent, and both within and without we quickly get more water by three feet. Without the bar the water gradually deepens and forms a very good road for ships to ride at low water, and is naturally protected from all winds except the north easterly and easterly, which blow right into the harbour's mouth, so that were there a little more depth of water over the bar, and this *certain*, this harbour would be capable of affording very good protection to merchant ships trading into these seas.

The

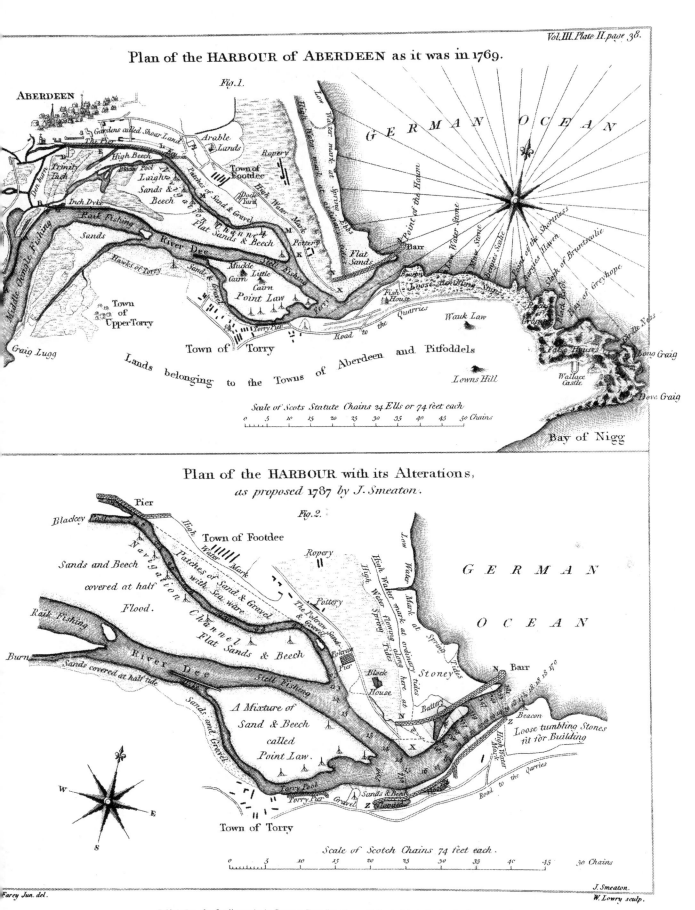

Plan of the HARBOUR of ABERDEEN as it was in 1769.

Fig.1.

ABERDEEN

Gardens called Shoar Land
The Pon
High Beech
Trinity Inch
Den burn
Inch Dyke
Raik Fishing
Sands
Middle Ghensle Fishing
Hawks of Torry
Craig Lugg
Town of Upper Torry
Town of Torry
Torry Pier
Point Law
Muckle Cairn
Little Cairn
Sands & Gravel
Laigh Sands & Gravel Beech
Patches of Sand & Gravel
Dock Yard
Town of Footdee
Arable Lands
Ropery
High Water Mark
Garron Channel
Flat Sands & Beech
Stell Fishing
River Dee
Pottery
Flat Sands
Point of the Hatom
Barr
Houpill
Loose tumbling Ships
Fish House
Low Water mark at Spring Ebbs
High Water mark at Spring Ebbs
Low Water Stone
Brine Stone
Pocras Scalp
Point of the Shortness
Torgies Haven
Bay of Greyhope
Inch of Bruntsalie
Castle Ness
Young Craig
Dove Craig
False Houses
Wallace Castle
Road to the Quarries
Wauk Law
Lowns Hill

Lands belonging to the Towns of Aberdeen and Pitfoddels

GERMAN OCEAN

Bay of Nigg

Scale of Scots Statute Chains 24 Ells or 74 feet each
0   5   10   15   20   25   30   35   40   45   50 Chains

Plan of the HARBOUR with its Alterations,
as proposed 1787 by J. Smeaton.

Fig.2.

Blackey
Pier
Town of Footdee
Ropery
Sands and Beech covered at half Flood.
Raik Fishing
Burn
Sands covered at half tide
Sands and Gravel
Navigation Channel
Patches of Sand & Gravel with Sea ware
Flat Sands & Beech
River Dee
Stell Fishing
A Mixture of Sand & Beech called Point Law.
Torry Pool
Torry Pier
Gravel
Town of Torry
High Water Mark
The Polorow Sand & Gravel
Pottery
Polorow Pier
Black House
Batter
Sands & Beech
High Water Mark flowing at Spring Tides
High Water Mark flowing along here at
Low Water
Stoney
N
Barr
Beacon
Loose tumbling Stones fit for Building
High Water Mark
Road to the Quarries

GERMAN OCEAN

W   E
S

Scale of Scotch Chains 74 feet each.
0   5   10   15   20   25   30   35   40   45   50 Chains

Farey Jun. del.

J. Smeaton.
W. Lowry sculp.

Published as the Act directs, 1812, by Longman, Hurst, Rees, Orme and Brown, Paternoster Row, London.

The caufe of the obftruction of the harbour's mouth appears to me to be this : The whole coaft, which ftretches away northerly, is apparently for miles a flat and fandy fhore, and I fuppofe from the harbour of Aberdeen till it meets with the point of Buchanefs, (which is at the diftance of feven leagues), continues of the fame kind ; confequently the wind at N. E. acting obliquely upon it, brings the fands and gravel intermixed coaftwife towards the fouth ; and as the coaft from the fouth fide of the entry of this harbour ftretches away nearly eaft for about three quarters of a mile, thofe fands would naturally be depofited in the angle of the coaft formed at the harbour's mouth, did not the land waters of the river Dee, in finding a paffage to fea, force themfelves a vent, which they maintain more or lefs clear according as the circumftance of winds, tides, and frefhes, balance one againft another. A hard gale of wind at N. E. as already mentioned, gradually brings the fands and gravel coaftwife fouthward, and puts in agitation that already lodged in the bank on the north fide of the harbour's mouth, at the fame time forcing it into the entry, and if at that time it happens to be fpring tides and little frefh water in the river, a ftrong tide of flood being the confequence, greatly co-operates with the wind and feas in carrying a large quantity of fand and gravel into the channel of the river ; and the frefh water in the river being fuppofed then very fhort, the reflux will be very languid, and being counteracted by the impetus of the fea, it cannot return ; and a continuance of weather and circumftances of this kind, will put the mouth of the harbour into the worft ftate, in which it muft neceffarily remain till by a contrary difpofition of circumftances a contrary effect is produced. On the other hand, a continuance of great land floods, either at fpring or neap tides, accompanied either with off fhore winds or moderate ones at N. E. gives the greateft advantage in fcouring away the fands and gravel from about the harbour's mouth. carrying it out into the road, from whence by degrees it gets round the point of *Girdlenefs* ; and if towards the clofe of the work there happen along with a ftrong land frefh, low fpring ebbs, which give the current the greateft fall to fea, and at the fame time run bare over the bar with a moderate wind at N. E. which will give the fand fome agitation without much impetus ; under thefe circumftances, the ftony body of the bar will be cleared of fand, and the harbour's mouth be put in its beft ftate, and fo will remain till the contrary caufes produce as before the contrary effects : and in this ftate of fluctuation muft the entry of the harbour of Aberdeen ever remain, till fomething is done to counteract the effects of that arrangement of circumftances whofe natural tendency is to do harm.

The only means by which I can fee that this is likely to be effected, is the erection of a north pier, as fhewn at N. N. in the plan, which will directly tend to the cure of the

evil

evil complained of; for it will not only keep the land freſhes more confined in a body till they come into deeper water, but what is of more conſequence, will in a great meaſure prevent the ſand and gravel from being driven in. It will not indeed ſtop the continual driving of the matter coaſtwiſe from the north, but after the back or outſide of the pier is filled up with ſand, &c. to a certain degree, it will then go round the pier head, and by the ſuperior action of middling freſhes and ſpring ebbs will be kept in deeper water, and ſo get round the point of Girdleneſs, without getting into the harbour's mouth, or at leaſt not in ſuch a degree as to obſtruct the navigation. By this means, as the bar will not only be kept clean down to the ſtone bed, but by lifting the larger ſort of ſtones by art, the remaining gravel will waſh out into deeper water, ſo as to make (as may reaſonably be expected) full two feet more water than there now is in its beſt ſtate.

I can ſee no objection to the putting of this work into immediate execution, ſave the expenſe of it. To reduce this as much poſſible, I have endeavoured to propoſe ſuch a conſtruction, as, conſiſtently with that ſolidity and permanency which a work of this kind ought to have, conſiſts of the ſmalleſt quantity of materials and workmanſhip that I can think ſufficient; and as the materials which nature furniſhes here are of the beſt kind for the purpoſe and are found near the place, I am in hopes that they will be raiſed and put together conſiderably cheaper than I have ſuppoſed in my eſtimate, which for that reaſon I deſire may be conſidered in no other light than a form or blank, comprehending ſpecies and quantities: for as my ſtay at Aberdeen was neceſſarily ſhort, the requiſite examinations immediately relating to the harbour prevented my entering into thoſe enquiries, which tended only to acquaint me with the price and value of labour and materials. I therefore can only fill up the blanks by compariſon with what has been done at other places leſs advantageouſly ſituated with reſpect to materials, and perhaps that of labour alſo: in order therefore to come at a real eſtimate ſuited to the place, I muſt beg leave to refer myſelf to the enquiries of the magiſtrates concerning prices, or to a committee deputed by them for that purpoſe; and which, from their particular knowledge of the country and of the workmen, will I apprehend be done to more advantage than I could have done myſelf, without a longer ſtay than my other avocations could poſſibly admit of.

One thing, however, I muſt beg leave to ſuggeſt, that though what I now offer is as I apprehend the complete thing, which I expect to anſwer in the beſt manner, yet I am of opinion that it will be found if the pier be not carried out ſo far by 200 feet, that it will in a great meaſure relieve the preſent annoyances, nay, that it will be of ſingular ſervice if carried out but juſt beyond the preſent pier, on the ſouth ſide of the entry, in which caſe it will be ſhortened by 350 feet. In the former caſe the expenſe will be reduced according

cording to my eftimate by £2,028, and in the latter by £3,549; and as the work ought to be begun from Sandnefs Point, where it is eafieft and cheapeft to be done, the workmen will gradually learn the way of doing it to the beft advantage, and will probably be able to contract for the more expenfive part, upon better terms than would feem to them feafible at the beginning; and as the work advances, the benefit and effect will be feen, fo that it needs not to be carried further than by experience fhall be proved neceffary.

I come now to the improvements that may be made in the internal part of the harbour. Here I cannot but lament that the courfe of the river is diftracted by fo many channels, and covers fo great a breadth of ground at high water, which want of confinement is not only detrimental to the procuring of a deep channel at the harbour's mouth, but within the harbour alfo, and it is particularly difadvantageous that the main current of the river does not fweep the face of the town's key at L: this I fhould without any hefitation advife to be done by art, were it not for the fifhing properties upon the main channel; but, as I muft fuppofe them irrevocably eftablifhed, it remains to point out what is the beft that can now be done, the fifhings remaining as they are. I obferve that within the Point of Sandnefs there is nearly the fame water as over the bar, till the main river channel and navigation channel divide, which part of the river is marked in the plan as the *Stell Fifhing*. In this part of the river, which is land-locked from all winds, veffels that will bear the ground, and whofe draft of water is fuch as not to admit of their going further up, may fafely deliver their cargoes, or fhelter themfelves when they come in by way of refuge; but after the aforefaid divifion the navigation channel M M becomes immediately fhallow, carrying however ten feet water till we arrive above the new pier called Pockraw Pier at K; when oppofite the ropery and dock-yard it falls a little fhallower to nine feet four inches, from thence it holds nine and a half and ten feet, till meeting with the town's pier or key E L, it again breaks into two channels viz. that which ftretches along the face of the pier, and that which is called the Blacky Pool. The navigation channel by the pier fide, from the aforefaid divifion, falls off at firft to nine, then to eight, and gradually to feven feet water.

The defirable improvement pointed out to me, and which feems of great confequence to the trade of the city, is to deepen the navigation channel quite away from the new pier to the weft end of the Town's Pier or Key.

I obferved when there, that little or none of the current water of the river Dee, in its common ftate at low water, goes down either the channel by the face of the pier, or the Blacky Pool, all the outlets from the main ftream that might be likely to take this courfe, being barricaded by ftone dykes, raifed from two to three feet, or thereabouts, above the ordinary furface of the Dee's water, fo that nothing worth notice, till the water is fwelled above thofe dykes, can go down the above-mentioned channel, fave the water of two fmall burns which empty themfelves by the navigation channel. With thefe helps, however, but principally by the current that paffes through it and the *Blackey Pool* during fuch fpeats as overflow the aforefaid dykes, the navigation channel is kept open.

It has been propofed to bank in the low grounds lying weft of the old pier marked D, fo as to pen in the fpring tides, and at that place to erect a fluice to be drawn at low water, and by making an artificial fcour to deepen the aforefaid navigation channel.

Great effects are capable of being produced by the operation and judicious management of fluices in fituations adapted thereto; that is, where there is a great command of frefh water, or a confiderable declivity in the part to be fcoured. Here as the water to be pent in is not confiderable in itfelf, muft be in a great meafure tide water, which, as the imbankation would in a great meafure prevent all currency through it, would be fubject to filt up the refervoir, and the length according to the navigation channel nearly a mile upon four feet only of defcent at low water, I fear, thefe circumftances confidered, the effect would not be found anfwerable to expectation. On the other hand, the imbankation preventing the fpeats which overflow the dykes from getting in at the head of the navigation channel, the principal natural agent would be prevented from operating, and which I fhould be forry to lofe. Were the long dyke, marked A A in the plan, broken down and removed, I make no doubt but that in the courfe of a few years the main ftream of the river Dee would make its paffage by way of the Denburn into the navigation channel, D E L M K, and by degrees of itfelf produce the effect defired, and with a little help would do it in a very few years. I ftate this not upon any fuppofition that the dyke is likely to be removed, but to fhew more ftrongly the ufe of fuch natural advantages as ftill remain. I do not fuppofe, however, that any ufe that can be made of the remaining advantages will of themfelves greatly deepen the navigation channel; but this I fuppofe, that after it is made deeper by art, thofe natural advantages may be fo applied as to keep it equally clean at a greater depth, as they now do at a leffer.

The whole channel from the new pier at K to the town's pier extending from D to L, does not need a great deal to make it good ten feet water, and it may be very fuccefsfully deepened by a ballaft lighter, conftructed like thofe ufed upon the river Thames for getting ballaft for the fhips. Thofe lighters work by direction of the Trinity Houfe upon fuch fhoals as are moft injurious to the navigation of that river, and all the fhips of that port are obliged to take their ballaft from them at a certain price. Perhaps much

ballaft

ballaft is not taken out from the port of Aberdeen, but fuch as is, may be fupplied by this lighter in aid, as far as it goes, of the expenfe of raifing it.

The channel for the whole length of the pier or key I would propofe to be deepened by the mattock and fpade at low water, which deepening being done two feet at a mean, will give ten feet water to the middle of the pier where now there is but feven feet fix inches; and this being done to a breadth of fixty feet will admit of two veffels to lie abreaft with fufficient paffage. This work will be attended with no extraordinary expenfe in proportion to the utility thereof, and will endure for feveral years before the ftate of the channel will return to what it now is; yet it would undoubtedly return by degrees to the fame ftate, unlefs fome counter-balance be applied to prevent it : what I would therefore recommend for this purpofe is as follows :—

To erect a ftrong ftone dyke beginning at the head of the Inch Dyke at F, in the direction of the dotted line' F E to the beacon E upon the Trinity Inch; or if it fhould anyways happen to interfere with the Raik Fifhing, to carry it from a lower part of the Inch Dyke to the faid beacon, according to the direction of the dotted line G E : this dyke to be made fo as to rife above and keep in the water of the high land floods at half ebb of the tide, and to be made firm, fo that the current may in great land floods at high water go over it, without hurting it; by this means the greateft part of the water that in time of fpeats flows over the prefent dykes, and makes its way partly by the Blacky Pool, partly over the furface of the higher lands, and partly by the navigation channel along the face of the pier, will all be conftrained to go through the channel alongfide the pier, and therefore will be as adequate to keep clear that channel at ten feet depth, as the prefent channel joined to Blacky Pool is to keep it at that mean depth from their junction to the junction of the main river at the Stell Fifhing: and as thefe operations are plain and fimple, and will be attended with no confiderable expenfe, I earneftly recommend their execution.

It is alfo very practicable in like manner to join the old pier D by a dyke acrofs the Trinity Inch to the elbow of the river, a little above the Inch Dyke B; and alfo by putting fluices upon the opening between the north end of the old pier D and the weft end of the town's pier or key to pin in the tides, in order to make artificial fcours : but as the building and maintenance of fluices would be expenfive, if made fo capacious as not to be an impediment to the current and action of the land flood waters, as above mentioned, and would have no confiderable effect, unlefs the water was pent up higher than the prefent fifhing dykes, (the banks whereof if fo would in a great meafure prevent the flood waters going that way); it therefore appears that fince fluices of any kind are

likely

likely to prevent more good than they will do, that the plain, natural, and simple method first described, of making a dyke according to the single dotted line E F as first mentioned, is the most eligible to be put in execution. Upon the whole, I am of opinion the methods pointed out to be pursued, without and within, will remedy the complaints the harbour is subject to, as far as is above specified, and possibly in a still greater degree.

*Austhorpe,*
19th *February* 1779.

J. SMEATON.

---

### ESTIMATE for the North Pier of Aberdeen, &c.

£  s.  d.

This pier N N being begun at the high ground of a point called Sandness Point, marked X, not subject to be overflowed with the tide, may be carried out for 400 feet gradually increasing, being 20 feet base, 12 feet top, and 12 feet high, will contain as follows, videlicet,

|  | Cube feet of solid rough blocks. |
|---|---|
| In the base for 2 feet high - - - - | 40 |
| In the side at 3 feet mean thickness and 10 high - - | 60 |
| In the platform at top - - - - - | 5 |

Solid rough blocks per foot running at 4d., £1. 15s; and for 400 feet length - - - } 105     700  0  0

To 87 cube feet of chiefly large and some small rough stones for filling, which reckoning 13 cube feet and a half to the ton (that is, 2 ton to the cube yard), makes 6 tons and a half per foot running, which at 2s. 6d. per ton laid in place amounts to 16s. 3d. per foot running, and for 400 feet to -     325  0  0

The parapet being at a medium 4 feet 6 inches base, 3 feet top, and 4 feet high, will contain 15 cube feet per foot running, and reckoning as before 4d. per foot for block stone, 2s. 6d. per ton for walling and filling stones, 2d. per foot superficial over all, for work in facing, and 2s. 2d. per cube yard for mortar and extra work in walling the parapets, will come to 10s. per yard, cube measure, and therefore for 15 feet to 5s. 6d. per foot running, and for 400 feet to -     110  0  0

The 1st stretch of 400 feet of the pier - - - £ 1,135  0  0

The

The 2d ftretch being carried out 400 feet farther, being a mean 28 feet bafe, 14 feet 6 inches at top, and 20 feet high, will contain as follows:

|  | rough blocks. | £ | s. | d. |
|---|---|---|---|---|
| In the bafe for 3 feet high | 60 | | | |
| In the fides 17 feet high and mean thicknefs 3 feet and a half } | 119 | | | |
| In the platform at top | 7 | | | |
| Solid rough blocks per feet running | 186 at 4d., £3. 2s. | | | |
| To 400 feet in length at £3. 2s. per foot running | | 1,240 | 0 | 0 |
| To 117 tons of large rough ftones for filling, at 2s. 6d. per ton, comes to £2. 4s. 3d. per foot running, and for 400 feet to | | 885 | 0 | 0 |
| The parapet will contain a cube yard per foot running, which for 400 feet at 10s. comes to | | 200 | 0 | 0 |
| The 2d ftretch of 400 feet of the pier | | £ 2,325 | 0 | 0 |

The 3d ftretch of the pier being carried out 546 feet beyond the former, and being at a mean 36 feet bafe, 24 feet top, and 24 feet high, will contain as follows:

|  |  | £ | s. | d. |
|---|---|---|---|---|
| The bafe of blocks 4 feet high | 114 | | | |
| The fides at a medium 4 feet and a half thick each | 180 | | | |
| The platform at top | 10 | | | |
| Rough blocks per feet running | 334 at 4d., £5. 11s. 4d. | | | |
| This for 546 feet in length comes to | | 3,039 | 8 | 0 |
| To 28-5 tons of large rough ftones for filling, at 2s. 6d. per ton, comes to £3. 11s. 6d. per foot running, and for 546 feet to | | 1,951 | 19 | 0 |
| The parapet will contain 2 cube yards per foot running at £1. per foot, and for 546 feet | | 546 | 0 | 0 |
| The 3d ftretch of the pier 546 feet | | £ 5,537 | 7 | 0 |

The pier head to be 60 feet diameter at bafe, 48 feet top, and 24 feet high, will contain as follows:

|  |  | £ | s. | d. |
|---|---|---|---|---|
| The bafe of blocks 4 feet high | 11,316 | | | |
| The outfide at 4 feet and a half mean thicknefs | 16,920 | | | |
| The platform | 707 | | | |
| Rough blocks in the pier head | 28,943 at 4d. | 482 | 7 | 8 |
| To 1,940 tons of rough ftones for filling at 2s. 6d. | | 242 | 10 | 0 |
| To 54 feet of parapet (being the length that the head is fuppofed to add to the 3d ftretch, making it in the whole 600 feet) at £1. per foot | | 54 | 0 | 0 |
| The pier head making 54 feet running | | £ 778 | 17 | 8 |

ABSTRACT

## ABSTRACT.

|  | | £ | s. | d |
|---|---|---|---|---|
| The 1st stretch containing 400 feet running | - - - | 1,135 | 0 | 0 |
| The 2d ditto - 400 | - - - | 2,325 | 0 | 0 |
| The 3d ditto - 546 | - - - | 5,537 | 7 | 0 |
| The pier head - 54 | - - - | 748 | 17 | 6 |

| Whole length 1,400 | Total of the pier - | 9,776 | 4 | 6 |

The taking up the bar so as to make two feet water more than at present when clear, with incidental charges, may be supposed - - 223 15 6

Total for the pier - £ 10,000 0 0

*Aufthorpe,*
19th *February* 1770.

J. SMEATON.

## ESTIMATE for the Interior Works proposed for the Harbour of Aberdeen.

|  | £ | s. | d. |
|---|---|---|---|
| The construction of a stone dyke in the direction F E, specified in the report and plans, being supposed at a mean 12 feet base, 6 feet high, and made rounding at top, will take about $3\frac{6}{10}$ tons of rough stones per foot running at 2s. 6d. per ton, and being in length about 1110 feet, will come to - - | 499 | 10 | 0 |
| To clearing the channel from the new pier K to the town's pier L D, in length about 30 chains or 2220 feet, and being supposed to be deepened at a medium one foot upon 60 feet wide, will contain 4933 cube yards, which I suppose may be done by hand at low water for 6d. per yard, will amount to - | 123 | 6 | 6 |
| To deepening the channel by the side of the town's pier L D, at a medium two feet upon 60 feet wide, for 1700 feet in length, will contain 7555 cube yards, which at 6d. will come to - - - - - | 188 | 17 | 6 |
| The interior works - - - | £ 811 | 14 | 0 |

*Aufthorpe,*
19th *February* 1770.

J. SMEATON.

To

To the Magiftrates of the City of Aberdeen.

The REPORT of JOHN SMEATON, Engineer, upon the *In-run* of the Seas into the Harbour of Aberdeen in easterly winds.

(See the Plan, Fig. 2. Plate 2.)

IN confequence of the memorial from the magiftrates of Aberdeen, dated the 9th of April laft, which I received in London from the hands of Mr. Profeffor Copland, and of your further requeft fignified to me by Meffrs. Carnegie and Black of Montrofe, I took the opportunity of vifiting your harbour, and made my obfervations upon the place, 1ft, 2d, and 3d of October laft, at which time the wind happening to be eafterly (though not very frefh) afforded me an opportunity of particularly viewing and confidering the mode of the action of the feas from this quarter, and had the pleafure to find that the deepening of the the entry into the harbour, and removal of the bar, that was the principal annoyance complained of when I viewed the harbour in the year 1769, has been effectually removed and cleanfed, and fo kept continually in that improved ftate by means of the north pier N N, which has been erected conformable to my report of the 19th February 1770. But though the main object has been anfwered, and the harbour and refort thereto very greatly improved, yet the very means by which this improvement has been effected have produced a caufe of complaint of a very different kind, which at the time I apprehend to have been altogether unforefeen, and that is from the increafe of depth and freedom of paffage, the fwell of the fea at high water meeting with nothing to controul it, makes its way through the clear paffage between the two piers, and meeting nothing within the great natural bafon or bay that forms the harbour to break or difperfe the feas fo brought in ; according to the nature of waves when paffing through a narrow into an expanded fpace, they turn round along the fhore and fpend their fury upon the neareft objects in the greater degree, and in proportion upon thofe more diftant.

While the Old Sandnefs Point remained, which had been formed in former times before the north pier was built, and which extended more than half way acrofs the fpace walled off by the new pier (as appears by the original Plan at X. Fig. 1.) this point of fhoal water while it fubfifted was capable of taking up and breaking the heavy feas in rolling over it, and thereby in a great degree difperfing them ; but in proportion as the bar was removed and the entry deepened, the feas falling more heavy upon this point, have gradually difperfed and removed it, and in confequence now pafs into the harbour without controul, as has been fully and clearly fet forth in your memorial above referred to.

The

The caufe of the prefent in-run in confequence of thefe alterations, being clearly owing in a great meafure to the lofs of the Sandnefs Point, it further appears that, if reftored, it would be inadequate to the difperfion of the feas that now enter, owing to a greater depth at the mouth. The remedy therefore is clearly pointed out to be a conftruction that fhall not only have the effect that it had in its original ftate, but one fo much greater as may be in proportion to the greater weight of the feas, which from the caufes ftated are now liable to fall upon it.

The caufe and the mode of cure being both perfectly clear, there can be no difficulty in judging that fomething fhould be done at the Sandnefs Point in order to its reftoration; but in what particular mode and degree, as alfo that it may be done in the moft œconomical as well as effectual way, have been to me matters of much reflection and ftudy.

The firft thing that prefents itfelf is, to begin from the fide of the pier near about where the Old Sandnefs Point was, and to depofit a body of rough ftones projecting gradually forwards towards the middle of the open fpace, rifing towards the pier and floping towards the low water; and as this work will naturally be done in progreffion, it may be gradually carried on till it is found to produce the defired effect; and this is doubtlefs the cleareft and eafieft way of doing the bufinefs, and no more needs to be done than what is fufficient; fo that in effect the *New Sandnefs Point* will now, inftead of fand and gravel, be compofed of a body of rough granite. This mode in its commencement would appear to be the moft œconomical, but it is very probable that before it is ended it will require fo great an area to be covered, as to contain fo confiderable a quantity of ftone, that the fame tonnage of ftone being put into a regular fhape would form a regular catch pier. The ftones compofing fuch a mound or bulwark muft not in general be of a fmall fize, becaufe if the feas breaking upon them remove them and wafh them into the navigable channel, they will produce an obftruction of a third kind; and if formed of large ftones, I apprehend that as the quarrying and carriage of fuch ftones will fo far exceed the coft of putting them in fhape when brought to the place, that they may juft as well be made to form a regular piece of work. I have, however, no doubt but that the mound now propofed would fully anfwer the end; and if the magiftrates are inclined to adopt the mode of conftruction, the outline of it, defcribing the fpread that I apprehend it may be needed to have when completed, is very well defcribed by the chain of dots that I left upon the plan that had been prefented to me along with the faid memorial, (thefe are feen in Fig. 2. at a a). In regard to any further directions they feem unneceffary, as the work is to prove itfelf, the fea being to break over it,

it

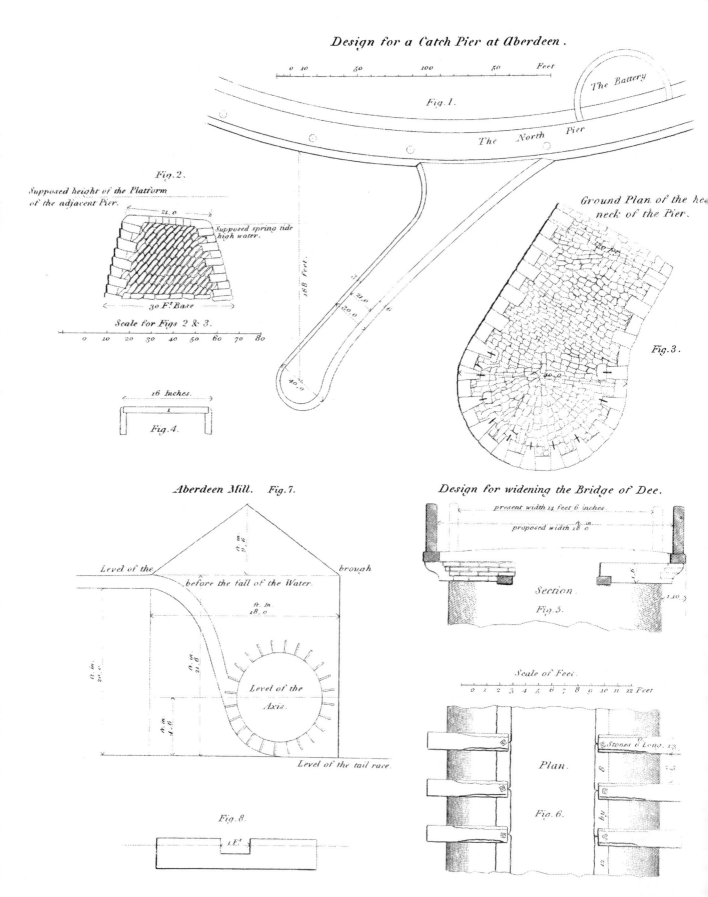

Design for a Catch Pier at Aberdeen.

0    10        50        100        50      Feet

Fig. 1.

The Battery

The North Pier

Fig. 2.

Supposed height of the Platform
of the adjacent Pier.

21. 0

Supposed spring tide
high water.

30 Ft Base

Scale for Figs 2 & 3.

0   10   20   30   40   50   60   70   80

16 Inches.

Fig. 4.

168 Feet

Ground Plan of the head
neck of the Pier.

Fig. 3.

Aberdeen Mill.    Fig. 7.

Level of the

brough

before the fall of the Water.

18. 0

Level of the
Axis.

Level of the tail race.

Fig. 8.

Design for widening the Bridge of Dee.

present width 14 feet 6 inches.

proposed width 18 0

Section.

Fig. 5.

Scale of Feet.

0   1   2   3   4   5   6   7   8   9   10   11   12 Feet

Stones 6 Long

Plan.

Fig. 6.

L.Farey Jun.del.    Published as the Act directs by Longman, Hurst, Rees, Orme & Brown, Paternoster Row.    Lowry

it is only neceffary to keep it conftantly higher towards the pier, and floping towards its bafe on all fides, as has been already mentioned, and was the cafe of the natural point.

I herewith fend a general Plan, (Plate 3.) fhewing the fhape, fituation, magnitude, and projection of a catch pier in refpect to the prefent works, and which if carried out to the extent defcribed, I have no doubt will have the effect of quieting the harbour; and as it will leave a clear breadth of full 300 feet for the navigable channel, I apprehend there can be no objection to its projecting the full extent of 180 feet from the fide of the prefent north pier, if fo required; but as it is not improbable, and indeed what I rather expect, that half of this projection may be found fufficient, when other things are done that I pointed out when upon the place, and fhall further enlarge upon, I would firft propofe to carry it out to the length of 90 feet, building up the head with a fquare return, in the manner that was done on quitting the work of the north pier at the conclufion of each feafon; and this being done, the effect of one winter will fhew what more will be neceffary, and in confequence the head as defigned may be immediately joined upon it, or it may be carried out to the full length, or to fuch extent as may then be judged requifite.

As this catch pier, fo far as I can fee, is not likely to be of ufe except as a break water, I do not mean to build it fo high as the other, but to about fpring tide mark; the feas being fuffered to break over it at high water, which will in fact have more effect towards their total difperfion, than that of folely diverting their courfe; for till they are broken and deftroyed they will always recoil and produce the effect fomewhere.

As I find by experience that there is nothing that fo much tends to quafh and difperfe a wave when it is raifed, as its breaking upon a floping beach; on this account the continuance of the fouth pier Z Z (Plate 2. Fig. 2.) fo far to the weftward, (which feems to have no other effect than as a wall to wharf up the floping fhore that lies oppofite to Sandnefs Point), has in reality a very bad effect in the prefent ftate of things, in keeping the feas from fpending and breaking as they would otherwife be inclined to do upon the *naturally* floped fhore at that place, as the flooded pools lying here behind the pier do teftify. For what reafon this wharfing has been carried fo far weft, it is not now eafy for me to conjecture; but doubtlefs the ill effect of it would not fo clearly appear before the erection of the north pier, as the north fide was then a floping beach entirely. But for whatever reafon it was done, the removal of it now is altogether neceffary, for the fhape and tendency of the catch pier now propofed, being to throw the feas more effectually over to the fouth fide, unlefs there is a floping beach for them to break upon, they will be again reflected back from that fide towards the north, and undoubtedly produce difagreeable effects in fome part of the harbour. I therefore propofe, from the place fpecified in the plan, where the weft end of the fouth pier is propofed to be terminated, to remove the prefent wall entirely, making it a floping beach quite away to Torry Pier; and to mend the flope artificially where wanted: the ftones will go in aid of the work to be carried on upon the north fide. Fig. 1. of Plate 3. is an enlarged plan of the catch pier as now propofed, accompanied with a fection, Fig. 2. and a plan, Fig. 3. of the ground courfe of the pier head, the two laft to a ftill larger fcale, and which, with the following explanation thereon, I apprehend will be fufficient for the guidance of the artificers.

There is not propofed to be any *work* upon the ftones of the outfide, further than fplitting at the quarries; and as in the fplitting of granite they can as eafily be cut into wedge-like pieces, as made of parallel fhapes, I have fhewn a method by which the ftones compofing the circular end of the pier can be retained in their places, as effectually or even more fo, than where the pier goes out near upon a ftraight line: for by making every other ftone a long ftone or *header* and more wedge-wife than the tendency towards the centre, the ftones lying betwixt them may be retained in the manner of a dovetail, and the header ftones themfelves being anchored at their tails, that is at their inward or fmaller ends, to an anchor or crofs ftone by means of an iron cramp to each, the whole will be held compactly together. Thefe wedge ftones may be of different fizes, as they will cut to the moft advantage in the quarry, as well as the intermediate ftones; all that is required, is, that they decently fit each other.

The wedge-like header ftones, employed to make the turn of the head, and hold in the intermediate ftones between them, are to be tied to the anchor ftones (fee Fig. 4.) at their tails by iron cramps of one inch fquare, turned down and rounded at each end to go into jumper holes, to be fixed therein if neceffary, with wood wedges without lead, as the weight of the courfes above them, bonded in the fame way, will be fufficient to keep them from ftarting.

I have fully attended to the contents of Mr. Carnegie's letter of the 24th ult. which only fhew that fomething effectual muft be done, even to preferve the internal works of the harbour; but when what is propofed is put in complete execution, I expect the *tongue*, and all its complaints, will vanifh of courfe by degrees.

*Aufthorpe,*
22d *March* 1788.

J. SMEATON.

## ABERDEEN BRIDGE AND MILLS.

(See Plate 3. Fig. 5. and 6.)

The REPORT of JOHN SMEATON, Engineer, concerning the Improvement of the Bridge of Dee, near the City of Aberdeen, and of the Common Milns of the faid City.

THE Bridge of Dee appears to be well built, and of very good materials, and the current of the river feems to fet fairly through the arches, fo that it promifes a duration for many years. Like moft bridges formerly built, it is too narrow, being but $14\frac{1}{2}$ feet wide within the parapets, which is too little for carriages to pafs with that eafe, freedom and fafety which are defirable. It is found by experience that 18 feet clear width admits of this convenience, and as this may be procured without deranging any of the more folid parts of the bridge, it feems a defirable improvement.

As the country affords an abundance of excellent granite, which, with little work upon it, is capable of being fplit into fquare blocks or pillars, I propofe to widen the bridge by laying out corbells of granite of one foot ten inches projection, and removing the parapets one foot nine inches further out on each fide on the corbells; the bridge will then be full 18 feet in the clear. The particular method of conftruction will be better underftood by the defign than by many words. Upon which I have only to obferve, that to prevent the leaft apprehenfion of the weight of the parapets overfetting the interior weight, I propofe that the fpaces between the corbell ftones be walled folid, and upon the ribband ftones which connect the tails of the corbells; fo that of mafonry and earth there will be more than double the counterpoife. The corbells are each to have a bolt to hold them down firm upon the ribband ftones; thefe are feen in the plan, Fig. 6; they muft have T heads in order more effectually to engage the ribband ftones; four inches in the T will be fufficient.

Having carefully confidered the fituation of the Town's Mills, I am of opinion that of the two mills within the city, now appropriated to the grinding of malt, if the machinery were made new, one of the mills may be made to do the bufinefs that is now done by both. This being the cafe, there will be one of the mills within the city that may be ap-

plied

plied to the grinding of corn, that is, to the fhelling of oats and grinding of oats and barley. I apprehend the malt mill may be altered for about £80. and the other to a corn mill for about £150.

On confidering the fituation of the Juftice Mill, I apprehend it will be beft to be fitted up for the purpofe of a wheat mill, for which it has been defigned. I judge, that when the full advantage of the fituation is taken, and the machinery properly adapted, the mill may be made to grind and drefs full four Winchefter bufhels of wheat per hour, which, when there is water in winter and wet feafons, may be continued for the whole 24 hours, and in dry feafons fuch as it was when I was there, about 12 hours in 24; fo that in the fhort water feafons, it will be able to grind and drefs at the rate of 12 bowls (fuppofing four Winchefter bufhels each) per day. The expenfe of erecting fuch a mill originally would be about £500; but as I expect the mill-ftones will do again, with the fmall machinery, thefe, with the larger materials and the buildings, will probably ftand inftead of £200. fo that to rebuild the mill and make it fit for bufinefs, will require a further outlay of £300. As I had not a fufficiency of time upon the place to take the neceffary meafures, I can only guefs at the quantity of bufinefs to be done, and the expenfe that may attend each of the mills; but in cafe the above idea appears eligible to be purfued, by receiving anfwers on the following points, I fhall be enabled more exactly to calculate what they will do, and form a proper defign for their rebuilding.

1ft. A plan and elevation of each of the mill buildings as they now ftand.

2d. The whole perpendicular fall of the water from the level trough, lander, or fhute, before it begins to fall towards the wheel to the bottom of the race at the tail of the wheel, noting upon the elevation where the horizontal lines cut the elevations of the refpective buildings, the fketch in Fig. 7. Plate 3. which is merely ideal, will be fufficient to fhow better than many words, what I want to have afcertained.

3d. Whether the water cannot without difficulty be brought to the place of its fall upon a ftill higher level, and whether the tail of the mill cannot be funk deeper, fo as to increafe the fall, an advantage in either of thefe particulars of half a foot is worth noticeing. Upon this head I muft obferve, I expect the lower mill within the town may have its tail race confiderably lowered; for that reafon I fhould be glad to know how much its bottom is above the ordinary high water mark, not only of fpring tides, but of the neap tides alfo. In refpect to the Juftice Mill, my propofition is to conduct the water to it, not from the tail of the upper mill, but upon the level of the head of the upper mill, as far as the height of the adjacent ground will admit, and then to take it into lander troughs to convey it over the declining ground into the wheat miln: and that there may

be

be no level unneceffarily loft in conducting the water, I fhould be glad to know the whole fall from the lead, after it has paffed the road next the mill-pond, to the tail of the prefent wheel, and from thence as far as the town's lands ftand, or they have power to let it fall, noting the refpective diftances, as alfo how far the water can be conducted on the furface of the ground in the lead from the mill-pond to the mill, and how far it will be required to be carried in a trough. According to this propofition, the upper mill may be left ftanding as it now is, to perform the fervice it is obliged to do: but as the water will then pafs by the wheat mill, without turning it, if this fervice will too greatly encroach upon the wheat mill, it will be the moft compleat way of all to fix up the mill ftones of the upper mill, to be turned by the wheat mill wheel, as then the upper mill fervice will be difpatched in one-fourth of the time; but then this will occafion fome addition of expenfe in new conftructing the wheat mill; yet the wheat mill may be fo conftructed as to admit of this addition afterwards, in cafe it fhall be found neceffary.

4thly and laftly. An experiment fhould be tried upon each ftream of water in the following manner :—Let a board of three, four, or five feet long have a notch cut in it as in Fig. 8. of one foot wide, and about fix inches deep, which fix in any convenient part of each of the watercourfes either above or below the mills, but fo that no water may go through it but what is employed in turning the mill referred to; this board is to be fixed in the manner of a dam, and to be made up with earth, clay, turf, fods or feal, fo that all the water turning the mill may be conftrained to go through this notch, and not to be interrupted by any water below it. The mill, in cafe it is not a time of fhort water, is to have its water paffing the notch fhortened, till it is as near as poffible, in the judgment of the miller, the fame quantity as they let down to work with in fhort water times; and the miller is to fpecify how many hours in 24 they can generally grind per day in dry feafons at that rate. While the water is thus going through the notch, it is to be noted, how much the dead furface of the water lying againft the board on the upftream fide of it is below the top of the board, and to be meafured at the diftance of at leaft fix inches on one fide the notch; this being done on each fide of the notch at the fame diftance therefrom, and if any difference taking the mean, this fubtracted from the whole depth of the notch, gives the thicknefs of the water flowing through it :—From this experiment I fhall be enabled to calculate exactly how much bufinefs can be difpatched by each mill.

Upon the whole, as ftreams of water run fcarce in the neighbourhood of Aberdeen, it feems of confequence to make the beft ufe of thofe fo advantageoufly fituated.

*Aufthorpe,*
19th *February* 1770,

J. SMEATON.

## DUNDEE HARBOUR.

(See the Plan, Plate 4. Fig. 1.)

The REPORT of John Smeaton, Engineer, upon the Harbour of Dundee.

IT is a very difficult thing to make an artificial harbour to be in all refpects complete, for the very means that tend to render it fafe and quiet for veffels, tend to make it lefs eafy of accefs, and more fubject to mud, filt, or fand, fuch as the coaft happens to be annoyed with. The harbour of Dundee by being too much inclofed is too fubject to mud, and in fuch a degree as greatly to diminifh the fpace inclofed, efpecially for larger veffels. I ob-ferve that the places where the mud chiefly lies are at A. and B. which are feveral feet higher than the entries and fpaces near the runs of the fluices, particularly that at B. The prefent means of clearing it are by drawing a complete quantity of water from one or other of the fluices upon the bafon at C. or D. and by a number of men fet to confine the current into a channel at pleafure, and throwing in the mud, whereby it is carried without the entries, and thence taken away by the run of the tides. This operation it is faid anfwers pretty well as to clearing the harbour, but it is attended with confiderable expenfe, and requires to be too often repeated. Notwithftanding what affiftance can be given by running the fluices, when unaffifted by men at the bottom, it is faid that before the pier E. was projected fo far, and before the Ground was made up at F. between the little bay G. and the harbour, that it was then lefs fubject to mud, particularly at B. and indeed this very well correfponds with the reafon of the thing, for I obferved when there, the 4th Auguft 1769, being the third day after new moon, and confequently at the height of fpring tides, that on tide of ebb a very ftrong current fet paft the end of the pier E. and was diverted thereby from the harbour's mouth; of confequence were there proper openings made in the neck at F. though they cannot be expected to clear away the mud at B. already depofited, yet being once done by other ways, this will be the moft likely means of keep-ing it clear ; I would therefore recommend that this neck be pierced by two fets of tunnels, in the directions fhewn at F. three and three together of 12 feet wide each. They may be arched over fo as to make good the platform at top, or by way of faving arching, may be fhortened according to the dotted line at F. Thofe tunnels fhould be made with a clear

passage

PLAN of the HARBOUR of DUNDEE with the alterations for improving it by J. Smeaton.

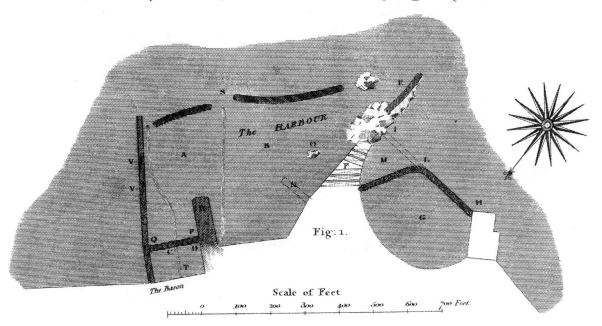

Fig: 1.

The Harbour

The Bason

Scale of Feet

0    100    200    300    400    500    600    700 Feet.

PLAN of the HARBOUR of DUNBAR, by John Johnson with a design for amending it, by J. Smeaton.

page 57.

Fig: 4. Face of the Pier

Fig: 3. Section

Low Water line

Scale for the Pier

50    40    30    20    10    0

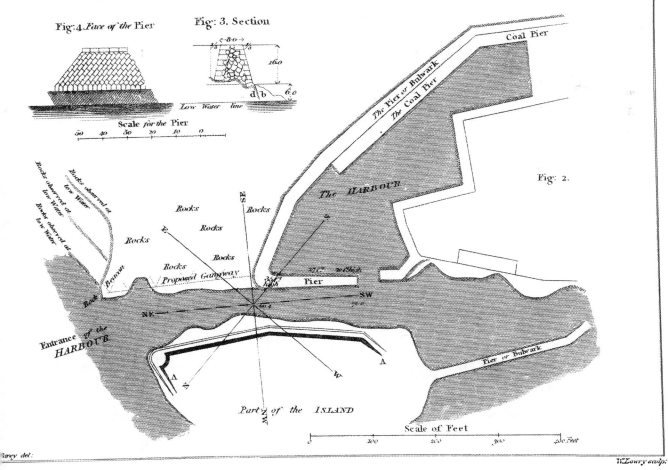

The Pier or Bulwark

The Coal Pier

Coal Pier

The HARBOUR

Fig: 2.

Rocks

Rocks

Rocks

Rocks

Rocks

Rocks

Rocks observed at low Water

Rocks observed at low Water

Rocks observed at low Water

Proposed Gangway

Pier

Entrance of the HARBOUR

Pier or Bulwark

Part of the ISLAND

Scale of Feet

0    100    200    300    400 Feet

Parey del:

W. Lowry sculp:

Published as the Act directs, 1812, by Longman, Hurst, Rees, Orme and Brown, Paternoster Row, London.

paffage as low as they can be made for the rocks, or to low water mark. The crofs wall H I. inftead of being carried from H. to I. muft be turned round at L. fo as to leave open the fpace M. of its natural depth. That the tunnels may be freely fupplied, the prefent little pier fhewn by dotted lines at N. fhould be totally removed, and the rock at O. fhould be levelled with the reft furrounding it, and the whole fo reduced as to give all freedom of water-way poffible through the tunnels, and in this ftate of things, if the pier E. had occafion to be carried out ftill further, it would bring the tide more forcibly through the tunnels.

The means above-mentioned, by promoting a more free circulation of water through the harbour, would tend to keep down the bank of mud at A. ; but I cannot think it would effectually keep it clear ; but I apprehend when this is once done by the help of men, as before, by a more judicious difpofition and application of fluices upon the prefent bafon, the whole harbour may be kept in a great meafure free. In the firft place, I apprehend the water-ways of the fluices are much too fmall, for though the bafon itfelf is much fmaller than were to be wifhed, yet the letting the water go off in a fmall body though continued a longer time is not making the beft ufe of it : a certain power of water will move that very expeditioufly, which applied in a lefs degree will let the fubject remain at reft. Confequently, the water being difcharged from the bafon in a large body, will do that in five minutes, which would remain undone, if the capacity of the opening were reduced fo as to play a quarter of an hour ; for this reafon I would advife the fluices to be full three feet fquare in the orifice, and to be laid rather lower than the bottom of the bafon. The fluice D. may ftill remain the place for one of them, but inftead of the fluice C. I would advife two upon that line at P. and Q. of the dimenfions aforefaid : That at P. will more directly tend to fcour the face of the pier R. than in its prefent pofition C. and alfo to keep down the growth of the mud at A. and more efpecially if a paffage be firft opened for it, to vent its water by the opening S. ; but the fluice P. being fo much nearer the pier R. it will be fo much farther from the eaft pier ; to keep the face whereof clear, it will be proper to have the other fluice mentioned at Q. which will alfo ferve a fecondary ufe ; for as the bafon itfelf is liable to fill with mud, which I fuppofe it is a confiderable expenfe to remove, this may be prevented by building a crofs wall according to the dotted lines, with a fluice upon it at T. of equal fize or rather larger than the others, conftructed fo as to pen the water either way ; this fluice may therefore be made ufe of by penning the water in one half of the bafon to fcour the other, and fo alternately ; and when open fuffers the whole bafon to act by any of the interior fluices ; to avoid lofs of water, it will be proper to have all the fluices, particularly thofe at Q. P. and D. executed in the moft fubftantial manner. I

cannot

cannot advife the doors to be cut, in order to make a valve to let the water in, but rather to fet one of them open by hand ; and to prevent accidents by neglect, a fmall tunnel with a valve in part to let in and effectually to keep in the water, may be fixed in the prefent opening at C. This I think will be the moft effectual way of applying the water of the bafon without the ufe of men, and I think it would alfo add, if to the opening in the eaft pier, which if I remember right is there already, there were another added, not lefs than 25 feet wide, fo that they may be nearly in the pofition V V. By thefe means, I expect the harbour may, when once cleared, be kept clear, or fo nearly fo, that with a little help from men occafionally, to direct the water to particular places, it may be effected in the moft eafy manner that the fituation will admit of.

J. SMEATON.

N. B. To prevent lofs of room in the bafon by the crofs wall, it may be a ftrong ftone wall, well built with lime, and aiflered on the two outfides.

## DUNBAR HARBOUR.

(See the Plan, Plate 4. Fig. 2.)

The REPORT of JOHN SMEATON, Engineer, upon the Harbour of Dunbar.

THE harbour of Dunbar appears from its form to be well adapted for the fecurity of veffels lying therein, and as it is fituated in the bottom of a bay, exclufive of the trade that is ufually carried on there, it becomes of confequence to the fecurity of veffels trading on this coaft; and its principal deficiency in both thefe views is the want of a good accefs, particularly in going into it; and this is not owing to an improper courfe or di‑rection of the entry from the bay, but from the narrownefs of the paffage, which is rendered in effect ftill narrower from the floping form of the rocks on the ftarboard or N. W. fide going in, to avoid which, veffels are obliged in prudence to keep nearer to the pier than they otherwife would do, and by this means, by the recoil of the fea from the rocks, are often driven againft the pier, or upon the floping rocks that are funken and hid. To remedy this inconvenience as much as poffible, it has been very properly propofed to cut off a certain part of the flope of the rocks, down even with the bottom of the reft of the paffage, and to build the face up above high water mark, (as fhewn at A A. in the plan), by which means, as the paffage will not only be actually widened full five yards in the nar‑roweft place, but by giving opportunity to veffels to fee every thing that can hurt them, they may fafely keep feveral yards further from the pier than they could otherwife have done; the paffage thereby in the narroweft part being rendered from about 45 to 60 feet wide; at the fame time the fea being prevented from breaking upon the floping rocks, t will recoil with lefs force than at prefent, and the veffels will be lefs fubject to be carried from fide to fide for want of fufficient way through the water. Laftly, this new pier or facing to the fide of the entry being carried up fufficiently above high water mark, it will enable affiftance to be given by throwing a rope on board from the moft projecting point thereof, fo as to tow in veffels when they do not come in with fufficient frefh way, to keep them clear of either fide of the entry.

With the fame view it would be well if a pier could alfo be carried out upon a ledge of rocks (which are dry at low water) on the S. E. fide of the entry from the north angle of the prefent pier, to the Beacon Rock. This would not only defend the paffage from the furge of the fea, but by its projection alfo enable affiftance to be given by throwing a rope on board on the larboard fide of the fhip, of which there is always an equal chance of its being preferable, by its being neareft; and ftill more effectual when it can be done on both fides. To this however may reafonably be objected the great expenfe that muft neceffarily attend the execution of a pier of fufficient bulk and ftrength to ftand in this place the full ftroke of the fea: however, to take advantage of that part of the utility of a S. E. pier, which confifts in giving affiftance by a rope, I have contrived a gang-way to extend itfelf in the fame direction as far as the Beacon Rock, as fhewn by the dotted lines in the plan, whereby any competent number of men will be able to give the fame affiftance by heaving a rope on board, as could be done from a ftone pier, and which is contrived on principles fo fimple that it may be executed at a very moderate expenfe. This will be fufficiently explained by a defign on purpofe, (fee Plate 5.) fhewing how the fame is to be executed, and indeed, had I not a good deal of experience in the of fixing temporary utenfils in the building of the Eddyftone Lighthoufe, I fhould have been very dubious in propofing a ftructure feemingly fo flight to ftand the violence of the fea. But there I learnt, that where the force of the fea is to be rather eluded than refifted, the lefs matter is oppofed to its action the better, provided that this be but fixed in the firmeft manner. I believe the whole of the defign will occur from the plan, fave the manner of fixing the iron bolts into the rocks, which are not propofed to be done with lead, as that is perpetually working loofe, but in the following manner:—Round holes about 18 inches into the rock more or lefs, according to the firmnefs thereof, are to be bored with a jumper of $1\frac{1}{2}$ inch diameter; to thefe the eye-bolts are to be forged a very fmall matter taper, and larger fo as to drive tight to their proper depth, with an iron maul; if they happen to be a fmall matter too fmall, then ftrips of plate iron put in along with them will make them to drive firm, and the ruft will fix them from drawing.

Refpecting the propofed pier or wharf (A A in the plan), as it will receive the full ftroke of the fea, with a S. eafterly wind, it will be neceffary to be built very firm, and I believe it will not fully anfwer the end if the ordinary feas break over it; I therefore propofe it to be raifed nine feet above high water at fpring tides, that is 22 feet above the low water line at the pier head, but it may decline in height as it runs S. weftwards towards the land. This height will carry it confiderably above the rocks where it is built, and therefore inftead of fo large a quantity of backing as will be neceffary to make the whole good to that level with the land, I have propofed it to be built upon the back fide as per fection, Fig. 3. The rocks b. at the foot of the pier, are to be taken away as above fhewn, before

the

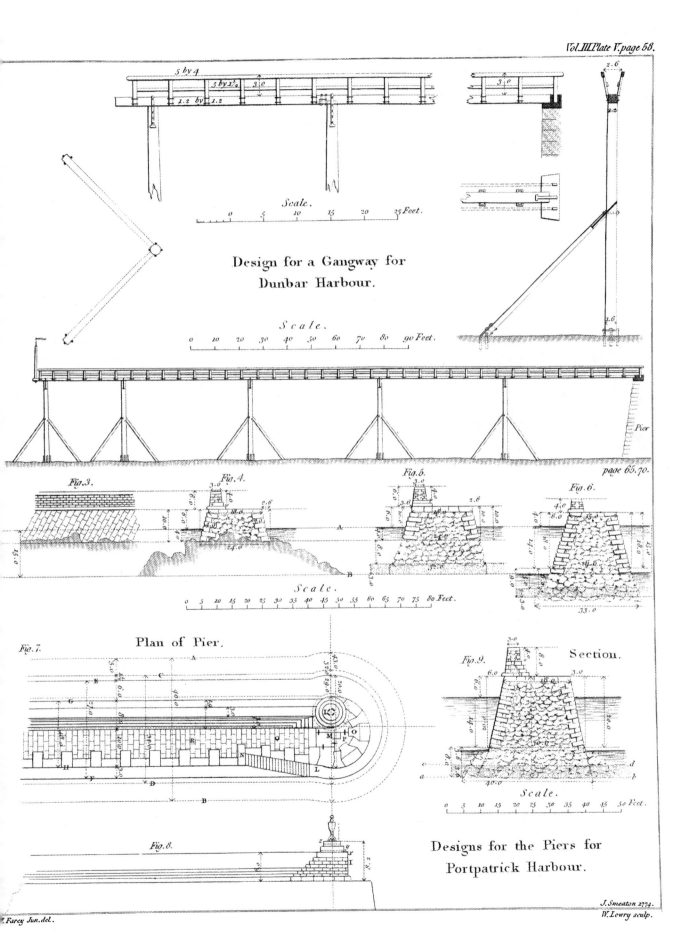

Design for a Gangway for
Dunbar Harbour.

Scale.

Plan of Pier.

Section.

Designs for the Piers for
Portpatrick Harbour.

J. Smeaton 1774.

Farey Jun. del.

W. Lowry sculp.

Published as the Act directs, 1812, by Longman, Hurst, Rees, Orme and Brown, Paternoster Row, London.

the pier is built; and when it is finished, they are to be cut off smooth according to the line a.

The bare stones of the pier (see the face view, Fig. 4.) are not proposed to be all on the same level, as there shewn, but the rock cut to give the stone its proper bearing.

The face (Fig. 4.) if you can procure the free-stone at a moderate price, will be best done according to the specimen; but if you work with rough stones, the more nearly they are laid conformable to the same idea, the better; and the platform at top should be laid on with the best mortar. As I cannot well judge of the prices at which the materials can be procured, and the work executed at Dunbar, I content myself with making an estimate of quantities as follows, which may be fitted up after proper workmen &c. are consulted.

## ESTIMATES.

**The Pier.**                                                                    £.   s.   d.

To cutting the rock according to the line in the plan 786 cube yards at -
To building in the whole pier 2,770 cube yards at                -        -        -
To free-stone in the face and platform, reckoned at a medium 2½ feet
 in thickness, reckoned extra in its value in the solid of the pier
 26,250 cube feet at          -          -          -          -          -
To posts of wood or stone (if thought necessary), as also for lime &c. -
To contingent expenses at 10 per cent on the above articles        -      -

    Total of the pier                £ _____


**The Gangway.**

To fir timber 942 say measured neat in place 1000 cube feet at          -
To iron work 19¼ cwt. say one ton at        -          -          -
To a large stone for fixing the end of the gangway to the parapet of the
 present pier, 30 cube feet at          -          -          -
Contingencies on the above at 10 per cent.                -      -

    Total of the gangway        £ _____


*Austhorpe,*
25th June 1772.                                    J. SMEATON.

## PORT PATRICK HARBOUR.

The REPORT of JOHN SMEATON, Engineer, upon the Harbour of Port Patrick, in the Shire of Galloway, with a Projection of Piers for rendering the same safe and commodious for Vessels of eight feet draft of water.

THE harbour of Port Patrick is generally esteemed the nearest port in Great Britain to Ireland, being, as it is said, but about seven leagues from the harbour of Donaghadee, upon the coast of Ireland, and which is almost right opposite : on account therefore of the shortness of the passage, it becomes a very desirable object that the harbour of Port Patrick should be rendered safe and commodious for such vessels as are best adapted to carry the pacquets, passengers, carriages, horses, cattle and goods between the two kingdoms; and on this head I am advised, that nothing less than vessels of 40 tons, drawing eight feet of water, and constructed upon proper principles for sailing, will be fully adequate to this purpose.

The harbour of Port Patrick is at present entirely in a state of nature, a small platform for the more commodious landing and shipping of passengers, &c. excepted ; and indeed it has many natural advantages, being very easy of access, and of sufficient depth to ride the vessels proper to be employed afloat at low water, and to protect them from storms coming from seven-eighths of the whole compass ; and had the remaining eighth been as well guarded as the rest, this harbour had been complete ; but the want of protection from those points, necessarily obliges the vessels employed to be of such a construction as unfits them for the general purposes. Those defects it is the business of art to remedy, which is the object of the present proposition.

This harbour is formed by two ledges of rocks running out almost parallel from the shore, so as to form between them a small bay of about 220 feet clear width, and about 550 feet in depth, that is in and out. The bottom is a clean sand, and the soundings gradually increase from the shore to 20 feet at low water in the mouth of the bay, and leaving from 9 to 10 feet at dead low water mark in middle of the harbour.

The

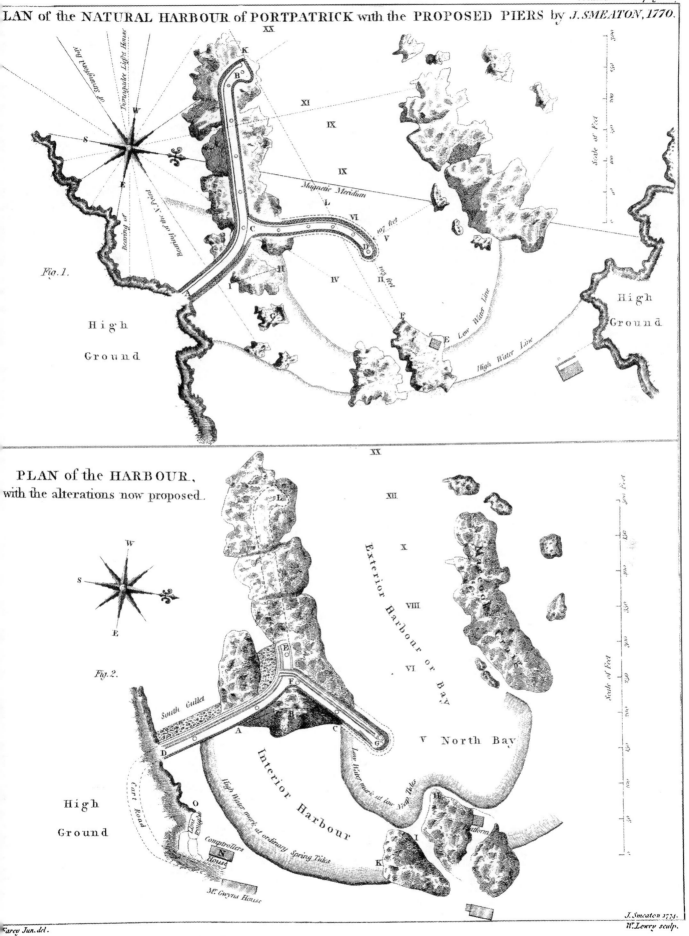

LAN of the NATURAL HARBOUR of PORTPATRICK with the PROPOSED PIERS by J. SMEATON, 1770.

Fig.1.

High Ground

High Ground

PLAN of the HARBOUR,
with the alterations now proposed.

Fig.2.

Exterior Harbour or Bay

North Bay

South Gullet

Interior Harbour

High Ground

Cart Road

Comptrollers House

Mr. Gwyns House

J. Smeaton 1774.

Farcy Jun. del.

W. Lowry sculp.

Published as the Act directs, 1812, by Longman, Hurst, Rees, Orme and Brown, Paternoster Row, London.

The coaft of Ireland lies right in front, extending from S. W. to N. W.,* and being fo near as not to admit of any confiderable fwell with the wind right in, it becomes naturally defended from the weftern quarter. From the N. W. & N. points, the fetch of the fea is not of great lengths, being in a manner land-locked by the Ila, Mull of Cantire, &c. and being laftly very well fkreened by the ledge of rocks immediately on the north fide of the harbour, which rife confiderably above high water, no confiderable violence is ex-perienced on that fide. The land lies from N. to S. fo that nothing can happen from the eaftern point; it is therefore only from S. to S. W. inclufive, that the harbour lies unprotected.

The ledge of rocks on the fouth fide of the harbour, which run out in a direction W. by S. and nearly point toward the lighthoufe of Donaghadee, would from their pofition afford a confiderable fhelter in all thofe winds, if they were higher ; but the Irifh fea being open from thefe points, and the rocks being in a great meafure covered at high water neap tides, and at three-fourth flood at fpring tides, the feas break over them with fo much violence in time of ftorms, that veffels lying there are obliged to be fuffered to beat up upon the fandy beach, at the bottom of the bay and to be retained by ropes as the only means of protection ; this in confequence not only creates a great deal of trouble in hauling them up, and launching them after the ftorms are over, but obliges them to be made fo ftrong and of fo flat a conftruction, that they will not fail except with wind on the beam or abaft. As therefore all the wefterly winds prevent their failing from Port Patrick to Donaghadee, and all the eafterly from returning, it follows that they cannot regularly go and return except the winds are foutherly or northerly ; nor indeed could they go very well with thofe, were it not for the very ftrong current of the tides which fet up and down this narrow channel, between the two kingdoms, twice each way in a day ; fo that by failing at a proper time of tide, they are prevented by the current from falling to leeward ; but could veffels conftructed upon proper principles for failing be protected here from all points, they would be enabled to turn to windward, and confequently make their paffage good in all winds when the weather is moderate, from the great advantage that arifes from the particular fet of the tides.

Now, as this harbour in its prefent ftate would be very convenient for the purpofe, being open and deep, and eafy of accefs and departure, were it not on account of the effects of violent winds from one-eighth part of the compafs only, as before mentioned, the place

* All the bearings in this report are according to the compafs, the north point of which is fuppofed to vary about 24 degrees to the weft.

itfelf

itfelf being fmall, it muft be a principal care, that while we are providing againſt the above mentioned difficulties we do not deſtroy its natural advantages. With this view I propoſe to run out a pier from the point of the rocks upon the main land at A. Fig. 1. Plate 6. and croſſing the gully between that point and the detached ledge of rocks hitherto called the South Ledge, to follow the general direction of the fame, and to terminate the pier at B. as particularly ſhewn in the plan. This pier to be raiſed ſix feet above the high water of a ſpring tide, with a parapet of ſix feet upon that; fo that the whole being raiſed 12 feet above high water, will effectually ſcreen the veſſels from the S. & S. weſterly winds, and greatly mitigate thoſe nearer the weſt; and fo far as this goes, the benefit to be obtained will be attended with no diſadvantages in other reſpects, that I can foreſee; and it is poſſible this being firſt executed, the internal pier may be found unneceſſary; but as the north point of Strangford Bay on the coaſt of Ireland is the ſouthernmoſt bearing from Port Patrick, viz. S. W. $\frac{1}{2}$ W. diſtant about nine leagues by the maps; when the wind s a little further out than this, viz. S. W. I am of opinion that fo much fea will get round the head as to reduce the ſpace naturally ſcreened thereby into fo narrow a compaſs, that veſſels properly built for ſailing will not ſafely ride in the harbour with the wind in ſuch poſition, as may be judged of by conſidering the poſition of the line upon the plan, marking out the bearing of the north point of Strangford Bay. It ſeems therefore neceſſary to provide a place of ſafety at ſuch times, which I propoſe to effect by means of the interior pier C. D. the poſition of which I have endeavoured fo to adapt, as to be ſufficiently capacious and deep, and without taking up any part of the external harbour, where I apprehend the veſſels uſually ride. This would have certainly been the caſe, had this internal pier been carried further out, fo as to have been capable of admitting veſſels drawing eight feet at low water; as it is, all the ſpace is preſerved where they cannot in ordinary float at low water, and veſſels will be capable of going in and out of the external harbour at all times, except from three-quarter ebb to one-quarter flood; and during this ſpace of low tide, the ſeas will be fo mitigated by the projection of the rocks upon the neighbouring coaſts, that a veſſel may ſafely ride in the external harbour, till the water flows ſufficiently to let her go into the internal. It is true, that as the veſſels in the internal harbour will be obliged to take the ground, they will be required to be ſtrong timbered; but as they are not intended principally to carry weight, that will not hinder their external form from being adapted in the beſt manner for ſailing, it being the caſe with few harbours the moſt reſorted to in Great Britain, that veſſels can ride afloat at low water.

The flow of the tides at Port Patrick is ſaid to be 15 feet at ſpring tides, and 12 feet at neap tides, but like other places variable according to the winds. When I was there, which

which was on the 11th of October 1768, being the day after the new moon, the perpendicular flow was only 10 feet, so that at a 15 feet spring tide the water would ebb out 2¼ feet lower, as appeared by a mark pointed out to me, and flow 2¼ feet higher; according to this state of the tides the internal pier will be placed at 4¼ and 5 feet water, at low water spring tides, and will have 19¼ and 20 feet at high water, and at such a tide as I saw (which was less than an ordinary neap tide), 7 and 7½ at low water, and 17 and 17½ at high water; so that it is evident vessels will be always afloat within the internal harbour, whenever it is a proper time of tide to go out, there being always 12 feet water at half tide.

This internal pier being at all times greatly sheltered from the violence of the seas by the external one, it is proposed to be considerably thinner on that account, and its body not raised so high by two feet, so that its top will be lower in the whole by four feet; that is, it will be only eight feet above high water spring tides; yet this it is presumed will always be fully sufficient to resist any seas that can get round the head of the external pier.

Since I have proposed the internal pier as a place of refuge, it may be reasonably enquired why this may not do without the external one? To this I must observe, that it is only so much of the external pier as projects beyond the internal one, that is the object in question; and this will be built with so much advantage to a proper height by being upon the rocks, that if omitted the internal pier will be required of greater base, height, and strength, and thereby the cost of the external one will in a great measure be incurred.

2dly. The vessels will be protected in the external harbour with wind at S. and S. S. W. and therefore can come in and find shelter, at a time when the tide would not serve to go into the internal harbour, and also will so far moderate the seas at S. W. as to enable the vessels to wait till they can get into the internal; and in all southerly winds will enable vessels to get under way going out, and to come to on entering, when they could not venture to do either from or to the entry of the interior harbour.

3dly. If the seas were not first moderated by the external pier, so much sea would get round the internal as not to admit vessels to lie there with sufficient quiet, unless it were so far contracted in the entry as to make it difficult of access and departure.

Within

Within the internal pier the veffels may occafionally take in their freight; but as I apprehend the place of the prefent platform will be found in general the moft commodious, efpecially about half tide, I propofe that platform to be extended, and the face of the rock to be extended, by blowing off the irregularities according to the line E F. and if the rock G. and part of the rock H I. to that line, be blown up to the level of the fand, it will be a means of preventing damage to veffels within, and of entertaining a greater number of veffels, for when this is done, the interior harbour will be capable of holding, on occafion, 10 or 15 fail, from 40 tons downwards.

Refpecting the conftruction of the piers, as the coaft is all rocky, I propofe to make all the ufe that properly can be done of the ftone that the place affords, but it happens to be of a kind not very well adapted for this fort of work, except for fillings and rough bafements of fuch parts of the piers as are not built upon the rocks, and to thefe ufes I propofe to confine it. The ftone compofing the two ledges of rocks on each fide of the harbour is of a very hard but very jointy nature, which feems a good deal the cafe with the ftone I fee thereabouts, fo that it can hardly be got without gunpowder, which would be apt to fhiver it into fmall pieces. I obferved that a confiderable quantity of large loofe pieces that have tumbled from the cliffs may be collected, enough to form the bafements of the piers as marked out in the fections, Plate 5.; but as it does not appear that a fufficient quantity could be collected to do the whole with rough materials, which would, if fo conftructed be required to be ftill more bulky, I propofe to face the upper part of the piers with quarried ftones, fcapelled to fomewhat of a fquare form, to be difpofed as fhewn in the fections, and to be filled with rough materials to be obtained from the rocks in the neighbourhood. I did not indeed fee any ftone in that part of the country fit for facings of the kind I propofe; but as this part lies open by fea to a confiderable extent of coaft of England, Scotland, and Ireland, it is not to be doubted but that convenient quarries of proper ftone will be found that can be brought by water, and for this purpofe I have made a confideration in my eftimate of 2s. 6d. per ton, that will be about 2d. per cube foot, which allowance I judge fully fufficient.

It is further to be obferved, that I propofe all the rough bafes and fillings to be compofed of as large ftones as can eafily be procured near the place, and no ftones of two tons or under be broken on account of removal; and that the fillings be at a medium, about a cube foot in a piece, the fmaller being difpofed in the interftices. The whole to be built dry, except what is particularly fpecified in the eftimate to be done with cement or mortar.

It

It has been obferved to me that a confiderable rivulet can be turned into this harbour; this may be of great ufe in cafe the interior harbour fhall be found inclined, from a greater degree of confinement, to retain more fand than it does at prefent, and will prove an effectual remedy to every evil that can arife from that circumftance. At prefent the harbour does not retain more fand than one would wifh it to do, for the fafety and eafe of veffels.

*Authorpe,*
18th *May* 1770.

J. SMEATON.

---

## EXPLANATION of the Plan of Port Patrick Harbour.

Fig. 1. (Plate 6.) Plan of the natural Harbour, with the projection of the propofed piers :—The Roman figures exprefs the depth in feet of the low water of a fpring tide, the perpendicular rife of which is faid to be 15 feet and the neap tides 12, confequently at neap tides low water, there will be 1½ feet more water than the foundings exprefs. The bottom is clean fand.

AB  The exterior Pier.

CD  The interior Pier.

EF  The line in which the face of that rock is propofed by blowing to be ftraightened, and to have timbers bolted to the face of it, to make the landing place there more commodious; the platform may alfo be extended to *e. f.*

G  A rock to be blown to the level of the fands; and

HI  Another to be blown to this line, in order to clear the interior Harbour.

The line KL fhews the bearing of the north part of Strangford Bay, upon the coaft of Ireland, being according to the map 9 leagues diftant.

Fig. 3. (Plate 5.) Shews the outward face of the Pier on the fouth rocks; and

Fig. 4. Do.  A fection of the fame, fhewing the manner of building it with quarried ftone, and filling up with rough ftones.

Fig. 5. Do.  Section of the Pier acrofs the gully, between the fouth rocks and the fhore.

Fig. 6. Do.  Section of the interior Pier :—All thefe figures are placed to fhew their relative levels to high and low water lines (A & B) at fpring tides.

ESTIMATE for building the Piers of the Harbour of Port Patrick, according to the defigns of John Smeaton Engineer.

## The EXTERIOR PIER.

|  | | £ | s. | d. |
|---|---|---|---|---|
| To form a bafement of rough ftones, to be collected from the adjacent coaft and cliffs in order to make up the gully between the main land and the ledge of rocks on the fouth fide of the Harbour, to be 32 feet mean bafe, 24 feet top, and 8 feet mean height, length 120 feet, will contain 1000 cube yards, which at 7s. 6d. per yard - - - - - | | 375 | 0 | 0 |
| The upper part of the Pier being fuppofed a medium 10 feet high, extending from the land 400 feet; the mean thicknefs of the facing on the outfide of the Pier being 3 feet 3 inches, and the infide 2 feet 9 inches, will contain cube feet - - - | 24,000 | | | |
| To make good the platform at top, being the fame length, 12 feet broad and 1 foot thick - - - | 4,800 | | | |
| Cube feet of ftone at 10d. per foot complete - - | 28,000 | 1,200 | 0 | 0 |
| To 2,044 cube yards of rough ftone in filling, at 5s. - - | | 511 | 0 | 0 |
| The circumference of the Pier head for 40 feet in length next the head, will be 112 feet, this at the mean thicknefs of 3 feet 6 inches, and 10 feet high, will produce cube feet - - - | 3,920 | | | |
| And to make good the platform to the above - - | 1,120 | | | |
| Cube feet of ftone at 10d. per foot - - | 5,040 | 210 | 0 | 0 |
| To rough ftone in filling 40 feet length, next the head, 332 yards at 5s. - | | 83 | 0 | 0 |
| The length of the front wall of the head being 40 feet, 15 feet in thicknefs, and 10 feet high, contains 222 cube yards; this being laid in cement, will, with extra labour and materials, at 7 fhillings, come to - - - | | 77 | 14 | 0 |
| To cutting footings for the bafes of the outfide cafings to reft on, 1,540 feet fuperficial, at 1 fhilling · - - - | | 77 | 0 | 0 |
| To 440 feet of parapet in mortar, at £1. per foot - - - | | 440 | 0 | 0 |
| | | £ 2,973 | 14 | 0 |

The

## The INTERIOR PIER.

|  | £ | s. | d. |
|---|---|---|---|

To the forming a bafement without the rocks of rough ftones, 33 feet bafe, 25 feet top, and 9 feet high, fuppofed to reach one foot above low water, and to fettle 3 feet into the fand; this, in a length of 180 feet will contain 1,740 yards, at 7s. 6d. - - - - - 652 10 0

The height of the cafing being 18 feet mean thicknefs, of the outfide 2 feet 9 inches and 1 foot 9 inches, infide 2 feet 4½ inches; this, in a length of 180 feet, will contain cube feet - - - 16,605

The height of the cafing over the rocks to join the former to the main pier, being 8 feet, will at the above thickneffes and 70 feet long contain - - - - - 2,870

To make good the platform in the whole length - - 2,500

21,975    915 12 6

To rough fillings in that part of the Pier without the rocks, cube yards   1,658

To Do. in the part over the rocks - - - 220

Cube yards of rough filling, at 5s. -   1,878   469 10 0

To cutting the footings upon the rock, 280 feet, at 1s. - - 14 0 0

To walling 5 yards, running next the head, in cement, containing 200 cube yards, extra labour and cement, at 7s. - - - - 70 0 0

To 240 feet of running parapet in mortar, at 15s. - - - 180 0 0

The Interior Pier - - £ 2,301 12 6
The Exterior Pier - - 2,973 14 0

Neat computation - - £ 5,275 6 6

To contingent expenfes, in iron work, fender piles, ftairs, blowing of rocks, machines, &c. being 10 per cent. upon the above - - 527 10 0

£ 5,802 16 6

*Authorpe,*
*18th May 1770.*

J. SMEATON.

     EXPLANATION

## EXPLANATION of the Plan for completing the Interior Harbour at Port Patrick.

THE work being now advanced fufficiently to the weft to break off from that part intended for the interior Harbour, the great feas that roll in with the foutherly and fouth weft winds  what is principally wanted to fecure the veffels lying under fhelter of the prefent work, is a flank Pier to prevent the feas from returning laterally after they have paffed the head, and then following the back fide of the Pier and ledge of rocks into the place of the interior Harbour.

For this purpofe, and by reafon that it is found expedient to complete the interior Harbour before the exterior Pier is carried further out, I have projected the Pier defigned in the prefent Plan in fuch a way, as that it may in the beft manner anfwer the end, on fuppofition that it may not be found expedient to carry out the external Pier further than it extends at prefent.

I have therefore defigned it in fuch a pofition that veffels, when coming from fea, may run into the interior Harbour under fail, leaving the bottom of the little Bay Road Stead or exterior Harbour as difencumbered as poffible, to be ufed in moderate weather, and when a greater degree of fhelter and protection is not neceffary for the fame purpofe as at prefent ; and at the fame time having the accefs to the interior Harbour as free as poffible, that time may not be loft in getting in and out of it.

The pofition of the propofed interior or flank pier, is nearly N. E. by the compafs, and extended from the main pier 175 feet, fo as to leave about 100 feet of opening between the Pier head and the neareft point of the platform rocks, which rocks will in effect anfwer the end of a counter pier. This pofition and length, by admitting of an opening of 100 feet, will enable veffels to get under fail in bad weather, that is, whenever they can venture in between the two ledges of rocks that compofe the exterior harbour, and will prevent the lateral return of the fouthweft feas into the interior harbour, and throw them into the bottom of the north Bay, fo as to fpend themfelves upon the beach. It is in conformity with this pofition of the Pier, and in order to afford room within the harbour, as well as to give the whole of the infide fuch a curve as may more effectually clear off the fands that may occafionally be brought in, that it is propofed to cut the rocks to the arch A B C. (Plate 6. Fig. 2.) It is alfo with this view that an opening of 100 feet is propofed to be left ; for though it will always hold true, that the lefs the mouh of a harbour is, the

quieter

quieter it will be proportionably within ; yet as the getting clear of fand, the getting fafely in during bad, and eafily out in moderate weather, are confiderations of the greateft confequence; and it being impoffible in fmall harbours to unite every advantage, it feems to me that the prefent plan will be likely upon the whole, to anfwer the end in the completeft manner poffible ; that is, for fheltering veffels from 30 to 40 tons and under, which it is fuppofed will be fully fufficient for the pacquet and paffage veffels attending at this place.

At prefent I underftand that the place of the interior harbour keeps itfelf as clear of fand as is neceffary, that the rocks may not become bare, and the veffels lie unfafely ; but when it becomes more inclofed by the erection of the flank pier, as it may not then clear itfelf of fands fo effectually, the bringing in of the burn occafionally in the way that has been already mentioned, by the garden of the collector of the cuftoms, I do not doubt will be effectual to keep it clear.

I have alfo endeavoured to make the pier fubfervient to fome purpofes, which, though fubordinate, will yet be found very convenient if not neceffary.

I have propofed to terminate the parapet by a round pillar of nine feet diameter, at the top of which may be planted a flag-ftaff in order to hoift a flag when there is a certain number of feet of water within the pier head, or by planting a light in a lanthorn or lamp fixed upon a proper iron ftanchion, veffels approaching in the dark will not only with more certainty make the bay, but turn the pier head in cafe they wifh to enter the harbour.

I have alfo defigned a ftair near the pier head infide, to which a boat can always go even at low water, and thereby paffengers be commodioufly conveyed on board or landed at all times of tide. Mooring pofts are alfo propofed and marked at proper intervals ; and as a poft near the extremity of the main pier may be very convenient for the faftening or fteadying veffels in the bay or exterior harbour, a paffage is purpofed to be made by an archway of two feet and a half wide, and four feet and a half high, at the joining of the flank pier's parapet to the prefent one.

The fpring tides are faid to flow here 15 feet, fo that there will never be lefs than eight feet water within the pier head at half tide.

N. B. The projection of the main pier weftward of the junction of the flank pier, will greatly contribute towards breaking off the feas from the harbour's mouth.

*Authorpe,*                                   J. SMEATON
6th *January* 1774.

P. S. It may be problematical till the flank pier is far advanced, whether it may be better to cut off the point of the rock marked I K. Fig. 2. by the dotted line or not. It will undoubtedly render the paffage in more perfectly clear if cut away ; but as it may tend to catch the fwell that will in fome degree roll in between the pier head G and the point of the rock H, this operation is left till the effect of the flank pier is feen.

---

### EXPLANATION of the Plan, Fig. 2. Plate 6.

A B C  The line to which the rocks are propofed to be cut away.

D E  The prefent pier as now terminated.

E L  The extenfion as at firft propofed.

F G  The interior or flank pier as now propofed.

G H  The opening between the pier head and the neareft point of the platform rocks, being about 100 feet.

I K  A line in which that point of the rock may be cut off, in cafe it fhall be found ufeful when the effect of the flank pier is feen.

      N. B.  The Roman figures fhew the foundings in feet at low water mark.

N  Comptroller's houfe.  O lime troughs.

Fig. 7. (Plate 5.) A plan of the intended pier F G in Fig. 2. Plate 6.

— 8. Elevation of the fame.

— 9. Section of the fame.

A B  Is the breadth of the pier upon the rough ground bafement 40 feet.

C D  The breadth of the rough ground bafement at about one foot above low water, 32 feet.

E F  The bafe of the pier at the bottom of the freeftone work, 27 feet.

G H  Shews the top of the cap or platform of the pier, 18 feet.

a b  Fig. 9. the fuppofed furface of clay or hard matter.

c d  Suppofed furface of the fand.

X Y Z  Fig. 8. The terminating pillar of the parapet capped with three circular ftones, the two upper ones are fuppofed to be each fingle ftones, the lower one of 2, 3 or 4 pieces, as moft convenient, the reft of the pillar to be built as the outfide of the parapet.

L N  Fig. 7. fhew the ftairs infide the pier head.

O P Q R  Shew the manner of forming the cap of the pier, being terminated with dovetailed ftones in the manner of the Edyftone, to bind the cap together.

M  A poft or pillar for warping veffels into and out of the harbour.

**DIRECTIONS**

## DIRECTIONS to Mr. GWIN, for the execution of the interior or Flank Pier for completing of the interior Harbour of Port Patrick.

THE middle line of direction of the base of this pier runs from a point about 45 feet from the westernmost termination of the top of the present pier, measured inside toward the east to another point about three feet from the southwest angle of the platform, measured along the south side, also toward the east, which line of direction will be nearly N. E. by the compass; and as the width of the pier at its base will be 40 feet, it will be proper to fix up leading marks from the present pier to the shore parallel to the above mentioned middle line of direction, at the distance of 20 feet on each side of it, so that a person standing at either end of these leading marks may be able to direct a person on board a vessel afloat, to drop stones in those lines, which are to form the extremity of the base.

It will be proper to have the stones so to be dropped (which are to be large pieces of rough rocks) suspended by tackles in slings, yet not to hook immediately by the slings, but by a loop made of as many turns of marline or rope yarn as will hold it, and then by cutting the loop when the stone is in its proper position, it will be sure to be dropped in the outline. The stones to form the internal part may be any how tumbled overboard, so as to be within the area formed by the outlines.

It is supposed that the first and perhaps the second set of stones so dropped will bury themselves in the sand, and the sooner or later according as there is more or less surge of the sea breaking upon the place where they are deposited; but notwithstanding the first, second, or third set of stones disappear, yet by repeating the dropping they will at last ground themselves upon the clay or gravel, and those dropped upon them begin to appear above the sand, and at last above the surface of the water at low water. But it is to be noted, that every subsequent sett of stones dropped to form the outline, must be nearer the centre line than the preceding, so that a slope somewhat like that shewn in the section may be formed. As to the interior parts they may be tumbled in promiscuously, and if they begin to appear in some places before others, the defective places may be afterwards supplied. The more rough weather you have in the course of making this basement, the more solid your work will be, as it will be the more firmly or the more deeply grounded upon the harder matter, and prevent after settlements, and in case any part of the work settles after it has appeared you must still fill up the addition by fresh matter.

The

The diameter of the circular part of the head being two feet greater than the common breadth of the pier, the bafement may be broader by one foot and a half each fide, that is, 43 feet over all; and having determined the extremity of the bafement of the head by crofs marks at or about 88 feet from the neareft point of the platform rocks, this crofs line with the others will be fufficient for dropping the foundation of the head in fomewhat of a circular form, correcting it as you fee it, for when you get the pier head to appear above low water, you then can mark off an actual centre and work by real lines.

As you have not faid at what depth below the furface of the fand you found the clay, gravel, &c. I have fuppofed in the additional drawing, Fig. 7 & 8, (Plate 5.) that the fand lie three feet and a half below the furface of the water at low water, and the clay three feet and a half below the fand, fo that from the folid matter to low water mark is fuppofed 7 feet; if it happens to be more in any or every part, the width of the ground tier of ftones muft be proportionably increafed; if lefs, you may diminifh it at difcretion, but in general it is good to keep a good breadth upon the ground.

It is probable that at low water you will find the ftones ftuck in the fand, fo that you may be able to borrow away your flings; but if you find the ftones to bury themfelves, you will fave expenfe by procuring old cordage to make them of.

On fuppofition that the ftones in ftill weather do not bury themfelves, if you can at low water give them fome affiftance it may not be amifs; but at any event, the piling a weight of ftones upon them will, with the aid of the action of the fea, caufe them to fettle, and at laft to ground themfelves.

Suppofing this to be an operation that would occafion the foundation to go on for more than one year, the beft way would be to carry on the work progreffively from the rocks towards the head, as by that means every part would have time to ground, while the part beyond it was advancing; but as the whole at leaft of the bafement is expected to be completed in one feafon, to give it all the chance poffible of ftorms to make it ground, I am of opinion, that you fhould begin with the head, and go on to complete the rough bafement between the head and the rocks, fo as to bring the whole to a level about one foot above low water, and then go on with the upper works (the outfides of which are to be of fcappeled blocks) to proceed progreffively from the rocks towards the head, for by this means the head, whofe firm eftablifhment is of the principal confequence, will have the greateft fhare of time to come to a firm bearing; and when you cannot on account of the tide work below, you may be employing the people in getting the work forward that is upon the rocks.

Your

Your firſt work will be during the winter to provide all the large rough pieces of rock you can procure, in order to make the rough baſement, and as far as may be convenient to get them brought to the place.

In regard to the maſonry of the upper works, as the turning of the head cannot be done otherwiſe than with ſtones upon their flat beds; and as I never intended this interior pier to be built otherwiſe than with the ſtones upon their beds, the diſtance between the head and the rock will be ſo ſhort, that it will be hardly worth while to change the maſonry; however, if you find any convenience in the diagonal way of working for the outſide, you are at liberty to put it in practice.

The whole I propoſe to be built dry till you come to lay on the platform or top of the pier, after which you will proceed with mortar as uſual. As the joints of the ſtones in turning the head of the pier will be radii, or pointed to the centre, and will have nothing naturally to retain them from getting outward but their own weight and the incumbent weight of the matter above, it will be proper to cramp them every third courſe, or otherwiſe to retain them to the more central parts by iron doggs. And by way of ſecuring the plat-form or cap of the pier head, which will be the moſt apt of any other to get looſe, and at the ſame time to make it a firm tie upon what is below, I propoſe the work thereof to be jointed dove-tail-wiſe in the Edyſtone faſhion, as ſhewn in the additional deſign, Fig. 7. Plate 5.

If you find it adviſeable by way of tying your work faſter together to begin with mortar above high water, for ſome of the laſt courſes next the platform, and can procure lime with pozzelana, I ſhall have no objection; but if this will embarraſs the work, I do not hold it to be abſolutely neceſſary.

*Auſthorpe*                                                              J. SMEATON.
6th *January* 1774.

N. B. As the high and low water lines are not certain determinate heights, you are to be guided by the pier already done; the cap or platform of the flank pier, and the top of its parapet, being reſpectively of the ſame height as the preſent pier.

## RAMSGATE HARBOUR.

(See the Plan, Plate 7.)

An Historical REPORT on Ramsgate Harbour, wrote by order of and addressed to the Trustees, by JOHN SMEATON, Civil Engineer, F.R.S. and Engineer to Ramsgate Harbour, 1791.

### To the Truſtees of Ramſgate Harbour.

SIRS,

THE following piece was begun laſt year by your order, principally with a view to inform the public of the improved ſtate of Ramſgate Harbour, and of the Improvements that were in a way to be further brought about ; as alſo of the unexpected difficulties that had occurred in the progreſs and actual execution of this long deſired eſtabliſhment.—Indeed the ſmall length of time that it had been a Harbour, capable of fulfilling the purpoſes for which it was begun, had ſcarcely given opportunity to the maritime part of this Nation, and more eſpecially to foreigners, to acquaint themſelves with the advantages they might derive from the uſe thereof.

It is now ſomewhat more than ten years ſince Ramſgate Harbour was ſo far cleanſed of ſand and ſilt, as to be capable of taking in ſhips of ſuperior draught of water and tonnage, to what appears to have been the object of Parliament in granting the act, as well as to the views of the original promoters of the undertaking; yet it was not till the winter before the preſent (January 1790), that the real practical utility of this Harbour appeared in full view, for it had ſo happened, that the ſame means that had been neceſſarily employed for cleanſing it, (that is to ſay, for conſtituting it a Harbour), had ſubjected it to that kind of agitation and inquietude, which in general rendered it more eligible for veſſels of burden, ſuch as might very well have come in as to draught of water, to ſubject themſelves to the wear and tear of their tackle, and the riſk of riding it out in the Downs, than come into the Harbour till they had received ſome actual damage.

The quietude of the Harbour has at length in a great meaſure been happily effected, along with the other advantages, by the progreſs made in the conſtruction of an advanced pier. This was only begun in the ſummer of 1788, and at Chriſtmas 1789 was run out the length of 120 feet, that is nearly one third of its propoſed length ; which ſo

ſenſibly

senfibly quieted the Harbour, that in January and part of February 1790, there were in it no lefs than 160 fhips and veffels that came in for refuge, and to fave the wear and tear of their tackle and furniture, all of which muft otherwife have crowded the Downs.— Almoft an equal number, for the fame reafon, came into the Harbour during the tempeftuous weather of the paft January.

It may however poffibly feem to fome, that the means now taking to quiet the Harbour may operate to render it lefs acceffible : but this, when properly underftood, will not appear to be the cafe, but fo far otherwife, that in reality a fhip will come in with greater facility ; for the Harbour's mouth is in effect as wide as it was before the advanced pier was begun.—The original width of the Harbour's mouth, as intended, was 200 feet ; the opening to be at S. S. W. The width ultimately fixed by the Truftees was 300 feet at S. ½ E. but that width and pofition appear to have been fettled at a time when the Truftees were apprehenfive of the Harbour entirely choaking up with filt, if more inclofed, and not from any neceffity of that width, merely for the fake of facility of entry; 200 feet clear opening being deemed a large fufficiency for the entry of an artificial Harbour, entirely raifed out of the fea by the hand of man.

When the Harbour was formed, it was found that during all the time of full fea, a ftrong current fets almoft right acrofs the Harbour's mouth, that is, from weft to eaft, which being a natural caufe, cannot be diverted ; a veffel therefore coming from the fouth, that is, from the Downs, were fhe to attempt to run into the Harbour, right acrofs the current, would be carried eaftward thereof, fo as to mifs it : it has therefore been an eftablifhed maxim, ever fince the Harbour was formed, and given out as a direction, for veffels to come in obliquely from the weftward, and as clofe as properly may be to the weft pier head ; and this courfe, as marked out many years ago upon the plans of direction, will carry a veffel right through the prefent intended opening, betwixt the weft head and the head of the advanced pier, which is full 200 feet in width, and fhe will always come in with the tide in her favour.

In confequence of your orders, in the courfe of laft fummer I made a confiderable progrefs with this Report, intending to have completed it at leifure this fpring, but if by being called upon more haftily than I had expected, it comes out lefs finifhed than I could have wifhed it, I have this only circumftance to plead in my excufe.

I remain, Sirs, &c.

J. SMEATON.

## SECTION I.

ACCOUNT of early attempts towards a Harbour for the Downs, and of proceedings, inclusive of obtaining an Act of Parliament, for establishing a Harbour for that purpose at Ramsgate.

THE expediency of a Harbour for the reception of Ships in the Downs, has doubtless subsisted as far back as the increase of our trade and shipping has rendered it important; but the want of a situation strongly pointed out by nature, was probably the reason why we have not heard of any attempt towards it in the earliest ages.

In the time of King Edward 6th it is said there was an attempt to make a Harbour from Sandwich into the Downs, and that the evident traces of a canal, which still subsist in the level grounds, between Sandwich and Sandown Castle, are the remains of that attempt. It is also said, that commissioners were appointed by Queen Elizabeth in 1574, for taking a survey of Sandwich Haven, and to give their opinion as to the making a better Harbour near Sandown Castle. Also that in 1705, a plan, report, and estimate, were delivered by persons appointed to survey and estimate the expense of a new Harbour from Sandwich into the Downs, accompanied with a certificate of the Flag Officers, and many commanders of the ships of her Majesty's royal navy, who then gave it as their opinion, that such a Harbour might be of general advantage to the public.

It seems also that petitions were presented to the Honourable House of Commons, praying for a new Harbour near the Downs, April 2d 1736; and that a committee appointed by the House heard evidence upon the matter thereof; and that in consequence a plan and survey of the Downs and coasts adjoining, were undertaken by Mr. Labelye, afterwards Engineer to Westminster Bridge, at the expense of Sir George Oxenden Baronet and Josiah Burchett Esquire, then members for the town and port of Sandwich; a copper plate of which was published in 1737-8, in which he exhibits a scheme for sheltering ships from the Downs by a navigable canal and bason, in the very direction of the Old Cut above mentioned, and by sluices to join the river Stour.

In April 1744, the House of Commons presented an address to the King, "that he would give directions to proper and skilful persons, to view the Haven of Sandwich, and examine whether a better and more commodious harbour may not be made into the Downs near Sandown Castle, fit for the reception and security of large merchant ships and men of war; and to survey the said ground and shore; and also the river Stour, neces-

sary

fary to cleaning and fcouring the faid Harbour when made; and to make an eftimate of the charges and expenfes thereof;" to be laid before Parliament the enfuing feffion.

In confequence of the above, an order was iffued from the Admiralty, appointing the following perfons to this bufinefs; W. Whorwood, John Redman, John Major, Thomas Slade, Charles Labelye, and R. Charles; who reported, that having made the neceffary obfervations, " a better and more commodious Harbour, than the prefent haven of Sandwich, may be made from the town of Sandwich into the Downs near Sandown Caftle."

They propofed to carry out two ftone piers, each 2,096 feet in length from the fhore, into twelve feet depth of water at low water; to have a clear opening between the heads of 300 feet; to narrow from that to 100 feet; and that the middle line fhould point S. S. E. $\frac{1}{2}$ E. by the compafs; that is nearly S. E. by the true meridian, or S. S. E. as the compafs now points. The eftimate for this work was £.389,168. 13s. 2d. exclufive of the value of the grounds to be purchafed.

The report being referred to a committee of the whole Houfe of Commons; after examining evidence of pilots and perfons beft fkilled in the navigation of the Downs, the committee came to a refolution, dated the 26th February 1744-5, importing, that it appeared to them, " that a fafe and commodious Harbour may be made into the Downs near Sandown Caftle, fit for the reception and fecurity of large merchant-men and fhips of war of fixty and feventy guns, and be of great ufe and advantage to the naval power of Great Britain.

Why this great work was fufpended, after being brought this length, does not now diftinctly appear; but if we confider the largenefs of the fum eftimated, for a work fuppofed to be undertaken by Government, at a time when we were at war with France and Spain, we perhaps need not be at a lofs to judge.

The whole affair feems however to have lain dormant for fome time, till the public was roufed by a violent ftorm which happened on the 16th December 1748, wherein a great number of veffels were driven from their anchors in the Downs, and being forced upon the fouth coaft of the Ifle of Thanet, feveral found fafety in the little harbour of Ramfgate.

This

This feems at once to have opened the eyes of the public, and caufed them to be turned upon Ramfgate as the proper place for the reception of fhips when in diftrefs from bad weather in the Downs ; and accordingly, the 8th February following, a petition was prefented to the Houfe of Commons, by feveral merchants of the city of London, owners and mafters of fhips, whofe names were thereunto fubfcribed, amounting to 131, which was referred to a committee appointed to fit the 13th following ; and on this day a petition was prefented from the mayor, magiftrates, freemen and inhabitants of Sandwich ; fetting forth the damage that would be likely to arife to the haven and port of Sandwich from the extenfion of piers into the fea at Ramfgate, which was referred to the fame Committee.

In fupport of the Petition of the merchants, owners, and mafters of fhips, a great number of witneffes were examined, and to a confiderable length : and the following points were fully and clearly proved, to the fatisfaction of the Committee.

1ft. That, in the faid great ftorm of December preceding, a number of fhips were actually forced into, and faved in Ramfgate Harbour, although then fo fmall as to be fcarcely capable of receiving veffels of 200 tons, at any time of tide ; that pier having been only built and maintained by the fifhermen of the place.

2d. That the winds in the Downs, whereby fhips riding there are moft apt to be annoyed, are from S. S. E. to S. S. W.

3d. That at Ramfgate, or near it, was not only the beft, but in reality the only place, where any harbour could be built, that could be ferviceable to fhips in diftrefs in the Downs, becaufe Ramfgate was right in the lee of that road, with fuch winds as produced that diftrefs ; and at fuch a proper diftance, that, after driving or breaking loofe, they had time to get under fail, fo that with a flender fhare of feamanfhip they could make a harbour if built there.

4th. That though this fhore was univerfally flat, yet as it gradually increafes in depth from the Cliffs towards the Downs, it was practicable, at a moderate expenfe, to carry out piers into fix feet and a half water, at the low water of a middling fpring tide ; and that, according to the rife of the tides, there would be water enough from three-quarter flood to one quarter ebb even at neap tides, to carry in veffels drawing fifteen feet water, which, if full built, was fuppofed to be full 300 tons burthen.*

* I underftand it can be made to appear that more than two-thirds of all the tonnage and value of fhipping is carried on in veffels not exceeding 300 tons.

5th. That

5th. That when veffels break loofe from their anchors in the Downs it is generally from three-quarter flood to one-quarter ebb,* during all which time the courfe of the current of the tide is to the North and N. E. which therefore would carry them rig'it into a harbour at Ramfgate, fo that by the time they got thither, it would be within an hour of high water.

6th. That the foil at Ramfgate being a chalk fufficiently firm to build upon, but yet fo yielding that the keels of veffels readily make a dock for themfelves therein ; this, with fometimes a flight cover of fand, forms a proper bottom to lay full built fhips aground upon at low-water ; and even if fharp built, will, in cafe of neceffity, fubject them to the leaft poffible damage; and indeed to little or none, if proper precaution be taken to lay them againft a pier ; nor can they fuffer in the leaft, if a proper bafon be conftructed to lay them afloat.

7th. That in time of war merchant fhips are built fharper than in time of peace ; but that at an average more of the London traders are built full than fharp.

8th. That the great fhips in the Downs are obliged to ride in a bad road to be out of the way of the fmall veffels, which commonly lie in the fmall Downs, and thofe fmall veffels being often ill-furnifhed with anchors and cables, frequently break loofe, and drive upon the large fhips, which then run foul of each other ; whereby fometimes a whole fleet is fet adrift; and in the opinion of Captain Conway (then an elder brother of the Trinity Houfe) if a harbour were only made for the reception of fhips of 200 tons and under, it would prevent nine-tenths of the damage that happens in the Downs ; as he fuppofes all fhips under 200 tons, waiting for a wind to proceed weftward, would take fhelter therein.

9th. That fhips in Ramfgate Harbour may fail out of it with any wind, that would carry them weftward out of the Downs ; and even with a ftrong wind at eaft, or with a fcant wind at S. E. by E. they can make good their courfe out of Ramfgate Harbour, by virtue

---

* This will appear clearly to be the cafe, when it is confidered that they are the Goodwin Sands that conftitute the Downs to be a road for fhips. At low water thofe fands may be confidered as a pier or break-water to all the eafterly winds ; and even at high water, it is too fhallow over them to admit the great feas to pafs, without being much broken and difperfed, efpecially in ftormy weather. From the fituation therefore of the Downs, thefe fands on one fide, and the coaft of Kent on the other, it is only the foutherly winds that can annoy them, which alfo are much moderated by the proximity of the coaft of France ; and ftill more fo, by the firft part of the flood tide running fouthward, and meeting the feas ; it is therefore not till the tide turns to the north (which is at or about three-quarter flood) that the combined force of wind and tide, makes the great effort to break the fhips from their moorings.

of the flood tide, under their lee, and sail westward, when ships in the Downs cannot pur-chase their anchors.

10th. That large craft might be constantly kept afloat in Ramsgate Harbour, at low water, such as might be able to carry out pilots, anchors, cables, and other assistance to men of war and large ships in distress in the Downs;* and the matter is so circumstanced that whenever they could not go from Ramsgate, boats may go out from Dover to ships in the Downs.

Upon this evidence, which I have carefully extracted from the Committee's report, I only beg leave to observe; that the tides, the sands, and the coasts remaining the same, as also the natural powers, what was true in the year 1749 will remain true in the year 1791.

During the whole of this investigation the great project of a harbour at Sandown Castle seems to have been altogether lost sight of; and perhaps at this we may the less marvel; if casting our eye upon the position thereof; vessels breaking loose from their anchors there, with wind at any point from S. S. E. to S. S. W. would be driven to the leeward of the harbour's mouth before they could get under way, so as to be under dominion of their helm; and therefore, after all, liable to be wrecked or run ashore upon the south coast of the isle of Thanet †.

Nor indeed, according to my judgement, could a more injudicious construction than the piers proposed at Sandown, be well imagined; for the heavy seas that would fall in between the heads (that is, all those producing distress in the Downs) would be so aug-mented by the gradual contraction of the distance of the piers, from 300 feet at their entry, to 100 feet near the bason; that vessels would not only be liable to be wrecked between the piers, but those augmented seas would infallibly destroy the gates of the bason; and

* Such crafts and pilots now actually station themselves at Ramsgate, and are the means of saving many lives and much property.

† While the investigation for a harbour at Sandown Castle was going on before Parliament in 1744, reasons were offered against making a harbour near Sandwich; to which at that time no regard seems to have been paid, viz. Sandwich having no convenient outlet for ships bound to the westward, they must remain wind-bound with many fair winds. 2dly. The inlet is inconvenient to receive ships bound to the westward, which are detained in the Downs by contrary winds and stormy weather; for if any ship's anchor start, or cables give way, they must drive past the harbour, before the ship can be brought into a position to put for the entrance.

probably

probably the fixed part of the ftone work alfo.—It hence appears moft evident, that nature has not in reality furnifhed any fituation for a harbour for the Downs preferable to that at Ramfgate; and therefore we need not wonder, that the projeft of the long-wifhed for harbour for the Downs was at that time promoted to be at Ramfgate, with a degree of eagernefs and even of enthufiafm.

The Aft paffed that feffion, fetting forth in its preamble, " Whereas frequent loffes of the lives and properties of his Majefty's fubjefts happen in the Downs for want of a harbour between the North and South Forelands; the greateft part of the fhips employed in the trade of the nation being under a neceffity at going out upon as well as returning from their voyages, to pafs through the Downs; and frequently by contrary winds being detained there a long time; during which they (efpecially the outward-bound fhips) are expofed to violent ftorms and dangerous gales of wind, without having a fufficient harbour to lie in or retreat into, or from whence they can receive any affiftance : And whereas a harbour may be made at the town of Ramfgate proper and convenient for the reception of fhips of and under 300 tons burthen; and from thence larger fhips in diftrefs in the Downs may be fupplied with pilots, anchors, cables, and other affiftance and neceffaries; and by the fmaller fhips taking fhelter in this harbour the larger fhips may take the anchorage which at prefent is occupied by the fmaller, and by that means their anchors will be fixed in more holding ground, and the fhips not fo expofed to the ocean; for carrying therefore a work of fuch public utility into execution, &c. be it enafted," &c.

The firft meeting of the truftees was appointed at the Guildhall London the firft Tuefday in July 1749, (that was then next following), at which a large number named in the Aft, to the amount of fixty-fix, appeared to qualify.

## SECTION II.

PROCEEDINGS of the Truftees from the Commencement of the Work in 1749, to the total Stoppage of the fame in 1755; upon Petition to the Houfe of Commons.

THE firft aft of the truftees regarding conftruftion was to appoint a Committee of their body to view the place at Ramfgate, where a new harbour was intended to be made; and report their obfervations and opinions to the truftees at their next meeting;

and that Mr. Robins and Mr. Turner of Gofport fhould be defired to attend them as engineers.*

The Committee accordingly met at Ramfgate, and on their return, reported, that in confequence of objections having been made before the Committees of the Houfe of Commons to Ramfgate, for want of a back-water, they having examined found from information, that in the year 1715 the pier then ftanding had been lengthened; and by infpection it appeared, that a bar of about forty yards breadth and length had been caft up, its higheft part in thicknefs two feet and a half; and they confidering this fmall quantity as the gradual effect of 34 years, thought it reafonable to expect, that when a greater depth of water was made by two piers inftead of one, the filling up by fullage or beach would become fo inconfiderable, that a fmall expenfe would continually prevent its increafe.

They further obferved, that the feaweed or fullage that drove in, came from the weftward; and that from the eaft there was a drift of large fhingle; which if it fhould take place, would be of advantage in backing the new piers. They alfo made and reported many other obfervations confirming what was given in evidence before the Houfe of Commons in its favour.

They endeavoured to fix their opinions as to one material point, which could not be fo conveniently fettled in evidence before the Houfe; and about which there was fome diverfity of opinion; and that was the pofition of the harbour's mouth, whether South, S. S. W. or S. W. were to be preferred. Towards the refolution of which queftion the committee premifed, " that the ftream of the tide in the Downs fets for fix hours to the northward, or at leaft between the north and the eaft, and then for the next fix hours the ftream turns and fets to the fouthward, or between the fouth and the weft; but the time of high and low water does not correfpond to the beginning and end of thefe ftreams; for it is high water about two hours after the ftream has begun to run to the northward; and it is low water two hours after the ftream has fet to the fouthward, fo that when the tide firft fets to the northward, more than two thirds of the tide has flowed; and high water happens about two hours after:" and that having feparately examined eleven captains or mafters of fhips of Ramfgate, they all unanimoufly agreed, that the moft dangerous winds in the

* It may at this day feem extraordinary, why Mr. Labelye, who appears to have made the original plan, upon which himfelf and affociates proceeded as mentioned above, fhould not have been called upon and confulted in this latter ftage of the bufinefs; but if it be remembered that in the year 1748, one of the piers of the new bridge of Weftminfter, built under his direction, moft unfortunately fettled, fo as to oblige two of the arches to be taken down, after the bridge had been opened to the public, we may be the lefs furprifed.

Downs

Downs were from the S. S. E. to S. S. W.; and that the time when ſhips run the greateſt riſk of being forced from their anchors, is when the northward ſtream ſets in; and that about the ſame time, that is the beginning of the northern ſtream, was likewiſe the moſt prudent for ſuch ſhips as ſhould intend to make for Ramſgate harbour, to ſlip their cables; for, that in either caſe, they would have both wind and tide in their favour, in ſtanding for Ramſgate; and that on their arrival, they would find it near high water, allowing an hour for their paſſage: and they unanimouſly agreed, that an entrance to S. S. W. was to be preferred: for if placed full ſouth, the tide near high water would run ſo ſtrong acroſs it, as to render it difficult to get in; and if at ſouth weſt, they feared there might be too great an indraught of ſullage. All the bearings then referred to were ſettled by the compaſs, which in the Downs was then $1\frac{1}{2}$ point weſt.

They alſo particularly attended to the point, whether the piers ſhould be built with wood or ſtone; but agreed unanimouſly ſtone was greatly to be preferred, had it not been for the great difference of expenſe, eſpecially as they had found (and brought up ſpecimens) that the worm bit conſiderably in the pier then ſtanding. On the whole they were ſatisfied, a harbour at Ramſgate would be as practicable, uſeful, and important, as they had before thought; and concluded with obſerving, that no other motive than that of avoiding ex-penſe ſhould have any weight in aſſigning to this harbour a form leſs perfect or exten-ſive than it was by nature capable of receiving. It does not, however, appear that the Committee, on this occaſion, were attended by Mr. Robins or Mr. Turner.

The next ſtep of the truſtees was to advertiſe to invite engineers to deliver plans ſealed up, to their ſecretary Mr. Elliot, for piers of ſtone or of wood, with the neceſſary ſpeci-fications, and explanations, againſt the 29th September following.

At this time, ſeveral plans and models were offered, and the 19th October appointed for taking the plans into conſideration: and alſo the ſecretary was ordered to invite en-gineers, or gentlemen acquainted with engineery, to aſſiſt the truſtees upon that occaſion. Accordingly, upon the 19th ſeveral perſons attended under that deſcription, and were deſired to appoint a meeting among themſelves to conſider the ſeveral plans, and give their opinion to the truſtees; and the ſecretary was ordered to invite the gentlemen who ſhould attend to dinner, at the expenſe of the truſt.

On the 26th of October the engineers delivered their report, which was ſubſcribed by Benjamin Robins, Thomas Innes, J. Leake, John Muller, and John Turner. The moſt material things that occurred therein were:

That

That it was improper to lay the foundation of the piers below low water mark, with loofe ftones thrown in at random.

That if the piers were not carried out into a greater depth of water than 6¼ feet, at low water of fpring tides, it did not appear neceffary to be at the expence of carrying on the foundation by caiffons.

That the propereft method of laying the foundation of the intended piers in the parts covered at low water, was by making a bafement of ftone fomewhat higher than low water mark, after the manner propofed by Mr. Turner.

At a fubfequent meeting, Mr. Defmaretz, Mr. Prat, and Mr. Mill, gave their opinions relative to the manner and materials neceffary for the building the piers of the intended harbour.

The truftees then ordered notice of a meeting upon the 15th December, for taking into their confideration all the plans and propofals that had been or fhould be laid before them, and to come to a determination thereon : and that all perfons having any thing further to offer, fhould be defired to deliver in their propofals before that time ; recommending a full meeting of the truftees on this extraordinary occafion.

At this meeting feveral plans and fections were examined, amongft which was a plan from Captain Robert Brooke of Margate, and a fection from Mr. Defmaretz chief engineer at Portfmouth ; and at a fubfequent meeting, three more plans and fections were delivered, amongft which was a plan and fection by William Ockenden Efquire, one of the truftees : and at a meeting after that, ten different plans, laid before the board in confequence of the original advertifement, were examined ; and an abftract ordered to be made by the fecretary, and laid before the next meeting ; at which, on the 12th January 1749-50, it was refolved, that the harbour fhould be proceeded with according to the plan figned by the chairman, fubject to the further alterations of the board. That the eaft pier fhould be proceeded upon with ftone ; and the weft pier with wood.

At a fubfequent meeting it was refolved, that the plan and fection propofed and delivered by William Ockenden Efquire, for erecting a ftone pier, fhould be the method of building the eaft pier ; and that the fection and model which was propofed and delivered for a wooden pier by Captain Robert Brooke, fhould be the method of building the weft pier to low water mark neap tides : and that Mr. Ockenden and Captain Brooke fhould be applied to, for their defcription of the proper materials for the piers refpectively propofed by them.

The

The 2d February 1749-50, it was refolved, that the eaft pier fhould be carried on by workmen appointed by the truft, and materials purchafed for 100 feet of pier; and Thomas Prefton was appointed mafon or foreman of the ftone work; who by the Board's order, fet out for Maidftone, Folkftone, Dover, and Ramfgate; and was to follow the in-ftructions given him; and Captains Conway, Stevens, and Bennet, were defired to affift Captain Brooke in purchafing materials and carrying on the weft head.

Mr. Prefton's firft report of the 23d following, chiefly contained an account of the ftone materials afforded, and ufed at the places fpecified; whereupon he was ordered to go to Purbeck, to make the proper enquiries concerning the ftone there: Mr. Ockenden, Sir Peter Thompfon, Mr. Fry, Captain Barker, Captain Hughes, Mr. Hyde, Mr. Norris, and Mr. Pole, were appointed a committee for carrying on the eaft head, and were defired to confer with the committee for the weft head, as often as they fhould find neceffary. Mr. John Scott was appointed foreman for carrying on the weft pier.

Things being thus fettled, nothing happened but what might be expected to be the refult of the difpofitions mentioned; and it is worthy of notice, that a quantity of Barrow limeftone was got, as recommended by Mr. Prefton.

In 1751, a committee being appointed by the board to make a furvey of the ftate of the works, made their report the 25th July. The moft material things to be obferved were as follows:—

That the ftone pier extended 390 feet, of which 104 feet were completely ready for the parapet, and the reft above the reach of a fpring tide; and that while the committee was there, the foundation was run on 83 feet further, in the whole 473 feet. They examined the work, and found it free from any kind of failure whatever.

The weft head was carried out about 460 feet from the cliff; and 540 feet thereof were propofed to be completed that year.

The commitee were attended by Mr. Turner of Gofport and Mr. Vincent of Scarbo-rough, and recommended the ufe of afhler inftead of backing ftone, fince a cube foot of fhell lime * coft more than a cube foot of afhler.

A locker or grind was then recommended, the committee being of opinion it would collect the fand and fhingle in the outward angle of the eaft pier with the land.

* At this time fhell lime was ufed for the backing of Ramfgate harbour.

By

By the feparate report of the engineers, of the fame date with that of the committee, it appears they were of opinion, that the part of the ftone pier then done, was but barely fufficient to refift the force of the fea; therefore recommended the future parts to be made with 40 feet bafe, and 30 feet at top; and an additional courfe of afhler, to the height of 18 feet above the foundation, both infide and outfide; and above that, one courfe only; the core to be leffened. The chalk to be well beaten, and mixed with gravel; and not mixed with mortar as then practifed. All the afhler to be fet with terras mortar, and the backing mortar made with brick-duft or fea-coal afhes.

In furveying, they found the bottom regular, and in the circumference of the harbour, the depth of low water not exceeding five feet fix inches; and that the foundation might be carried out without caiffons, fo as to anfwer the defign, in a manner by them defcribed; which was that of laying large Portland blocks in the ground courfes, fo as to reach above low water.

The following Board agreed to the ufe of afhler for backing, for the reafons given by the committee in their report. A locker or grind to be made as propofed. Captain William Read was appointed haven-mafter, with inftructions to hoift a flag when there was ten feet water at the old pier head.

6th December 1751, a general meeting was appointed to confider of the method of founding the piers beyond low water mark, when Mr. William Etheridge produced a model of a caiffon; and a model of a tool, and method of working it, for making a trench, and levelling the ground under water for fetting the caiffon upon. Captain Robert Brooke fent a model for laying the foundations of ftone, without caiffons, accompanied with a defcription. In confequence of which the truftees ordered an advertifement for contractors to deliver plans, propofals, and eftimates, for conftructing the piers beyond low water mark.

The 3d January 1752, Mr. James Morehoufe prefented a plan, propofals, and eftimates; which however were not approved; but it was refolved, that a foundation, not exceeding 200 feet, fhould be carried on with large blocks of ftone, agreeably to two plans given in by Mr. Prefton the mafon.

24th January; it was refolved that the faid 200 feet of ftone pier fhould have its foundation laid in a channel or trench, dug into the chalk ten or twelve inches deep; and a motion was made, and agreed to, that a furveyor fhould be appointed; and that a proper

advertifement

advertifement fhould be confidered for the purpofe: Friday the 21ft was appointed for the election of one, from fuch perfons as fhould offer, to infpect and direct the careful and ex-peditious carrying on of the building ; and the mafon's report was poftponed till a furveyor fhould be chofen; but Boulogne lime and lime kilns were ordered to be fet about immediately.

21ft January. It was refolved that fuch furveyor as fhould be chofen, fhould have a falary of £200. per annum, and refide at Ramfgate.

On this occafion, feven perfons applied ; amongft whom were Mr. Etheridge, and Mr. Vincent the Engineer of Scarborough Pier, who both produced ample teftimonials ; but on holding up of the hands, the majority was declared for Mr. Etheridge, who was ftrongly recommended by Mr. Ockenden ; and who received on this occafion the thanks of the Board for his plan, care, and attention ; and which were defired to be continued.

28th February. Inftructions were given by the truftees to the furveyor, and were in fubftance as follows :—

" The furveyor is to refide in Ramfgate, and not be abfent without leave firft ob-tained from the truftees at one of their meetings ; and to have the infpecting and directing of all the works that have already or fhall be hereafter carried on, under the direction of the truftees of Ramfgate Harbour ; and alfo to have the direction of all the perfons em-ployed therein, at Ramfgate, except the clerk of the cheque. He is weekly to tranfmit to the fecretary an account of the progrefs of the work, and of the tranfactions relative thereto, that they may be laid before the Board at their feveral meetings. He is truly and faithfully to infpect the feveral materials which fhall be delivered, and report to the Board thofe which are not according to agreement or contract ; and frequently to examine the lift of the workmen employed, and fee that they perform the labour they are paid for ; and as foon as he conveniently can, he is to take an account of the abilities of the feveral work-men employed, and report to the fecretary, whether they are deferving the wages therein charged ; but no alteration of wages fhall at any time be made by him, without firft mentioning his intention to the truftees, and receiving their approbation. The increafing or leffening the number of workmen to be left to his difcretion. But he is from time to time to acquaint the Board of his reafons for whatever alterations he fhall make among them. Nor is he to make any material alteration of any kind, without previoufly ac-quainting the fecretary thereof, and taking the directions of the truftees thereon. In general he is to infpect and direct every thing relating to the works, fo that they may be carried on in the moft expeditious and frugal manner."

25th

25th March. Mr. Etheridge delivered his firſt report, which deſcribes the ſtate of the works at Ramſgate on a general view; and having inſpected a cargo of backing ſtone brought from Purbeck, he thought it very good, and capable of making very ſufficient work, without building the wall entirely with aſhler; but for the benefit of the work, he recommended to ſend a proper perſon to Purbeck, at the expenſe of the truſt, to ſee that the ſtone be good, properly worked, and the courſes ſhipped as wanted; which was afterwards ordered by the Board.

Here it may be neceſſary to obſerve, that Mr. Etheridge appears to have been a perſon of a truly mechanic genius, and having been brought forward by the celebrated Mr. King, carpenter of the works of Weſtminſter Bridge, as his foreman, and after Mr. King's death become his ſucceſſor in completing this branch of thoſe works, might be preſumed to be a man of much experience in the carpentry line: but being here appointed director of the whole work, and looking upon himſelf as competent to the maſonry as to the carpentry branch, there very ſoon aroſe a difference of opinion betwixt the ſurveyor and Mr. Preſton the maſon, that afterwards turned out of great detriment to theſe works. Mr. Preſton, though by no means equal to Mr. Etheridge in general mechanical knowledge, yet was an excellent maſon, and well informed in the nature of the materials proper to his trade:—the making of mortar being rather of a chemical than a mechanical nature. Mr. Preſton, before the appointment of Mr. Etheridge, not being ſatisfied with the uſe of ſhell lime under water, which was in part uſed at Ramſgate Harbour, recommended to the truſtees the trial of various lime ſtones; viz. Aberthaw, Barrow, and that of Boulogne, of which cargoes had been ordered; but their merits had not then been ſufficiently inveſtigated. Mr. Etheridge unfortunately had adopted ſhell lime for waterworks, and accounted that of Maidſtone, and St. Vincent's Rocks at Briſtol Hot Wells, as preferable to the above. †

Mr. Etheridge, as reported, being of an auſtere temper, not readily giving up what he had once advanced, a ſhyneſs took place between theſe two officers; which though it did not prevent either of them from punctually doing his duty, that is, did not prevent Mr. Preſton from ſcrupulouſly purſuing his orders from Mr. Etheridge; yet it prevented that interchange of ſentiments and confidence, which is ſo eſſentially neceſſary among the principal officers of a great work or enterprize, that it may be carried on to the beſt advantage.

* Mr. Etheridge afterwards deſigned and built the famous wooden bridge at Walton-upon-Thames; the middle arch of which ſpanned 120 feet.

† See the chapter on water cements, Smeaton's account of Edyſtone lighthouſe.

Mr.

Mr. Etheridge now proceeded to put in practice his proposed method of laying the foundation of the piers, in cases or caissons, and shewed that method of digging a trench under water, and levelling it, which, being attended with certainty, and every necessary degree of dispatch, has ever since been the method put in practice here, and so continues to this day.

Every thing appeared to go successfully on during the year 1752; and the committee of the 29th September, reported that 68 feet of stone pier, of the same dimensions as the east pier, had been added to the 550 feet of timber work of the west pier, and 138 feet of foundation carried out five feet high.

The 26th January 1753, produced an order of Board, that Mr. Ockenden, and any other gentlemen of the trust, should be desired to go to Ramsgate to consult with the surveyor there, and give directions for a proper plan to be drawn of the extent and manner in which the work should be carried on and finished; and that the said plan should be laid before the Board for approbation.

14th May 1753, the Board read Mr. Ockenden's report of his survey made at Ramsgate, and considered the plan for carrying out the work, which is dated the 21st April, and signed by him and the surveyor, but came to no resolution thereon; however Mr. Ockenden received the thanks of the Board, for his great care and pains in making his late survey and report.

At a general meeting, 14th December 1753, for considering tne plan of the Harbour, a motion was made, that the Harbour be contracted to 1,200 feet in width, according to the plan this day laid before the Board by Mr. Ockenden; which, being debated, was carried on a division of 28 to 15. And on an adjournment to that day se'nnight, the resolution was confirmed; 26 to 7.

Upon this contracted plan, which appears to have originated with Mr. Ockenden, Mr. Etheridge seems not only to have vigorously proceeded, but even to have pushed the execution; for, by the Committee's report on their visitation, the 9th October 1754, after declaring their opinion, that the contraction ordered by the Board would leave the Harbour large enough to contain more ships than would ever have occasion to lie there, at the same time; and " that the curve at the west pier continuing the same as it was in the original plan, would give sufficient room for ships to bring up;" they reported, that at the west pier, the 138 feet of foundation mentioned last year to be laid, was taken up, and a temporary crofs wall of backing ftone had been carried out eaftward 278 feet; but that from the eaft end of this, nearly 300 feet had been carried out foutherly, to five feet high and upwards; and 111 feet to 19 feet 8 inches; and furthermore, that the piles had been driven for founding 80 feet more.

At this meeting the furveyor pointed out a communication to be neceffary by land to the weft pier at high water, and recommended a ftair-cafe from the top to the bottom of the cliff; which work being foon after executed, was called Jacob's Ladder, which name it ftill retains.

This appears to be the laft work of Mr. Etheridge at Ramfgate, for while the works of this fummer were going on, remonftrances to the truftees came from feveral ports, declaring ftrongly againft the contraction, and which ended in an application to Parliament in 1755; when the petition of merchants, owners, and mafters of fhips, whofe names were fubfcribed, fet forth; " That from the works carried on under the direction of the Truftees of Ramfgate Harbour, for fome time the petitioners had great reafon to hope, that a fafe and commodious Harbour would be made there, for the reception of fhips and veffels; and thereby the intention of the Legiflature anfwered in impofing the duties; but that, in or about December 1753, the faid Truftees had refolved to alter and contract the faid Harbour; and during the laft fummer, fuch works had been carried on towards altering and contracting the fame, that if further proceeded in, and finifhed according to the faid refolution of the faid truftees, would, in the petitioners' apprehenfion, render the faid Harbour in a great meafure ufelefs; and that the expenfe thereof would be loft to the public."

" That as the petitioners were fully convinced a fafe, commodious, and ufeful Harbour might be made at Ramfgate, to the great advantage of the commerce of this kingdom, under fkilful and proper directions;

" The petitioners therefore humbly prayed the honourable Houfe to take into their confideration this matter, in which the public was fo greatly interefted, and grant fuch relief therein, as to the honourable Houfe fhould feem meet."

The matter of the above petition having been fully heard by the Houfe of Commons, in the early part of the year 1755, that Houfe thought proper to addrefs his Majefty, " that he would be gracioufly pleafed to appoint proper and fkilful perfons, to make a furvey of the works carrying on at Ramfgate Harbour, and give their opinions thereon; and what might be the moft proper plan for finifhing and completing the faid Harbour; to make

an

an eftimate of the expenfe thereof; and that his Majefty would be gracioufly pleafed to caufe fuch plan or plans and eftimate, together with an account of all the proceedings of the faid perfons, to be laid before the faid Houfe in the next feffion of Parliament."

In confequence, Sir Piercy Brett and Captain Defmaretz were appointed to this fervice; who, the latter end of the fame year, delivered their plan, report, and eftimate; and in the feffion of 1756, a bill was brought into Parliament, which was much agitated, and canvaffed; but it did not ultimately pafs into an act. Thefe proceedings however had the effect of putting a total ftop to the works at Ramfgate.

## SECTION III.

OCCURRENCES from the ftoppage of the works in 1755, to their recommencement in 1761.

IT may in this place be proper to give a fhort account of this fcheme of the furveyors appointed, as nothing was in reality done upon it.

The report of Sir Piercy Brett and Captain Defmaretz is addreffed to the Lords of the Admiralty, and was printed in 1756. It almoft fets out with faying, that " according to the beft of their judgement the works already made, and every plan hitherto impofed, feem liable to very material objections;" they go on to fay, " and we think it proper that the work done upon the contracted plan fhould be taken up, and the materials made ufe of in carrying out the Weft Pier."

It is however to be noted, that the whole of the work that had been done, exclufive of what concerned the contraction, was propofed to make a part of their own fcheme, being no other than the ftraight walls or piers from the fhore; but they propofed, by engrafting upon them, to carry out the pier heads further by full 400 feet to feaward, than according to any of the plans that had been before exhibited, and which they fuppofed would place them in eight feet water. They alfo propofed to make the opening of the Harbour's mouth right towards the Downs; whereas according to the advice of the eleven Captains, mentioned before, all former plans had exhibited that opening to be S. S. W. that is, about three points more wefterly than the fituation recommended by thofe gentlemen. The width of their opening, as fhewn in their plan, is 275 feet; which is 75 feet greater width than any of the former fchemes.

They

They further propofed, " for the better prefervation of fuch fhips and veffels which are fharp built, and liable to receive damage by laying a-ground, to have a bafon at the eaft fide of the Harbour, to be funk fixteen feet from the low water of a fpring tide, and partly inclofed with a ftone wall, as defcribed in the plan; where 50 or 60 fhips of 300 tons might always lie afloat in fmooth water: and they humbly conceived, that without fuch a bafon the Harbour would not anfwer the good purpofes intended by Parliament."

|  | £. | s. | d. |
|---|---|---|---|
| Eftimate for finifhing the Harbour, according to the plan - - | 153,548 | 18 | 10 |
| Add for completing the propofed bafon - - - | 50,567 | 14 | 0 |
| Total expenfe of the works propofed - - | £. 204,116 | 12 | 10 |

N. B. From the above the furveyors deduct for old materials to be drawn from the contracted walls, the fum of £.8,210 5s. 4d. fo that the total, at the foot of their eftimate, ftands £.195,906. 7s. 6d.

Such was the plan or fcheme of Sir Piercy Brett and Captain Defmaretz, upon which I fhall have a few ftrictures to offer when they come in courfe.

How it happened that this contracted plan of Mr. Ockenden's, which was productive of fo much mifchief and delay, came to be fo ftrenuoufly efpoufed by him, is not now eafy to fay; certain it is, that on a calm review it amounted to no more than to make a lefs Harbour at a greater expenfe than a larger; and that without any apparent inducement, unlefs what is alledged in the Committee's laft report, p. 89. is to be looked upon as fuch; viz. that the contraction will leave the Harbour large enough to contain more fhips than will ever have occafion to lie there at the fame time. How it happened that Mr. Etheridge fhould come into the fcheme is equally wonderful, as Mr. Etheridge appears to have been a very ingenious perfon, and which therefore can fcarcely be accounted for, but from jealoufy between him and Mr. Prefton, who it feems was againft the contraction. Finally; how it came to be adopted by the truftees at large, many of whom were able and experienced feamen, can only be accounted for by fuppofing Mr. Ockenden, having had occafion to confider works of civil engineery in his private affairs, became the leader of the majority; but ftill, why Mr. Ockenden, who was a man of fortune, and then was, or had been in Parliament, and does not appear to have had any immediate intereft in this bufinefs, fhould have wifhed in that ftrenuous manner to lead the truftees into a fcheme fo unneceffary, is to me at this day equally wonderful and unaccountable;

countable; but in this I am firmly perfuaded in my own mind, that had Mr. Etheridge fuffered Mr. Prefton to have had his due weight, this fcheme of contraction had never been brought forward; which would have faved the truftees much trouble, the truft much expenfe, and the work itfelf great lofs of time.

In this ftate the Harbour was at the time of this ftruggle, the eaft pier being carried out 757 feet from the fhore, and the weft 849. There however appears not the leaft apprehenfion from the public that the fituation of the Harbour at Ramfgate was improperly chofen; the effort was, that it fhould not be contracted, and its utility thereby diminifhed; nor did any feem to wifh it enlarged, Sir P. Brett and Captain Defmaretz excepted; and how muft thefe gentlemen have ftood aghaft, fuppofing their projected bafon (fixteen feet under low water mark) could have been at once executed, if they had afterwards found, on turning the curves of the piers, that the firft operation of nature was to fill it up, not only level with, but much above the level of the ground wherein it was dug.

After this total ftoppage of the works, a confiderable paufe followed; for we do not find the record of any tranfaction, till 30th April 1760, when the Board came to a refolution to proceed with the works, and gave an order for ftone and other materials neceffary for the purpofe. It may indeed feem extraordinary, that in the laft four years nothing was done. But I have been informed that the truftees were totally at a lofs what to do, as the works had been entirely fufpended, and no new act of Parliament made to chalk out what was to be done. In this interim, I underftand, the truftees applied to the Houfe of Commons, and alfo to the Lords, and were told verbally, that as nothing had been done to fufpend or alter their original powers, the beft way would be to proceed according to their beft difcretion. Accordingly the 20th June 1761, a Committee, by appointment of the Board, affembled at Ramfgate, and reported that it would in the firft place be neceffary to take up the contracting walls; of which they ordered a beginning.

## SECTION IV.

TRANSACTIONS of the Truftees from the recommencement of the work in 1761, to the end of the year 1773; when the Committee preffed the Board to confider of more effectual means for cleanfing the Harbour.

A COMMITTEE, by appointment of the Board, met at Ramfgate the 25th June 1762, and reported that a good deal had been done in laying frefh foundations, and taking up the

works

works of the contraction. Little had been then done at the eaft head; and they recommended the works to be for the prefent carried on at the weft head only

The feveral fucceffive committees of vifitation chiefly contented themfelves with reporting the progrefs of the work. That of Auguft the 25th 1766, particularly mentions that the work of the fourth kant or flexure of the eaft pier towards the weft, was then going on, and the fecond kant of the weft pier was in hand. "That the Committee obferved a collection of fand in the harbour, increafing under the eaft head, near the ftairs; and recommended, as the principal work for the next year, to finifh the weft head as foon as poffible; which they hoped would prevent a further increafe."

Till this vifitation, the works of Ramfgate Harbour (the ftoppage from the contracted plan excepted) appear to have gone on with alacrity, and in full confidence of every degree of fuccefs that had been expected; but it was now a mortifying fight to the truftees to find, that no fooner had they begun to bend the -walls towards one another, fo as to afford protection as a harbour for fhipping, than fand began to collect, threatening to choak it up. This ftruck the truftees with fo much chagrin, that the year 1767 had no vifitation; and in July 1768, after reporting the progrefs of the walling, they only recommended the work to be continued at the weft head, without mentioning the ftate of the fand †.

The fucceffive accounts from Ramfgate contained fo little of confolation, or hopes of amendment, that the years 1769 and 1770 paffed without a vifitation from the truftees; but the next year the Committee's Report, which bears date Ramfgate, 25th Auguft 1771, after ftating the progrefs, and that the eaft head was founded, acquaints us, that the Committee, in order to fix the entrance of the harbour, directed three poles with flags upon them, to be erected one at 200 feet, one at 250, and the third at 300 feet from the eaft head, which laft was then found to be twenty-five feet diftance from the work of the weft pier; and that the Committee, after taking feveral different views of the harbour, and upon mature confideration, recommended that the entrance be 300 feet; and Mr. Prefton was then directed to prepare a proper finifhing for the weft head, agreeable to thofe dimenfions.

* In this report mention is made of a contract for getting ftone upon the fea fhore under the cliffs at Lyme Regis in Dorfetfhire, for burning into lime. For the qualities of this lime, fee Smeaton's account of Edyftone.

† The moft ftriking incident of this report was, that a veffel with lime-ftone from Lyme had run foul upon the works, and done them confiderable damage, which fhews this lime to have been in ufe at Ramfgate.

The

The Committee conclude their Report with obferving, that though a hopper and two lighters had been employed fince January 1770 to empty the harbour, yet they were forry to inform the truftees, that the fand had much increafed therein; but that they had given directions to continue, with the utmoft diligence, to do every thing in their power to make the harbour ufeful; and recommended to the Board to confider if they could find out a more effectual method.

The Committee's Report of 1772, comprehends only the report of progrefs; the two heads being now in great forwardnefs.

Ramfgate, 31ft Auguft 1773, the Committee, after reporting progrefs towards finifhing the heads, and that an engine had been fitted to a lighter for taking up fand, and another Engine for throwing up fand had been repaired, add that they could not help expreffing their great concern in finding a vaft quantity of fand and fullage lodged in the harbour, notwithftanding that fince January 1770 upwards of 52,000 tons had been taken out at an expenfe of £1,100, and it was feared it was rather increafed than diminifhed; and furthermore that the men employed herein refufed to work without an increafe of price.

The Committee defired to remind the Board that in 1771, the gentlemen expreffed their apprehenfions on the fame account; and recommended it to the Board to confider of fome more effectual method of clearing the harbour; that in confequence thereof the Secretary had directions to write to Mr. Smeaton, that the truftees would be glad to have his advice upon the clearing and deepening of the harbour; but as the Committee were informed that he could not attend them that year, on account of a pre-engagement of his to go to Ireland, the Committee therefore advife nothing more to be done, till the opinion of that gentleman, or fome other able engineer, could be had. The Committee however obferve, that of the fand thrown over the eaft pier, confiderable part returns with the tide, and fettles in the harbour: therefore fhould the Board think it advifeable to continue taking up the fand, &c. they recommend it to be taken 200 feet from the eaft head, and 400 feet range along that fide; and that it be carried up in the harbour within the 119 feet of crofs wall now building; and to prevent its wafhing out again to build another wall joining that, right in towards the cliff.

SECTION

## SECTION V.

PROCEEDINGS of the Truſtees from the beginning of the Year 1774 to the delivery of Mr. SMEATON's Report, and Tranſactions upon it, to the end of the Year 1778.

RAMSGATE, 6th April 1774, Mr. Smeaton attended a committee of nine truſtees and the ſecretary.

The ſtorekeeper and harbour-maſter were ordered to prepare an account of all the ſand and ſullage that had been carried out of the harbour, for the information of Mr. Smeaton. The harbour-maſter, foreman maſon, and boatmen to attend him, in taking ſoundings, &c. In conſequence of which he proceeded to make experiments upon the quantity and nature of the ſilt brought in, and to inveſtigate every matter that appeared to him neceſſary to make a full report upon the ſubject of cleanſing the harbour now before him.

The Committee further report that Mr. Preſton laid before them a plan for cleanſing the harbour, by harrowing up the ſand, ſo that it might mix with the water, and be carried out with the ebb tide ; and he was ordered to prepare a model of the machine for that purpoſe, as ſoon as poſſible, and ſend it for the inſpection of the Board.

The 27th Auguſt the ſame year, the Committee again met at Ramſgate, and mention that a quantity of ſtone had been laid to ſecure the outer end of the eaſt head ; and that 260 tons of Portland had been taken out of the foundation of the contracted wall.

Mr. Preſton now produced a model for cleanſing the Harbour ; in conſequence, the Committee gave directions for making a number of rakes for ſtirring up the ſand in the Harbour ; and the 29th the Committee ſtationed fourteen men with the new rakes at the weſt ſide of the Harbour, at two-third ebb, and attended their working till flood ; and ordered them to continue working every tide for ſeven or eight days ; and that ſtakes properly marked ſhould be driven where they worked, and a marked ſtaff to be got by the harbour maſter, to take the depth of the ſand at the centre, within and without the Harbour's mouth, to aſcertain the effect of the rakes in reducing the ſullage of the Harbour.

24th October, Mr. Smeaton tranſmitted his report on Ramſgate Harbour, in conſequence of his ſurvey of the 6th April before mentioned ; an abſtract and extracts of the moſt material parts thereof, are as follows :—

" Having

" Having been confulted by the truftees of the Harbour of Ramfgate, upon the beft and moft effectual method of clearing the harbour from filt gathered therein, having carefully viewed the faid harbour in April laft, in the prefence of a Committee of truftees, and having taken fuch foundings, admeafurements, and other obfervations as appeared neceffary ; having alfo confidered the feveral plans and papers fince put into my hands, by the Board, it appears to me as follows:—

" That a large mafs of filt, confifting partly of mud but chiefly of very fine fand, has been brought into the harbour by the tide, the tide water upon this part of the coaft being charged with a confiderable quantity of mud and fandy matter, whenever it is agitated by the wind, accompanied with a quick flowing tide. This filty matter being thus carried into the harbour along with the water that contains it, and there finding a place of repofe, fettles to the bottom ; and as there is nothing to raife the mud upon the reflow, the water quietly ebbs out of the harbour, leaving the filt behind. And as the fame caufes conftantly operate to produce the fame effects, a continual increafe of filt muft be expected to take place, till fome caufe is brought to operate in a contrary way.

" This is the natural tendency of all harbours ; for wherever there is mud or matter to depofit, an addition to the foil is the natural confequence of a place of repofe ; and a depofition and increafe muft take place, unlefs there are powers either natural or artificial to produce a contrary effect."—" The common natural power is a frefh water river, which continually tending towards the fea, and often, in time of floods, with great impe-tuofity, makes an effort to carry out whatever oppofes it. The fand and filt therefore brought in by the tides, is carried out by the torrent of frefh water. Harbours therefore that have no land water or back-water cannot naturally keep open for a large courfe of years *. Thefe being the effects of the powers of nature, we muft by no means wonder that the harbour of Ramfgate, into which and through which not the leaft rivulet or runner of frefh water takes its courfe, has obeyed this general tendency. For, in proportion as the work of the piers has advanced, the fpace being inclofed, and the water rendered more quiet, and in that refpect more fit for the purpofes of a harbour, in much about the fame proportion has the filting taken place, and muft continue to increafe till the area of the harbour becomes dry land, and inftead of a receptacle for fhips, exhibits a field of corn ; that is, unlefs recourfe be had to fuch artificial means as have the due efficacy.

" How far thefe effects were, or might have been, forefeen before the harbour was built ; or being forefeen, how far it might have been proper to build a harbour there, is

* Large natural harbours or arms of the fea will neceffarily be a long time in filling.

not now the queftion. The fact is, that a noble piece of mafonry has been erected, at a very confiderable expenfe, inclofing a large area in a place where a harbour muft doubtlefs be of very great utility; in cafe the ground fo inclofed remained as clear of filt as it was before its inclofure. The queftion therefore now is, what in effect you put to me; how to make it as ufeful as poffible, and at the moft moderate expenfe?"

In the courfe of this report, it is ftated from actual computation, that at this time there was not lefs than 268,700 cube yards of filt in the harbour: that the two barges then employed by the truftees, with ten men each, got about feventy ton of filt per day: and fuppofing them capable, from weather, regularly to work at this rate, which is fcarcely poffible, and that a ton of filt will be a cube yard; of which, in reality, it is much fhort; yet the harbour at this rate would be above twelve years in clearing, even fuppofing that no frefh filt were to come in during the time.

It is further fhewn, that the whole harbour contains forty-fix acres; and that the area of the external harbour where the filt chiefly lies, being thirty acres and a half, one tenth of an inch in thicknefs over this whole area, would amount to 410 cube yards, or tons; and this at feventy tons per day, would take a week to clear. Now fuppofing the mud to have come in at this rate only, the prefent mafs, independently of what had been carried out, would have taken twelve years and a half to collect: but as it has been chiefly collected fince the inclofure of the harbour, by the curves having been got above half tide *; the increafe of filt could not be reckoned at lefs than double that quantity, or one fifth of an inch per week; which would afford a continual employ for four barges; and therefore that this, with the clearance of the prefent accumulation, and that their work muft in reality fall fhort of the calculation, would render the whole fo tedious a bufinefs, that it by no means appears to be the cheapeft, or moft effectual means of clearing the harbour; which was the queftion before Mr. Smeaton. He therefore propofed, " a method " of procuring an artificial backwater by means of fluices."

" Where no frefh water is to be procured, as is the cafe at Ramfgate harbour, the only refource is to conftruct a pool or bafon to take in the fea water; the tide having there a confiderable rife and fall. This has been done in many cafes abroad, and particularly in the low countries, and in fome cafes in England; but the method has fallen into difrepute here, by its having been found, that the bafon into which the tide water has

* It was only at the vifitation of Auguft 1766, that the growing of the fand was firft noticed by the truftees.

been

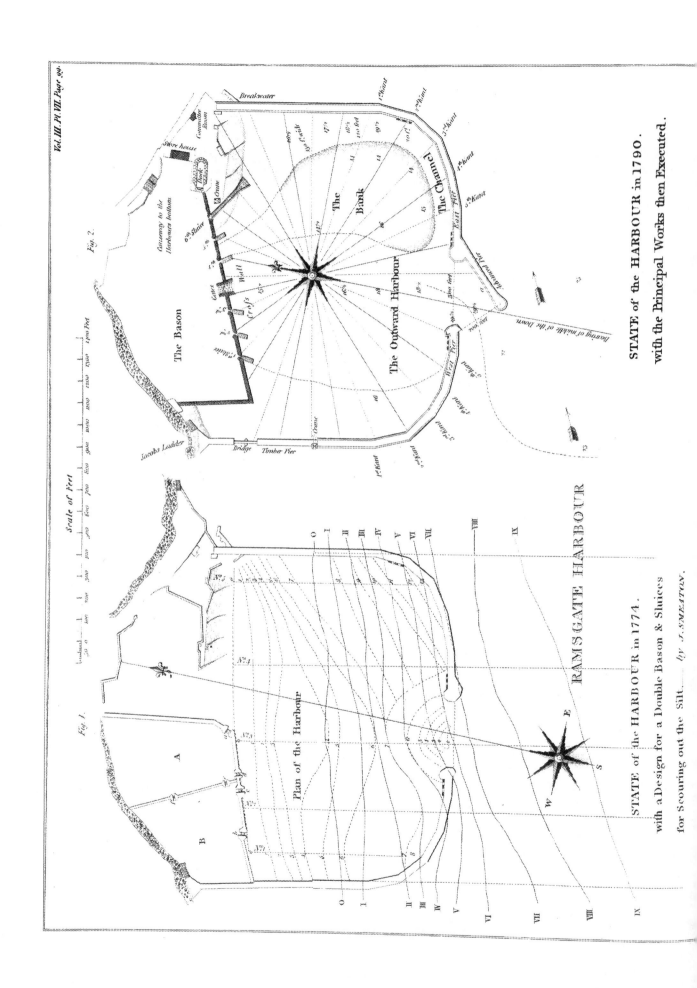

Fig. 2.

Breakwater

Commodore Room

Store house

Causeway to the Harbour bottom

Dock

Crane

The Bason

The Bank

The Channel

East Pier

Old Roadway

The Outward Harbour

West Pier

Jacobs Ladder

Bridge

Timber Pier

Crane

Bearing of middle of the Down

Scale of Feet

Fig. 1.

A

B

Plan of the Harbour

N°.1.

N°.5.

RAMSGATE HARBOUR

STATE of the HARBOUR in 1790.
with the Principal Works then Executed.

STATE of the HARBOUR in 1774.
with a Design for a Double Bason & Sluices
for Scouring out the Silt.    by J. SMEATON.

been received, has itself in a moderate space of time silted up and become useless.—
A method however suggests itself to my mind, though I do not remember to have any
where seen it put in practice, of keeping the bason equally as clear as the harbour intended to be cleaned thereby; and that is by dividing the bason into two parts, by a partition, with a sluice, or sluices, capable of retaining the water in either while the other is
empty; for by this means they can reciprocally be made a bason for clearing each other;
and be both united for clearing the harbour."

The harbour of Ramsgate is very well adapted for the execution of this scheme. It has
every where a found bottom of chalk, upon a regular decline from the cliffs towards the
harbour's mouth, and from thence out to sea. The set of the tide, which is pretty brisk
at particular times, runs crossways upon the harbour's mouth, so that as soon as the sand
is washed to the outside of the heads by the artificial current, the natural current of the
tides will wash it away, and effectually prevent any bar from being formed before it.

" The sand itself is of such a nature as to give the least resistance possible to a smart
current; for the grains being small, though hard, and specifically heavy, yet not being
united by any loamy or tenacious matter, will give way to the impression of a
current."

This report was accompanied with a plan, (see Plate 7. Fig. 1.) distinctly shewing the
mode of executing the scheme; which was by carrying forward the cross wall, already
begun for another purpose, in an eastern direction to a certain point.

A space thus marked out was shewn to contain eight acres of water; which was
proposed to be divided into two basons of four acres each; which it was shewn would
produce a very powerful effect in cleansing the harbour, insomuch as in six months to
be likely to make from fifteen to sixteen feet water, at a common neap; and from eighteen
to nineteen feet at common spring tides.

Those basons were designed to have nine draw-gates, four upon the westernmost, and
five upon the easternmost bason; the whole being pointed in three different directions;
two towards the curve of the western pier; four towards the harbour's mouth, and three
towards the curve of the eastern pier.

To give the sluices all the effect possible, it was proposed to construct a caisson, shaped
something like the pier of a bridge; which being floated to its place, and there sunk, after

being

being put in a proper direction, might be used to divert the current to the right hand or left, as might be wanted. This report concludes with observing, that after the silt is carried by the sluices without the harbour's mouth, there is not any danger of its lodging there; because, having particularly examined the outside, and found all clean of sand, there is no doubt but that the same cause that now operates to keep that ground clean, will continue to do so. Furthermore, that the same means that will clear the harbour, will keep it clear: and that it might be expected that the playing of the sluices eight tides at each spring tide will preserve it.

This report, transmitted to the Board in October 1774, lay before them till August 1775, when the committee at Ramsgate took into consideration Mr. Smeaton's plan, for making the upper part of the harbour into a bason or basons, for scouring the harbour by means of sluices. The committee likewise conferred with Mr. Preston, and considered a plan of his for cleansing the harbour, by making all the upper part into a bason, for scouring it with sluices; and Mr. Preston now declared his opinion that it would clear away the silt to low water mark, wherever the power of the sluices could reach *.

After further consideration of the matter before the Committee, they being cautious in adopting a scheme of expense, gave directions in order to try an experiment what effect an artificial back water might have, for a lighter of fifty tons to be scuttled, seventeen inches deep, and fourteen inches broad on the starboard bow; this to be placed near the end of the cross wall, and filled with water. The sluice being opened at low water, it ran out in a few minutes, and made a cavity in the sand of several feet in diameter and depth. But the ground here being higher before the head of the lighter than at its stern, the water spread a stern, without making the impression expected.

The Committee now ordered the lighter to be removed, and placed between the cross wall and the harbour's mouth; and there repeated the experiment. The first tide the water being let out, made a greater impression than before; but yet spread itself without making any channel.

A channel was then ordered to be made and guarded with planks, and in the afternoon (the lighter remaining in the same place) the experiment was tried again; and the cavity

---

* Mr. Preston seems here to have lost sight of his scheme for harrowing up the silt, proposed to the Board in April 1774; and which was tried by the Committee in August following: in short, ten months after the delivery of Mr. Smeaton's report, Mr. Preston seems to have come to a different view of this business to what he had before.

made

made by the difcharge of the fluice, was about feven feet wide at the furface of the fand, and fix feet deep to the bed of the chalk; and a channel was then forced of near 100 feet long. The Committee now feeing the effect of the confinement of the water, and of repeated difcharges, tried the experiment of the barge twice more, when the hole was rendered ten feet wide upon the furface of the fand, fix feet deep to the chalk, and three feet wide upon the bed of chalk; the channel being full 100 feet long; and the ftream carried fome clods of fand to a confiderable diftance. Thefe experiments were tried in the prefence of Meffrs. Barker, Slater, Aubert, Laprimaudaye, Bennet and Bennet junior, who all became fully fatisfied of the effect to be expected from fluices when made in large.

Mr. Barker and the two laft mentioned gentlemen continuing at Ramfgate another week, in that time further confulted with Mr. Prefton, upon the eligibility of carrying the wall already begun, acrofs the harbour to the old pier; in order to make *all* the upper part of the harbour into a bafon for fcouring it with fluices: and after conferring with feveral old men of the town, refpecting the moft practicable method of clearing the harbour, Mr. Prefton was ordered to prepare eftimates of the ftone and timber for the crofs wall and fluices, and Mr. Stead the carpenter, to prepare an eftimate of the expenfe of making one fluice, and the time neceffary to conftruct it; and to make a model of one, to be fent up to town for the infpection of the Board. The refult was, that the Committee finally recommended to the Board, that Mr. Prefton's plan for cleanfing the harbour by means of fluices fhould be carried into execution with all poffible difpatch.

In the year 1777, Mr. Prefton the mafon being then dead, Mr. Hurft, who had been foreman under him from the beginning of the work, appears as mafter mafon, and Mr. Henry Cull as his foreman. A confiderable progrefs with the crofs wall and fluices, as well as other works of the piers, was reported on this vifitation as the produce of this year's work; as alfo, by the Committee of July 1778, a ftill further progrefs in the works of the crofs wall; the floor of the fifth fluice being laid; and the great gates, and five fluices framed. The Committee now perceived a greater agitation of the fea within the harbour, than before the crofs wall was built fo far forward; and recommended a range of piles and planking to be erected on the eaft fide of the harbour to abate that agitation; and be a means of making that part of the harbour, which is the defirable ftation for the fhipping, more fafe and quiet. This to be done with all poffible difpatch, and the crofs wall to be carried out with all expedition.

SECT.

## SECTION VI.

The firſt Starting and Effects of the Sluices in 1779, with the Account of their Operation, to the end of the Year 1781.

1779.—THE Committee of the 9th Auguſt report the range of piles driven, and in part planked, the works of the crofs wall much advanced ; Mr. Barker, earneſtly defiring to fee a trial of thefe works, though far from finiſhed, or the walls got up to high water, yet, the 1ſt, 2d, 3d, and 4th ſluices being fixed, made a ſtrenuous effort to get the great gates hung, and the fifth ſluice made capable of penning in the water while he ſtaid at Ramſgate. Meſſrs. Aubert and Laprimaudaye at this time arrived, and Mr. Paris attending : the gates being hung, and all the ſluices put down at high water, the Committee attended at low water, and having ordered all the men to be ready, and placing them in proper proportions at the gates and ſluices; the men applied themſelves to the handles to ſtart the ſluices, when the ſpindles upon which the wheels were fixed, broke upon the firſt attempt of every ſluice ; fo that at prefent there was no poſſibility of raifing any more than the two ſluices in the gates, and which indeed were drawn by tackle blocks ; but the force and power of the ſtream iſſuing through thefe two ſluices only, was fo amazingly great, that in its immediate action it forced up the Chalk Rock to the depth of fix and feven feet, and carried pieces of it of three to four hundred weight, to the diſtance of 60 or 70 feet ; and in its courfe, cleared away the filt and fullage down to the chalk, to low water mark ; the ſtream continuing ſtrong 2 or 300 feet, without the Harbour's mouth. Some defects were found in the wall and other parts, which being remedied as well as time would permit, the water was ordered to be pent up in the bafon again; and the tackle blocks to be applied to the ſluices, as well as to thofe of the gates ; yet though all the people were properly placed, there was no poſſibility of raifing more than two ſluices, excluſive of thofe of the gates, notwithſtanding 30 or 40 men heaved at each ſluice. However, with a great deal of labour and difficulty, the Committee afterwards got thofe ſluices ſtarted twice again, and were happy to inform the Board, that the effects produced exceeded the general expectation : the ſtream of water carrying the fand a great way beyond the entrance of the Harbour, in fuch quantities, that the fea at the diſtance of a mile and a half, was obferved to be exceedingly thick and foul. The deep channels through the fand in the harbour, appeared fimilar to the beds of rivers ; and the general voice feemed unanimous in teſtifying their firm belief, that the back water would effectually cleanfe the harbour.

To

To devife fome proper method to ftart the fluices, feemed a difficulty not eafy to be got over, as the people could not contrive any purchafe adequate to the purpofe. But Mr. Aubert obferving that the planking of the fluices were put to draw crofs ways of the plank, which operating in a rough groove of ftone muft neceffarily caufe a great friction; orders were immediately given for the trial of one made in a different way, and they then found it to work with fufficient eafe; and the fpindles being repaired and made of wrought iron, the water was again pent up, and it was difcharged in the prefence of the before mentioned gentlemen, and of Mr. Sibley and Mr. Lutkens, who were then arrived, and things were now found much better adapted to do their duty; but the Committee finding that the rapidity of the ftreams of water iffuing from the fluices had, by its great power, forced up the chalk rock to a confiderable depth; infomuch that the crofs wall would be in great danger of being undermined; they gave orders to the mafon and carpenter to conftruct proper aprons for preventing thofe effects in future.

The Committee further report, that during the laft winter it had been a prevailing complaint that the Harbour had been greatly agitated, and at fome particular times rather unfafe for fhips to lie in; which was not the cafe before building the crofs wall, as the fea then broke and fpent itfelf upon the fhore: and as it had been obferved that the fea moftly ranges along the weftern pier, and that the crofs wall ftopping and repelling the fwell, it returns on account of its not having any vent or outlet, and caufes that great difturbance and agitation now complained of. Under thefe circumftances, the Committee recommended that about 2 or 300 feet of the weftern end of the crofs wall (or what might be neceffary) fhould be taken down; and that from thence a wall fhould be built up towards the cliff; and furthermore, that about 80 or 100 feet of the timber pier fhould be taken away, beginning about the end of the crofs wall, and the opening to extend towards Jacob's Ladder; the Committee having great hopes that this plan would render the Harbour more fafe and quiet.

The Committee further mention it as a general opinion, that another fluice would be found neceffary to fcour the upper or northern angle of the eaft pier; and therefore recommended that a fluice fhould be made from the angle in the old pier, to be carried through the work yard to the end of the carpenter's fhop.

It may here be proper, before we go on further, to advert to the ftate in which the Harbour really was previoufly to the trial of the fluices: indeed, it was fo bad that I do not find any memoranda have been preferved of its condition. I muft therefore recur to the ftate in which I found it at the time of my furvey in 1774. At that time, in the very centre of the outward Harbour the fand was accumulated to an elevation four feet above the

the level at which the thresholds of the present gates are laid; and this being then the best, that is the deepest part of the Harbour, vessels drawing above ten feet water could hardly be said to get into it even at spring tides. At low water there was no water to be seen in the Harbour, except a small roundish area reaching a little within the pier heads at neap tides, and at spring tides none but what was immediately between the heads. Under the curve of the east pier, which was the proper birth for large vessels (could they have occupied it), the sand lay considerably higher; so that in the third angle, which is naturally the best birth, there was no less than thirteen feet in depth of filt, lying upon the chalk bottom, which would be seven feet above the level of the present threshold of the gates; and if this was the condition of the Harbour in 1774, we must conjecture how much worse it had become in this year 1779, when the fluices were first brought into action, as has been described, notwithstanding that the barges had been all the time employed in getting the sand out as fast as they could. In this forlorn state the Harbour of Ramsgate had become justly reprobated by the public, as a work not having the least appearance of utility, or likelihood of being made useful.

1780.—The Committee's report dated 7th August, notices that 150 feet of the cross wall had been taken down, and a considerable progress made with the returning wall towards the cliff; and that 100 feet of the timber pier had been removed to make an opening; which was then made and paffable by a wooden bridge, with a stone pier in the middle. Aprons being also laid to all the fluices, and examined by Mr. Barker, the water was ordered to be shut in, and Messrs. Slater and James arriving, took their station at the pier head, and found that the stream ran exceedingly strong beyond the mouth of the Harbour, carrying the sand into the tide.

The fifth fluice having not yet been put in action, it was ordered to be forthwith completed; and Mr. Barker finding a great bank of sand extending eastward of the fourth fluice, directed the harbour-master to place two barges to turn the current of the water to run upon the point of the bank. The fifth fluice being completed during Mr. Barker's stay, he ordered it to be run without the rest, and found the effects of it to be amazingly powerful; infomuch that in a few times running it entirely cleared away the fullage within a few feet of the old pier, and carried out of the Harbour's mouth great quantities of sand. Mr. Barker being defirous to have all the fluices completed as soon as possible, and as there was the greatest reason to believe that the fifth, with an addition of a sixth fluice, would operate very powerfully in cleansing the east side of the Harbour, which, as already noticed, is the principal birth for shipping, he ordered the digging for the foundation of the conduit of the sixth fluice to be begun; and before he left Ramsgate, saw a considerable

able

able length of the foundation dug, and the ſtone work itſelf begun. He alſo left direc-tions to run the ſluices once in 24 hours, and ordered the chalk that had been forced up by the running of the ſluices, to be laid under the croſs-wall to ſecure its foundation.

Obſerving now, that a prodigious quantity of ſilt accumulated in the baſon or upper harbour, and that the clearing it would make it vaſtly more commodious for ſhipping; by way of experiment he ordered the ſluices to be put down at low water, and that one ſluice only ſhould be opened about an hour before high water; for it was ſuppoſed that the action of the water iſſuing into the empty baſon, would ſtir up and looſen the ſilt ſo as to facilitate its getting out when the ſluices acted in the contrary direction. The expe-riment was twice tried and found to have ſome effect; but the power of the water ſeeming to endanger the foundation of the wall, he found it more adviſeable to order that ſix or eight men ſhould attend at the running of the ſluices, to ſhovel down the ſilt into the current, when the water was ſufficiently down to admit thereof.

1781, Committee's report, Auguſt 6th.—The work at the fifth and ſixth ſluice being now completed, a new channel was ordered to be dug through the ſand from the ſixth ſluice, and a couple of barges to be laid, ſo as to direct the water thereof through the channel. In a few times running, it was found that the bank was conſiderably decreaſed, and that the water of this ſluice flowed ſo high as to overtop the conduit wall; from the whole of the operation of the ſluices there appeared the greateſt reaſon to believe that the fifth and ſixth ſluices, and particularly the laſt, would be effectual in cleanſing the eaſt ſide of the harbour.

From this era Ramſgate Harbour began to put off that forlorn appearance of a repo-ſitory of mud, and to aſſume one more reſpectable than it had done for 15 years before; and the truſtees ſeeing that they were now become competent to the cleanſing of the out-ward harbour, and finding the great utility of the baſon, turned their mind towards the clearing of that from the fullage gathered therein; for which various propoſitions preſented themſelves; viz. Firſt, the barges to be filled at a price per ton, as had been done in the outward harbour: Secondly, the fullage to be taken up and lead out with carts: Thirdly, a premium of three-pence per load to be offered to the farmers to take it out as manure: Fourthly, the barges to be filled by the labourers, and carried out by day works: Fifthly, the barges to be filled with water, and being ſcuttled, to play upon the bank of fullage at low water, in the manner the experiment was conducted; labourers being employed to throw down the bank of ſand. The two laſt of which ſeemed then the moſt practicable ſchemes.

* Neceſſary alſo for the purpoſe of a backwater.

## SECTION VII.

TRANSACTIONS from the Propofal of a Dock in 1782, to and inclufive of SMEATON's third vifit to Ramfgate in the Summer of 1787; with the Tranfactions to the Death of John Barker Efquire, in the Autumn of that Year.

1782, COMMITTEE's report, dated Ramfgate, 5th Auguft.—On this vifit Mr. Barker had the higheft fatisfaction in finding the great improvement that had been made in the harbour by means of the fluices, and particularly in the middle part; and that in the channel under the eaft pier, there was full nineteen feet water at a fpring tide. This improvement was greatly owing to a temporary fence having been made, for turning the water of the fifth fluice to operate along with that of the fourth; but which having been carried in an irregular line, fo that the water went over it, did not produce all the effect it might have done; and therefore, by way of expedient, he ordered the barges and lighters to be laid along-fide the fence, the more effectually to confine the waters and grind a channel. He had alfo high fatisfaction in finding, that by the means propofed laft year, a great part of the bafon was now cleared of fullage down to the chalk; and that from the gates to the crane there were fourteen feet water at fpring tides, and he fuppofed that at leaft 80 fail of veffels might be fheltered therein.

He found in the bafon a Venetian veffel of 300 tons, that had been hove down, and her keel repaired; and alfo a Swedifh fhip of 340 tons meafure, that had been unloaded of 1,700 quarters of wheat in the bafon, and there undergone a thorough repair; and with pleafure obferved that veffels brought in from the Goodwin Sands, made directly for the gates, and failed into the bafon without difficulty. Therefore, taking all thefe things ferioufly into confideration, and finding that this harbour was not only a place of fhelter for fhips in diftrefs, but alfo for the repair of their damages at fea; two more articles feemed effentially neceffary to make the harbour of that complete utility to the public, that now appeared near in profpect; and thefe were, a ftorehoufe contiguous to the bafon, for the reception of the goods that were thus obliged occafionally to be put on fhore while the veffels were repaired; and a dock for occafionally taking in a veffel to be thus repaired. Thefe feemed, indeed, matters fo important and of fuch immediate concern, that Mr. Barker defired to have Mr. Aubert's advice, and that he would alfo bring down with him Mr. Smeaton, to advife with refpect to the practicability of the dock, &c.

Meffrs. Aubert and Smeaton immediately attended this fummons; and all agreed that the ftorehoufe would be of great utility as well as the dock; but that as the latter

was

was likely to be a work of expenfe, it would be proper not to lay more burthen upon the truft, in the whole, than neceffary ; and therefore that it would be eligible to let an area of ground for a ftorehoufe on leafe to private adventurers, under certain reftrictions ; but that with refpect to the dock, it would probably require that care and caution in conftruction, as not to make it eligible to be trufted to private execution.

In confequence therefore of my orders, I proceeded to infpect the ground, with a view to the building a dry dock for the graving and repairing of veffels ; and to inform myfelf of every circumftance relative thereto, that could in its nature be then known.

On this vifit, it was not with a fmall degree of fatisfaction that I was eye-witnefs to the very great difference that there was in the condition of this harbour, to what it was in the year 1774, when I was laft there : and that the hints fuggefted in my report of that year being carried into execution, had been attended with every profpect of fuccefs that could reafonably be expected.

The gates having been fhut in, and having feen the operation of the fluices, what feemed to be now moft wanting, was an increafe of power fo as to widen the channel under the eaft pier. And as I obferved that the bringing of the ftream of the fifth fluice to act along with the fixth, was but very imperfectly executed, to prevent lofs of time I then ftaked out a curve for the direction of a fence or turnwater ; which being temporarily executed with ftakes and boards, and found to anfwer the end, has, with repairs, remained there to this day.

It was then complained, that the gates were much too narrow ; but finding them thirty feet clear width, and therefore wide enough for the largeft veffels then expected into the bafon ; I obferved, that veffels any thing near that width could never venture to run in under fail. Nor did it appear to me, that thofe gates had ever been defigned with that idea : for I found the entering angles were very little rounded, and the entering walls themfelves upon a parallel ; fo that the leaft fwerving of the veffels, to the right hand or left, would make them ftrike the ftone work, unlefs in the cafe of fmall veffels where there was a good deal of width to fpare. To remedy this inconvenience, I recommended to build flue walls, to project forward thirty feet, and in that length to fplay out, or widen, five feet each, fo that at this new entry the width would be virtually forty feet ; and which might be deemed wide enough for fecuring the entry of veffels of thirty feet. I alfo noticed, that if the fharp angles of the ftone work of the fluices were alfo properly rounded off, they would give a more full bore of water.

P 2

The

The gates themfelves, alfo, being of too flight a conftruction, were already got into fuch bad condition, that I was defired likewife to make a proper defign for a new pair of gates.

I afterwards took an opportunity of viewing the docks at Liverpool, and early in the year 1783 delivered, not only my defigns for new gates for the bafon, with the flue walls, which were immediately ordered into execution; but for a graving dock with a timber bottom, like thofe I faw at Liverpool, and which indeed appeared to me abfolutely neceffary in a fituation like Ramfgate. I knew that the natural bottom would be the fame as that of the whole harbour, which was univerfally a chalk; and which, though it had never been proved by any excavation under low water mark that had been attempted to be kept clear of water, yet it might be reafonably pre-fumed, that though the foundation was likely to be free from grofs fprings, yet it would be fubject to fuch tranfpirations as could not take place in clay or folid ftone, and would therefore need a wooden floor; and as in works of this kind it is ufual to call upon the defigner, in cafe any thing in the nature of the ground or foundation turned out different from what was originally to be fuppofed, there could therefore be no neceffity to give directions for every poffible cafe and contingence.

1783, Ramfgate, 4th Auguft.—Mr. Barker found that the ftorehoufe recommended had been built; and that the fharp corners of the fluice walls had been rounded off, in order to give more water way; that a new fence of piles and planks had been executed, to unite the waters of the fifth and fixth fluices; that a hill of chalk in the bafon had been ploughed up and removed; that the eaft flue wall of the gates had been carried up to the height and dimenfions fpecified by Mr. Smeaton's plan; and that the weft flue wall was in great forwardnefs, and he found it to be the general opinion, that thefe walls would greatly facilitate the entrance of fhips into the bafon; and that the graving dock, already refolved on by the Board, would be very ferviceable in the repair of fhips damaged at fea, and brought in for that purpofe. During Mr. Barker's long ftay here at this vifi-tation, the new gates were actually hung, the water fhut in, and every thing found to anfwer; fo that, after feeing the fluices run feveral times, he concluded that they would in every refpect fulfil the intended purpofe. He expreffed himfelf as particularly happy in acquainting the Board, that the vaft accumulated body of filt which formerly choaked up the bafon, was now entirely cleared away, and that fhips from 200 to 400 tons might

* It was afterwards intimated to me, that, as the dry docks of Liverpool were efteemed of the moft complete kind for merchants fervice, the truftees would wifh me to take an opportunity of feeing them.

lie

lie in the bafon with the greateft eafe and fafety, and that in a confiderable number : and fays, " that in the courfe of laft winter and fpring, property to a very large amount had been preferved, as feveral large fhips with valuable cargoes were brought in and taken care of. The cargo of one large fhip only being eftimated at £35,000. ; and that under Providence, it has been the means of preferving many ufeful lives to their families, and the community.

1784, Ramfgate, 26th July.—From Mr. Barker's report upon this vifitation it appears, that a new dock being determined upon by the truftees to be built according to Mr. Smeaton's plan, it was immediately begun ; and Mr. Barker laid the firft ftone 31ft July, and the workmen were directed to proceed with all poffible expedition *.

On this vifitation, Mr. Hurft the mafon, having been found incompetent to this work, was fufpended : this was attributed to a fit he had had two years before ; but being an old fervant, from the firft commencement of the work, under Mr. Prefton, and always confidered as a man acquitting himfelf well, Mr. Barker recommended it to the truftees to allow him fomething by way of fubfiftence. Mr. Cull therefore, his foreman, was recommended inftead of Mr. Hurft, as mafter mafon, and Mr. Speers to be his foreman.

Mr. Cull being confulted by Mr. Barker, recommended that the thicknefs of the north wall of the dock fhould be increafed by two feet ; and to build both the walls two feet higher than they had been fet out in Mr. Smeaton's plan, to which there could be no objection. With refpect to the floor, they apprehended no difficulty in making it of ftone like that of the gates ; thinking it more eligible than of wood, as being more durable, and not fubject to the worm.

Mr. Barker had the great fatisfaction to find that both the outward harbour and bafon had been greatly improved ; as the channel under the eaft pier had been confiderably widened and deepened : and the fullage of the bafon, being now entirely and conftantly kept down, betwixt four and five feet depth had been gained at the upper end thereof.

During this vifit at Ramfgate, in a hard gale of wind which happened at N. E. Mr. Barker had an opportunity to be eye-witnefs to the more quiet and preferable ftate that.

* Here it may be proper to obferve, that had Mr. Smeaton's plan been in reality complied with, it would probably have taken the whole of this feafon to have completed, and laid down the timber floor ; before a fingle ftone could have been laid : but it was thought proper to begin the walls according to Mr. Smeaton's lines, exclufive of the floor.

the

the harbour was now in, to what it was before the wefternmoft part of the crofs wall was taken down, and the opening made through the weft pier, as has been defcribed. The new ftorehoufe now built by direction of the truftees, he found to be a ftrong and com-modious building, convenient, and every way adapted for the reception of goods ; as fhips of 400 tons could range up clofe under it, and unload their cargoes with the greateft facility.

1785, Ramfgate, 5th Auguft.—On this vifit Mr. Barker reports that the dock was in great forwardnefs ; for the walls were built up twelve feet high, and that 3065 feet of pavement had been laid at the bottom. That the gates were made, and ready for hanging ; and that the whole feemed to be an exceedingly good piece of workmanfhip ; and that it was likely to be ready for ufe by February next. He notices that the worm prevailed much here, and ate the wood work of the fluices to fuch a degree, that he recommends every part to be of ftone that can be.

He had the pleafure to find every thing in an improving ftate ; and that the encomiums which he thought it only juftice to beftow upon the harbour the laft year, had been more amply confirmed by the appearances in this. He fays the whole body of fullage in the outward harbour was certainly confiderably diminifhed ; and the channel under the eaft pier improved and widened, but ftill not to that extent to fatisfy his wifhes ; or that was ex-pected from the united waters of the fifth and fixth fluices ; however, as the widening of this channel was of important confideration, he ordered two of the barges to be laid clofe in the channel ; and ten or twelve men to fill them, as had formerly been done before the fluices were brought to act. It was however obferved to him by the ableft pilots and feamen beft acquainted with the harbour, that it would not in reality be well to have the whole bank of fand taken away from the eaftern fide of the harbour ; and they put a cafe, that fuppofing the harbour quite freed of this bank, and a fhip coming in in a hard gale of wind, without anchors and cables ; if there was nothing to bring her up, fhe would be liable to be driven againft the walling of the piers, and be knocked to pieces ; they therefore ftrongly urged the neceffity of having a bank on the eaft fide of the harbour upon which fhips might bring up.

He notices that it had been (and not long fince) the reproach of this harbour, that it never could be made capable of receiving any thing but fmall veffels ; but that now it was capable of fheltering almoft any merchant fhips (Eaft Indiamen excepted) ; and as a proof of it, a fhip which was faid to be of 800 tons, took refuge in the channel of the eaft pier laft winter, and lay in the faid channel for two months ; and a large Norway fhip, called

the

the Ebenezer, with upwards of 30,000 deals on board, got upon the Goodwin fands, where her mafts and rigging were deftroyed ; yet fhe was neverthelefs got off by the Ramfgate boatmen, and brought into the channel of the outward harbour ; though fo water-logged that fhe actually drew eighteen feet and a half water ; and in a gale of wind that happened lately, about 30 fail of veffels, Englifh and foreign, with valuable cargoes, the greateft part of which having loft their anchors and cables in the Downs, had been preferved by being brought into the harbour and bafon, where they had every neceffary affiftance ; fome of which were then unloading their cargoes into the ftore-houfe, being leaky and much damaged. Indeed it was, he fays, now generally admitted, that within a very few years, the harbour had been improved, to a degree, even beyond what its warmeft friends and promoters ever expected.

1786, Ramfgate, 5th Auguft.—" Mr. Barker thinks it proper to acquaint the Board, that the dock is built, and is an exceeding neat piece of architecture ; but the water unfor-tunately rifes up through the pavement, a great part of it having been heaved up and loofened. A difagreeable circumftance, equally unforefeen as unexpected.—It is now perceived that there are natural fprings which rife in the bed of the dock ; that thefe, with the great weight of water in the bafon, when it is fhut in, united, have been fo ftrong in their effect as not only to break through the cement, but have in many places broken the paving ftones ; a circumftance very mortifying, after fo much pains have been beftowed upon it." Mr. Barker, in confulting with the mafter mafon upon the moft probable means of preventing the waters rifing, was informed that twenty-three years ago a new dock had been built in the King's yard at Plymouth ; and when fhut in, it was found that the water rofe up in the bottom, and loofened the pavement fimilar to what had happened here ; and that no other remedy could be found but new laying the floor with heavy Portland blocks, which method Mr. Barker recommended to the Board to be taken at Ramfgate.—He fays, " It is propofed to take up the prefent pavement, and lay down large blocks of ftone three feet by four, and two feet fix inches deep ; each ftone to be one ton and a half ; and that the mafter mafon is firmly of opinion, the weight of thefe blocks will effectually prevent the water from rifing, and keep the dock dry."

The ftone merchants at Purbeck on being immediately written to by the fecretary, returned anfwer, that they were not able to make ftones to the dimenfions required, but recom-mended ftones of a lefs bulk ; but as that would introduce more joints and other difad-vantages, Mr. Barker took the mafon's advice, in having them ordered from Portland.

Mr. Barker had the higheft fatisfaction in informing the Board, that every year more and more demonftrated the utility of the harbour, as was manifeft from the great number

of

of ſhipping that in the courſe of the laſt winter had taken ſhelter in it, and the number of lives preſerved. He mentions the following facts; a large Swediſh ſhip ran upon the Goodwin and parted; two Ramſgate boats, at the hazard of their lives, brought the crew, conſiſting of the captain and twenty men, ſafe on ſhore from the wreck *. In December laſt, eleven ſhips, all of which had loſt their anchors and cables off Dungeneſs, and ſeveral of them otherwiſe damaged, were brought into Ramſgate by Dover pilots; and the day before Mr. Barker left Ramſgate, a Hamburgh ſhip ran upon the Goodwin, and the captain and crew deſerted her; but ſhe was got off and brought into Ramſgate by two boats, one of Ramſgate and the other of Deal. She came into the baſon and delivered her cargo into the ſtorehouſe (eſtimated at betwixt 8 and £9,000), till ſhe repaired.

1787.—The Portland blocks being got with readineſs the work of the floor was vigorouſly puſhed during the winter and ſpring; ſo that it was completed early in the ſummer, which being notified to the Board, orders were ſent to the maſter maſon to put the work to a trial. In conſequence, Mr. Cull's letters of the 3d and 7th Auguſt, acquaint the Board, " that agreeably to his orders he had ſhut in the dock, when to his great concern and ſurprize, and that of every one preſent, by the time it was high water, the greateſt part of the pavement was disjointed, and hove up; and what was yet more aſtoniſhing, nearly 100 feet in length of the north wall was hove up alſo.

Ramſgate, 22d of Auguſt.—Mr. Barker on his viſit, taking the affair of the dock into ſerious conſideration; that the harbour being now become very uſeful to commercial navigation, and that numbers of veſſels took ſhelter therein in winter, for the purpoſe of repairs; the dock muſt certainly be not only an uſeful, but a neceſſary accommodation; and conſidering that a large ſum of money had been already expended thereon, he judged it proper to engage Mr. Smeaton to come down with Mr. Aubert to examine it, and if poſſible, to contrive ſome method to render it uſeful.

In conſequence, Mr. Aubert and Mr. Smeaton arriving, the gates were ordered to be ſhut at low water; and Meſſrs. Barker, Aubert, and Smeaton attending the riſe of the tide; Mr. Smeaton's report to the Board of truſtees contains the following account: " The tide roſe that day to the height of thirteen feet four inches upon the apron of the gates of the dock; but before it had riſen two feet, it begun to ſpring through ſeveral joints of the ſtone-floor, which had been laid with ſolid Portland blocks of two feet and a half in thickneſs, in faſhion of an arch; and which, to all appearance, had originally been ſufficiently

* It is proper in this place to remark, that the Ramſgate boats lie conſtantly afloat, ready to put off at a minute's warning; had this not been the caſe in that critical moment theſe unfortunate men would have periſhed.

well

well jointed, and indeed the whole building in point of workmanſhip, as a piece of maſonry, had been done in a very maſterly manner.

" As the height of the tide increaſed upon the apron, the leakage through the joints of the floor gradually formed a greater depth of water upon it, ſo that when it was high water upon the apron, there was a depth of five feet three inches upon the floor ; and we obſerved, that while the tide was riſing, the joints of the ſide wall, on the north ſide next the baſon, apparently opened, ſo as in ſome places to let water through the wall ; but in far leſs quantity than appeared to riſe through the floor *. Every thing being left ſtanding when the tide had ebbed, ſo as to be upon a level with the water within, its depth upon the floor was ſeven feet two inches."

" The cauſe of theſe derangements, was doubtleſs owing to the preſſure of the water under the bottom, endeavouring like a veſſel ſwimming in the water to buoy it upwards ; and which, in the circumſtance of only eight feet difference of preſſure (which in this ex-periment was the greateſt) would amount to 1000 tons upon the area of floor. This power acting upwards, would indeed be the ſame, whatever material the floor was com-poſed of ; but from its conſtruction, as an arch of ſtone laid very flat, its lateral preſſure would, in this caſe, be much greater than the abſolute weight of the wall upon its baſe ; and therefore no wonder that it ſhould ſhove it outward ; and that the effects thereof ſhould be perceived by breaking the joints of the ſide wall, as has been mentioned."

" He thinks it neceſſary to ſtate to the Truſtees, that this failure has not been owing to bad materials, or bad workmanſhip, or to taking a method in itſelf bad, but only become ſo, by the conſtruction not ſuiting the ſituation and ſoil ; for had the ground whereon the dock was founded been a ſtiff or moderately compact bed of clay, or a rock, either hard or ſoft, that would not have ſuffered ſprings to percolate through its pores ; the dock built as it was, or even in its firſt ſtate, would doubtleſs have ſucceeded ; but it was the circumſtance of there being ſprings iſſuing from the area of chalk, on which the floor was laid, that has occaſioned the miſchief †. And in this reſpect a wooden floor, according to the original plan, would have been the eligible expedient, as it would have been ſubject to the upright preſſure only, and not to a much greater lateral preſſure as in ſtone arches."

* The Committee's report mentions that the joints of the pavement, or more properly the caping of the wall, had opened one-third of an inch, though the wall itſelf was fourteen feet thick at its baſe.

† As I obſerved a conſiderable ſtream of water iſſuing from under the apron at low water, I ordered this water to be drained therefrom by a chain pump, and found that this ſpring (which was ſalt) vented at the rate of 160 barrels per hour, which vent expoſed the bottom to the action of the tides waters.

The Committee's report ſtates that it was Mr. Smeaton's opinion, that the whole bottom of the dock, as alſo the greateſt part of the north wall, are ſo greatly deranged and diſjointed, that it will be neceſſary to take up all the pavement and take down a great part of the north wall; and that, to render the dock ſecure and dry, the bottom muſt be laid upon a different principle, and with a different material; and the wall itſelf made a ſolid piece of maſonry, ten or twelve feet thicker than it is at preſent, for which he was deſired to make the proper deſigns. Accordingly, the men were immediately ordered to begin to take the wall down.

Every thing elſe was found in a way of doing its buſineſs, and anſwering its end; and in attending to the running of the ſluices, it was ſeen that the water ran with ſo great a current and body, as to make its way into the ſea, apparently half a mile beyond the mouth of the harbour; but yet Mr. Barker's favourite wiſh of widening the channel under the eaſt pier to a yet greater extent, induced Meſſrs. Barker and Aubert to take the opportunity of conſulting Mr. Smeaton upon that head alſo; and he recommended it to the truſtees, to build another ſluice to the eaſtward of the crane, and by uniting its water with that of the fifth and ſixth, the whole would have an increaſed effect, not only in cleanſing, but in widening that channel.

Meſſrs. Barker and Aubert, attentive to every thing that might conduce to the improvement of the harbour, made uſe alſo of this opportunity in having Mr. Smeaton's opinion upon a point ſo very material and critical, that before the erection of the ſluices, it could not have even been thought of; and that was, that though by the opening of the weſt pier, much good had been done towards quieting the harbour; yet as in very hard gales of north-eaſterly and eaſterly winds, the harbour was ſtill liable to a greater agitation than could be wiſhed, a complete remedy was greatly deſirable. And finding that the ſeamen and pilots of Ramſgate had formed an opinion, that a wall or pier extended out from the eaſt head to 350 or 400 feet in a proper direction, would greatly tend to keep out the ſea, and quiet the harbour; this propoſition was laid before Mr. Smeaton, and he was ſtrongly and clearly of opinion, that ſuch a work would render the outward harbour ſtill more quiet; and that if another ſluice were built it would, together with the good effects from it, before propoſed, countervail any difference that might ariſe from the more weſterly direction of the harbour's mouth, to take in ſilt, which might (not without reaſon) otherwiſe be apprehended.

It appeared to the Committee, that in the courſe of laſt winter, no leſs than ſixty ſail of ſhipping had been ſheltered in the baſon at one time, many of them brought in without anchors and cables, and otherwiſe much damaged, and obliged to unload their cargoes to

be

be repaired; and that Mr. Rowe the pilot had actually carried a ship of 500 tons directly into the bafon.

The preceding proved to be the laft report of John Barker Efquire, who, with great affiduity, attention, and perfeverance, had prefided over the execution of the Ramfgate works for nearly twenty-eight years.

## SECTION VIII.

THE unanimous choice of Alexander Aubert Efquire, Chairman, in Autumn 1787; and the fubfequent tranfactions down to the prefent time.

IN confequence of Mr. Barker's death, which happened the 1ft day of November, the truftees were unanimous in requefting Alexander Aubert Efquire, to take the chair; which having been done, the gentlemen obferved, that from the fituation of Ramfgate har- bour, and the very great ufe it has already been to fhipping, there was no doubt but it might be made of the utmoft utility to commercial navigation; but as yet there remained a great deal to be done, and many works to be carried forward to bring it to that ftate of ufefulnefs and perfection it was capable of; and confequently it being an affair of great mag- nitude and importance to the public, it required the conftant care and attention of a gentle- man of abilities and refpectability, therefore the Board in the fame unanimous and earneft manner, requefted Mr. Aubert to take the lead in the management and direction of the bufinefs and affairs of the harbour, and every thing relative thereto; to which requeft Mr. Aubert politely affented, at the fame time requefting the gentlemen to affift and fup- port him in every meafure tending to the benefit and public utility of the harbour.

Mr. Aubert being elected chairman of the Truftees of Ramfgate Harbour, a vifitation was appointed, and the chairman to be attended by the fecretary and Mr. Smeaton, and they arrived at Ramfgate at Chriftmas this year.

The firft object of Mr. Aubert's attention was to render the Harbour perfectly quiet; the utility of it in its prefent ftate could not be more evidently proved than by the great number of veffels that had taken fhelter therein *. It proved a hard gale of wind at eaft on Chriftmas day, fo that Mr. Aubert was himfelf witnefs of the effects thereof, and faw that a great degree of agitation prevailed in the outward Harbour, beyond what before he had a conception of; he therefore affembled and confulted fome of the ableft and moft experienced pilots, upon the propofition mentioned at his laft attendance at

* - - - - - - They were informed that above 70 veffels had been in the harbour at once, fince they were there in Auguft laft, the greateft part driven in by diftrefs of weather; fome of them were of 350 tons.

Q 2

Ramfgate,

Ramſgate, and they all agreed that there was an abſolute neceſſity for ſuch a work being carried into execution ; and that if a pier of 350 or 400 feet were carried out in a proper direction, it would not only keep out the heavy ſea that now tumbled in in hard gales of wind, and make the Harbour more ſafe and quiet, but that the coming into the Harbour would in reality be more ſafe and eaſy. For they obſerved, that at and near high water the tide runs briſkly from the weſtward acroſs the Harbour's mouth ; which obliging the ſhips and veſſels that intend to make the Harbour to come down from the weſtward ; if a pier were extended in a proper direction, they would then come in right along with the tide, and with more facility than at preſent. The only thing that appeared to them in the ſhape of a doubt or a difficulty, was, whether the Harbour could be as effectually cleanſed from ſilt, that is continually brought in, as it now is * ? To which Mr. Smeaton made anſwer, that before the eſtabliſhment of the baſon and ſluices, ſuch a work could not have been thought of; but there was now ſo great a power of backwater, which could be increaſed as already intimated, if there were found occaſion ; that if the work pro-poſed would not leſſen the facility of the entry, which in their judgement it would not, he would be anſwerable to keep the Harbour to as great a depth, and as clear of ſilt as it now is.

Mr. Aubert concluded his report to the Board, with obſerving, " that the works of Ramſgate Harbour have, for a ſeries of years, been carried on without the aſſiſtance or direction of any engineer, or even a reſident ſurveyor †, one or two occaſional con-ſultations excepted ; and in thoſe particular caſes the plans of the engineer were only in part adopted. It therefore ſeemed to him, that in works of this nature and magnitude it was of great importance to have an able and ſkilful engineer ; who uniting the powers of mechaniſm with a thorough knowledge of the materials, ſo as to apply them to the beſt advantage, might rationally be expected to be the means of ſaving many thouſand pounds, and give durability to the works ; and is conſequently deſirous of being guided by ſome profeſſional man in that line."

" Upon theſe conſiderations Mr. Aubert has no doubt but the Board ſee the neceſſity of employing conſtantly ſome able engineer in the deſign and conſtruction of their future works ; and therefore as Mr. Smeaton has already been employed occaſionally by the truſtees, and by ſeveral of the executive Boards under government, and is well known to ſtand high in profeſſional character, Mr. Aubert is perſuaded the Board will approve

* Mr. Smeaton's ſketch of an advanced pier was then laid before them.

† Mr. Etheridge was the only perſon who had been employed in this capacity, except occaſionally, and whoſe ſervice terminated with the year 1753.

of

of his being requefted to take upon him the guidance and refponfibility of the engineering part of the works. But although Mr. Smeaton may be induced to go down to Ramfgate periodically, and to furnifh every neceffary plan, as he cannot upon any confideration be refident upon the fpot; it appears neceffary, that he may be made refponfible for the per-fect execution of his plans, that he fhould have fome experienced refident furveyor, in whom he can place confidence; and fuch a one, in cafe the Board fhould think proper to employ him as their refident furveyor, he is enabled to recommend in the perfon of Mr. John Gwyn, who has been employed twenty-feven years under Mr. Smeaton's direction as deputy furveyor; and who during that whole period has given him the greateft fatisfaction in the execution of feveral capital works.*"

In confequence of thefe recommendations Mr. Smeaton was unanimoufly chofen engineer, and Mr. Gwyn refident furveyor.

Being now entered upon my office of engineer of Ramfgate Harbour, it may be fuppofed that in carrying forward the account of the tranfactions thereof, I am no longer under the neceffity of drawing my materials chiefly from written documents, but that I may in my own perfon carry on the narrative; and this I am the more inclined to do, as I can ftate my own remarks on perfons and things with much greater brevity.

1788.—In April I attended the chairman on his firft vifitation, accompanied by the fecretary. At this time the whole of the north wall of the dock was taken down and the area cleared, and a confiderable progrefs made with the timber new floor. Upon this vifit the chairman fully opened and explained his intentions to us, the principal officers and artificers; viz. That though the re-conftruction of the dock was with him a great object; as when completed it would be of very great utility and accommodation to thofe who are driven into the harbour by ftrefs of weather; yet he could not confider it as the primary object to which we ought to bend all our powers. It appeared to him, that the rendering the harbour a place of quiet and fafety for the fhipping, was to be the thing aimed at, and endeavoured to be atttained, in preference of all other confiderations. That therefore the works of the propofed advanced pier were to be pufhed, at the fame time that thofe of the dock might be carried on whenever it could be done without retarding the primary object; and with this view he ordered an additional number of workmen.

* Mr. Aubert fpecified the works executed by Mr. Gwyn, under Mr. Smeaton; viz. the Calder Navigation, the Bridge of Perth, the Harbour of Portpatrick, the Pier of Aberdeen, the Pier and Harbour of Cromartie, and the Drawbridge of Hull.

On

On this occafion I took the liberty to recommend, that myfelf accompanied by the two principal artificers, Meffrs. Gwyn and Cull, fhould vifit the ftone quarries of Purbeck and Portland, to fee and examine their prefent ftate and produce; and alfo more efpecially to examine the limeftone quarries of Lyme in Dorfetfhire, as I had been informed that the true Lyas limeftone was to be had there; and which, if fo, was likely to prove a valuable acquifition to thefe works *. This journey having been made, in confequence, the ufe of this valuable lime has been eftablifhed here and continued ever fince.

The chairman on examination found that the bafon was kept clear of filt, by the ufe of the horfe-dredges or drugs, and the channel under the eaft pier improved; there being now good eighteen feet water at a middling tide from the harbour's mouth, up to the ftairs neareft the town, and the fullages of the whole harbour vifibly decreafed fince Chriftmas, by the frequent running of the fluices.

Our fecond vifit was in June, at which time the labourers were diligently employed in getting up a large quantity of ftones that had been thrown in to fecure the pier head, as mentioned in the year 1774, which it now appeared were neceffary to be removed in order to clear the foundation for the advanced pier. But as it feemed dubious whether they could all be got up in nine and ten feet water by the ufual method of tongs from the barges; this occafioned me to turn my thoughts upon a diving machine I had formerly made ufe of with fuccefs in doing works confiderably under water.

I had fcarcely returned from this vifitation before a requifition came to defire the expedient I had mentioned might be got ready, which was done with fuch expedition, that I fet forward for Ramfgate the 6th July in order to put it in ufe; and the 12th left Ramfgate, after a full trial of the diving cheft, and the certainty of fuccefs †.

With this machine, which enabled the workmen (or divers) therein to ftay under water any length of time at pleafure, that is when the wind was moderate, that the

---

* I did not at this time imagine this lime had ever been ufed in thofe works, as appears to have been done; but without underftanding its merits, was now grown into difufe.

† Inftead of the ufual form of a bell or of a conical tub of wood funk by weights (externally applied), this for convenience, was a fquare cheft of eaft iron; which being 50 cwt. was heavy enough to fink itfelf; and being 4¼ feet in height, 4½ feet in length, and three feet wide, afforded room fufficient for two men at a time to work under it. But it was peculiar to this machine, that the men therein were fupplied with a conftant influx of frefh air, without any attention of theirs; that neceffary article being amply fupplied by a forcing air pump, in a boat upon the water's furface.

boats

boats could attend, in the courfe of that and part of the following month the foundation was cleared; and the tools for levelling of the ground, the fame that were originally invented and applied by the late ingenious Mr. Etheridge, were now put in ufe, under the management of Mr. Cull the mafter mafon, who had formerly been employed in that part of the bufinefs; it being the Chairman's wifh, as beft for the work, that every thing fhould go on in the fame method as originally practifed.

Soon after our return from the June vifitation, Mr. Aubert having had the misfortune to break his leg by a fall off his horfe, could not attend the third vifitation of this feafon; but I attended the fecretary to Ramfgate; and the third of September we were prefent at the finking of the firft caiffon of the advanced pier, when every thing was carried on with regularity and fatisfaction; as much additional ftone being built therein that tide, as would prevent its being afterwards moved by the fea; and the next day it was got up to its height; that is, above the low water mark of ordinary neap tides.

The piling of the dock's bottom was alfo now confiderably advanced, and feveral of the double beams of the floor were laid, and every thing going on to fatisfaction.

The Chairman being recovered, attended the fourth and laft vifitation of this year. His report thereon bears date 15th November 1788. It notices that the getting up of the ftones (fome of them in $12\frac{1}{2}$ feet under water), which from the circumftances had become a work of abfolute neceffity, had been fuccefsfully performed; and alfo the finking of the caiffons of which he had the pleafure to find four in their places, and built up to their proper neap tide low water height. He alfo found that the fifth and fixth caiffons with their materials were all in readinefs; yet obferving the effects that the fea had had in a gale of wind the night preceding, it having torn away one of the wooden fides of the fourth caiffon, he confidered that the feafon was far advanced; and confulting the engineer and principal officers, it was unanimoufly agreed, that it would be imprudent to attempt to fink any more caiffons at prefent, but rather to do what would be likely to fecure what had been done as foon as poffible; and then poftpone the further progrefs till the fpring.

The floor of the dock being in a ftate of forwardnefs, above half of the ground timbers being laid, the Chairman ordered that the men fhould be employed in that work as much

---

* It was computed that about 160 tons of ftones had been got up in clearing the foundation; and that about 100 tons thereof had been raifed by the diving machine, many of above a ton each; but the want of the machine would doubtlefs have been the lofs of the feafon.

as they could be during the winter feafon ; but that the primary object might be kept in view, Mr. Aubert recommended, that carpenters be employed in preparing and framing more caiffons ; and that a proportionable number of mafons be preparing the ftone for the fame, fo that the works of the advanced pier might commence again, and be carried on with vigour, as early as weather would permit the next fpring.

The Chairman further reported to the Board as follows : " that there is always a light put upon the head of the weft pier at night, as a direction and guide for veffels and boats, coming in and going out of the harbour ; and alfo that a flag is hoifted on the north weft cliff, when the water rifes to ten feet at the gates of the bafon ; and continues flying (in the day time) till the ebb reduces it below that depth, as a fignal for veffels in the Downs ; but as bad weather and gales of wind naturally happen in the night, as well as in the day, and confequently fhips and veffels may meet with accidents from ftrefs of weather, or from other incidental caufes, and be defirous of running for the harbour for fhelter, fuch veffels perhaps not knowing the depth of water in it, are reftrained from attempting to make for the harbour, when they might do it ; Mr. Aubert therefore thought it advifeable to call in fome of the moft experienced pilots ; who unanimoufly agreed, that a double light always put upon the head of the weft pier, in the night, when there fhall be ten feet water at the gates of the bafon ; the additional light to be elevated about eight or ten feet above the prefent light, and fufficiently ftrong and luminous to be feen in the Downs, would be of great utility, as the double light would be a fignal, denoting to mafters of veffels that they might make for, and enter the harbour with fafety."

Mr. Aubert (whenever the tides ferved) directed the water to be fhut in the bafon, and the fluices to be run ; and had the fatisfaction to obferve, that they continued to operate with their ufual force and power ; as he perceived the water, with a rapid and ftrong current, carried the filt out a great way beyond the tranfport buoy : and that by working the dredges, and running the fluices occafionally at the fpring tides, the bafon is kept clear of fand. He had likewife the pleafure to inform the Board, " that the channel under the eaft pier is confiderably widened, that it can receive and fhelter a great number of fhips of large tonnage, that the depth of water is confiderably increafed, and the whole harbour is in fo improving a ftate, that there is the greateft probability in the courfe of a few years, it will become a place of the greateft confequence to the commercial navigation of this kingdom."

" Mr. Aubert embraced the opportunity of again affembling and afking the pilots their opinion refpecting the advanced pier, and they profeffed the fame belief of its beneficial confequences,

confequences, in quieting the harbour; they likewife obferved, " that the prefent ufual and fafeft track, for veffels making for the harbour, is to come in from the weftward, and that when the advanced pier is completed, veffels will then (as now) come in right before the tide; and in all probability, with more eafe and fafety."

1789.—The Chairman's firft vifitation to Ramfgate this year, was in March, and we arrived there the 12th. The 14th the fheet piling inclofing the circumference of the dock floor, was clofed in; all the beams having been previoufly laid.

With refpect to the works of the advanced pier, the caiffons were in fufficient for-wardnefs, but it was yet too early in the feafon to attempt any thing at that: it had been built upon and fecured, fo that no damage had happened to it during the winter. The Chairman therefore contented himfelf with examining the ftate of the works, and giving the neceffary directions, that every thing might be in forwardnefs as foon as the feafon became favourable.

The fecond vifitation was the latter end of May. We then found the floor of the dock laid, and the mafon-work begun: but that Mr. Gwyn the deputy furveyor, being afflicted with an abfcefs, was unable to attend at the yard *.

The works of the advanced pier were now going on, and four caiffons had been funk this year (making eight in the whole), and others forwarding. Upon this journey we vifited and landed upon the Goodwin Sands, to have a view of them and examine their nature; and found that though of the nature of a quick fand, clean and uncon-nected, yet the particles laid fo clofe, that it was difficult to work a pointed iron bar into the mafs, more than to the depth of fix or feven feet.

On this vifitation I took feveral levels of the relative heights of different parts of the works compared with the bottom; alfo, particularly examined the lighthoufe, as relative to the Chairman's laft propofition to the Board.

In Auguft the truftees affembled at Ramfgate, on a general annual vifitation, and the 24th, reported as follows:

" Your Committee proceeding in their furvey, ordered the water to be fhut up in the bafon, and have the pleafure to inform the Board, that by running the fluices three

---

* Of this he died in the courfe of the next month; a real lofs to the public, as well as lamented by his family and friends.

or four times at fpring tides, and working the horfe drags, the bafon is kept clear of fullage; your Committee likewife obferved, that the channel under the eaft pier is confiderably wider, and in its prefent ftate capable of receiving a number of fhips of 500 tons and upwards; and it is acknowledged by all perfons acquainted with the harbour, that it has been wonderfully improved within a few years paft, as in the moft ufeful parts of the harbour there is an increafe of between five and fix feet water; and whoever contemplates it in its prefent ftate, and perufes the evidence given at the bar of the Houfe of Commons, prior to the paffing of the Act of Parliament, and alfo the preamble to the Act, will be convinced that Ramfgate Harbour now exceeds the hopes ever entertained by its projectors and friends; or, that the legiflature had any idea of, at the paffing the Act, which fuppofed it might be made a receptacle for fhipping of and under 300 tons, but could not imagine it ever would be capable of receiving fhips of much larger burthen. That the harbour is of great importance to commercial navigation, the well attefted documents of the office fufficiently prove; videlicet, That the harbour (under Providence) has already been the means of faving property to the amount of between three and four millions fterling, and between eight and nine thoufand valuable lives, to their friends and fociety; yet notwithftanding the truth of thefe ftriking facts, the prefent ufeful ftate of the harbour, through prejudice or want of proper information, not being known to the public as it ought to be, your Committee think it neceffary to recommend, that Mr. Smeaton your engineer be defired to draw up a proper account of the harbour in its prefent improved ftate, to be publifhed at a convenient time."

The walls of the dock were then up feven or eight courfes high; the whole defign of it was fully apparent to the Committee.

The feventh caiffon was now funk, while the Committee were there; and every thing being in a ftate of going on fuccefsfully and fatisfactorily, the Committee fignified their wifh to fee the operation of the diving machine or bell (as in conformity to cuftom, it had now acquired that name,) for it appeared to the Committee, that by means hereof it would be very practicable at any time to examine the ftate of the foundations that had been laid a courfe of years; and particularly, whether the timber bottoms of the caiffons, that laid under the piers, and immediately upon the chalk, had not greatly fuffered by the worm? and further, that by the fame means, it would be practicable to make in future any repairs to the foundations that might, on account of the above or any other caufe, be found neceffary.

On

On this occafion, I had the honour to attend the Chairman down to the bottom of the fea, upon which we could ftand, and work dry. In this fituation the Chairman finding himfelf perfectly at his eafe, and very comfortable, ftaid full three quarters of an hour. We found that the ground work, that is the bottoms of the caiffons, was now fo deeply buried in fand and filt, that it would be a work of confiderable labour to rid it out and clean it; and as this is doubtlefs the beft defence againft the worm, he concluded it inexpedient to attempt to difturb it: having fully fatisfied himfelf of the practicability of ftaying any length of time, and of performing any kind of work, that the bounds thereof admitted, with a capability of removal almoft at pleafure, we gave the fignal for afcending, and were received with great joy by our friends, whom we found furrounding us in boats, and who by this time were beginning to be apprehenfive that fomething might have happened to us.

This vifitation I took an exact plan of the weft pier head, to enable me to make out a defign for a lighthoufe for the approbation of the Board, for erecting a double light upon that head.

The Chairman's laft vifitation of the year 1789, was November 7th. At this time, though the weather had lately been but indifferent, yet eight caiffons had been funk, and fixed this year; which, with the four funk the laft year, completed twelve in two years, as we originally had in view. The Chairman confulting the engineer and principal officers, confidered that thofe twelve caiffons would need to be fecured; and that the beft way to do this was to build the fuperftructure upon them, with all poffible diligence. And furthermore, that it would be eligible, rather than completing any particular part, to go on with the whole length, which being now 120 feet, that is, nearly one third of the propofed advanced pier, if weather permitted, that it could be got up to high water mark by Chriftmas; this would be fufficient to demonftrate the effect that was likely to be produced, and would be of great fervice, in affording an addition of fhelter to the fhipping the enfuing winter; in fhort, it would give us an experimental proof of the good or ill that was likely to arife from the farther profecution of this work; and the harbour having been founded, nothing fo far appeared but what was in favour.

It alfo had happened in the late autumnal winds, that the break-water or jettee in the external angle of the eaft pier with the land, had been wafhed down and totally demolifhed. This was a work that had been erected feveral years before, chiefly of old fhip timber, by way of experiment, for the defence of the pier, and the eftablifhments there. This having, with frequent repairs, anfwered the end ever fince, it appeared

R 2

therefore

therefore very defirable and neceffary to be rebuilt. This being the cafe, the Chairman thought it ultimately for the greateft benefit to the harbour, to rebuild it with ftone, and in the moft fubftantial manner, being greatly expofed : and I was ordered to prepare a proper defign for a ftone break-water or bulwark, for the approbation of the Board. Here therefore were two objects that wanted to be profecuted both at once ; and both with our whole force ; but as the break-water could not be more than totally deftroyed, which it then was, the Chairman judged it of the moft importance, both in point of fafety and ufe, to carry on the advanced pier till it was got to its height ; which was done in the courfe of this year.

1790. This year commenced with the building of the break-water. In January many heavy gales at Weft came on, which of courfe detained the fhipping in the Downs, many of them therefore fought fhelter in Ramfgate Harbour, and found it in a degree never experienced before ; fo that in the courfe of this month, there were in it at one time 160 fail, the greateft part of which lay in the bafon, and there was room in the harbour for many more. This affemblage of fhipping afforded a fpectacle fo new, that the people all round the country came to fee it.

The chairman's firft vifitation this year was the beginning of March ; but the getting up of the advanced pier, and after that of the break-water, having totally occupied the mafons, the dock was obliged to remain as it had been left the laft fpring. The break-water was now in a ftate of proceeding very vigoroufly ; but it being too early in the feafon for any thing to be done at the advanced pier, the chairman could only give the neceffary orders, that nothing might hinder the proceedings when the feafon was more advanced.

The beginning of April a very hard gale of wind happened at E. S. E. and the capping courfe of the break-water not being quite clofed in, the violence of the fea wafhed ftones of one ton and a half out of their places, though every thing that had been completely fixed, ftood faft. This, however, fhewed us that we had not beftowed on this work more folidity than neceffary, as fome had been induced to imagine. The violence of the fea alfo fell fo heavy upon the advanced pier, that the extreme unfinifhed termination became underwafhed, and caufed a fettlement of the exterior angle, which immediately obliged Mr. Cull to guard it with rough ftones, in the manner the original pier-head had been before the removal thereof ; and likewife to take fome of it down, in order to ftraighten the wall.

Thurfday

Thurſday, 27th May, the Chairman arrived at Ramſgate, upon his ſecond viſitation ; and then found the break-water completed ; the damage at the advanced pier rectified ; and two caiſſons, viz. the 13th and 14th ready to be put down, and which were ſunk while we were there. We ſaw the ſluices run ; examined the channel, and found it with-out any ſenſible alteration from the prolongation of the advanced pier.

Thurſday, the 12th Auguſt, the general viſitation of truſtees took place, at which time the eighteenth caiſſon was ſunk, the works examined, the ſluices run, and every thing found ſatisfactory.

Six caiſſons having been laid this year, and finding ourſelves too late in the laſt ſeaſon, by laying eight, it was thought beſt by the Committee to terminate the work of this ſeaſon, as to the laying down of caiſſons ; and to employ the remainder of it in getting the pier built up, ſo that every part might be ſeaſoned before the winter and heavy gales come on : and as there would now be full 180 feet of the advanced pier laid, this, when raiſed to its height, would be an effectual trial of the utility, effect, and validity of this work.

The Committee had the ſatisfaction to obſerve a very viſible alteration in the width of the channel under the eaſt pier, which had been chiefly brought about by the running of the ſluices, during the crowd of veſſels in the harbour in January laſt.

Upon the Chairman's fourth viſitation of this year, we arrived at Ramſgate the 21ſt Oc-tober, and then found the preſent ſtretch of advanced pier in general up to high water mark.—During our ſtay here, we had a hard gale of wind at eaſt, for three days together ; and then had the great ſatisfaction to find the harbour, which when the wind was at that point uſed to be much agitated, to be moſt remarkably ſtill and quiet ; ſo great a change in this reſpect could have ſcarcely been expected, till the pier was car-ried out to its full length, of which what was now done was ſomewhat more than half. On this occaſion we thought it neceſſary to compare the harbour-maſter's ſoundings, taken in May laſt, and alſo the preſent month, with our own now taken in his preſence ; and on comparing them, had the ſatisfaction to find the channel of the harbour, notwith-ſtanding the advancement of the pier, to be in an improving ſtate in depth as well as in width.

*The*

## *The present State of Ramsgate Harbour.*

THE operation of the fluices, as has been defcribed, has gradually cleared out a broad fpace or channel through the middle of the outward harbour, from the gates to the pier heads; and the bottom lying upon a gentle flope, makes above fix feet more water in that material part now, than in the year 1774, fo that veffels drawing from ten to eleven feet water, can go into the bafon in neap tides; and in fpring tides thofe drawing from fourteen to fifteen feet.

Under the curve of the eaft pier, the fluices have now cleared a channel capable of taking two fhips abreaft, with clearance for paffage, where at neap tides there is from fixteen to feventeen feet water, and at fpring tides from twenty to twenty-one feet, and often twenty-two; fo that not only veffels of 300 tons, the primary objeft of this harbour, may come into it in all tides; but at fpring tides as large fhips as are ordinarily employed in the mer-chants fervice. It is here in reality no material objeftion, that a veffel cannot come in from the Downs at low water; becaufe fhe is not in diftrefs there till the tide is rifen to that point of height that it begins to run northward, and then it has been amply fhewn, that there is always water to go into Ramfgate; and that with every wind whereby fhe can be annoyed in the Downs, fhe will be right before it into Ramfgate; and every wind that will be fair for fhips to proceed upon their voyages from the Downs, will be alfo fair for their failing from Ramfgate.

If therefore it be really eligible to have a harbour for the reception of fhips in diftrefs from the Downs, it muft be upon the flat fhore of the Ifle of Thanet; and no place has yet been pointed out fo proper as Ramfgate.

It probably will be thought by many who curforily view the place, and are not fully ap-prized of the requifites of an artificial harbour, to be a defeft, that this harbour is not en-tirely covered with water all over its area at low water; but the bank is really of the greateft utility, as will appear when the pilot's reprefentation above noticed is fully confidered. However, notwithftanding that for the reafons already mentioned none of the fluices have been brought to play upon the bank, yet it has in reality fo much wafted, that the higheft part of what now remains is lower by five feet than the middle of the harbour was

in

in 1774; and indeed it is fo far wafted and wafting, that probably it will not be many years before expedients will be found neceffary to preferve it. There have already been complaints, that it is grown fo low, that at neap tides the veffels cannot get their ballaft therefrom; and the expedient of filling barges in readinefs has been ordered by the truftees for a remedy of that defect. At a fpring tide there is now thirteen feet water ove it, fo that a number of the fmaller veffels may occafionally lie upon it.

Befides the completion of the advanced pier and works now in hand, there is obvioufly a number of articles of confiderable expenfe, that will greatly tend to improve, ftrengthen, and confirm the whole work; and which may very well be expected, after the various councils, turns of fortune, and changes this work has undergone, are confidered. And after all, a harbour that muft fubfift by the artificial power of fluices, muft be fubject to a continual expenfe, and great care to keep every thing in repair and in order; but when all thofe things are duly, properly, and attentively performed, I doubt not to fee the time when it will be faid, notwithftanding its misfortunes, and the obloquy that has been occafionally caft upon it, to be a work worthy of the expenfe it has incurred, at leaft by the attempts to recover it from the condition it was in in the beginning of the year 1779. I will conclude with faying, that according to my information, 130 fail of veffels were at one time in the harbour, driven in by ftrefs of weather in the late winds of January 1791, among which were four Weft Indiamen richly laden, from 350 to 500 tons; and if we are to fuppofe that the whole or the greateft part of thefe 130 veffels would have been riding in the Downs during this ftormy weather, we need not be at a lofs to judge what a number of additional dangers and difficulties would have been in the way of thofe which actually did ride there. I underftand that the number of veffels in the Downs at one time has rarely ever exceeded 300 fail; but in the bad weather in the beginning of the year 1790, and in the prefent year, the Downs were in a great degree cleared, there being in reality few fhips left riding in them.

A Lift of the number of Ships and Veffels that have taken fhelter in Ramfgate Harbour in ftormy weather:

| In 1780 | - | 29 | In 1786 | - | 238 |
|---|---|---|---|---|---|
| 1781 | - | 56 | 1787 | - | 247 |
| 1782 | - | 140 | 1788 | - | 172 |
| 1783 | - | 149 | 1789 | - | 320 |
| 1784 | - | 159 | 1790 | - | 387 |
| 1785 | - | 213 | | | |

Among the above were feveral from 300 to 500 Tons burthen, and upwards.

Within

Within the laſt ſeventeen months upwards of ſix hundred ſail of ſhips and veſſels have taken ſhelter in the harbour, of which above three hundred were bound to and from the port of London.

Evidence can be produced, that the harbour has been this winter the means of ſaving a great many ſhips and veſſels, and property to the amount of between two and three hundred thouſand pounds, with a great number of valuable lives, which otherwiſe would have been driven upon the flats and rocks, and in all probability loſt.

---

As an addition to this Report I have the pleaſure of informing the public, that, on the 17th July 1791, at a high ſpring tide, the new dry dock built in the baſon for repairing ſhips was tried in the preſence of the chairman, for the firſt time ſince it was found neceſſary to build it with a timber floor, which is of a new and peculiar conſtruction, on account of the ſprings from the chalk riſing ſo powerfully under it as to force up the ſtone floor with which it had before been twice tried. The experiment anſwered in the completeſt manner ; the dock remaining perfectly dry till low water, when the ſluices of the baſon were opened for ſcouring the harbour, ſo that this very deſirable object, that has been ſo much deſpaired of, is now fully obtained, and muſt prove of great utility to the public.

## SANDWICH HAVEN.

The REPORT of JOHN SMEATON, Civil Engineer, upon the State of Sandwich Haven.

To the Mayor and Magiftrates of Sandwich.

GENTLEMEN,

THAT according to hiftorical accounts Sandwich was once a famous and flourifhing fea-port, and that the Ifle of Thanet was completely furrounded with navigable water, there is not the leaft reafon to queftion ; for the change from that to its prefent ftate, without fuppofing any particular convulfion of nature, or neglect of particular perfons, is conformable to the regular change that is conftantly experienced in all fimilar fituations; the mud wherewith the fea-water is always charged in a greater or a lefs degree, being ready to fubfide and fix wherever it can meet with a place of reft. The great caufe whereby it is kept fufpended, is the movement and agitation of the waters that con- tain it. Wherever the quantity and agitation of the water are too little, or the quantity of mud is too great to be kept in motion by the agitating powers, its natural tendency to reft will operate ; and the effects of want of motion increafing in proportion to the progrefs made, a more rapid change enfues towards the completion of nature's operation, than in the more early ftages. Every creek, inlet and bay, that has not a fufficiency of frefh water rivers to keep it open by being difcharged through it, has a tendency to become land. While fuch creek or bay remains deep, a quantity of tide's water flowing in and out twice a day, tends to keep the mud in agitation and from fettling; but as the tide of ebb is naturally weaker than the flood, the ebb will not carry out all that the flood has brought in ; and when the depofition is fo far advanced as to contract the breadth of the water, and render it to a certain degree fhallow, the quantity of water flowing in and out being leffened, its power is weakened. The natural means whereby an inlet is kept open, is the difcharge of a frefh water river through it, which oppofing the influx of the tide, and add- ing to the force of its ebb, will always maintain a certain channel in proportion to the quantity of land-water that requires to be difcharged. The tendency of nature therefore is to contract the channel to fuch a fize, that the natural power of the ftream can juft maintain it ; and in this ftate the wide extended arm of the fea, anciently flowing by Sand- wich, and up the general vallies, as now called, feems to have been at the period that the new cut at Stonar was projected and executed.

A river such as the Stour, is by no means adequate to keep open so large an arm of the sea as that through which it flowed in former ages; and the powers of its current having become considerably weakened by the meandring course it took, (which in effect lessens its declivity) the difficulties of draining the marshes and adjacent low grounds became greater in proportion as the depth and width of its channel grew more contracted. Thus, while so obvious a remedy presented itself as the turning of the river through the narrow neck of land at Stonar, where, in the space of a furlong, there was a fall of more than a fathom; it is not to be wondered that the landed interest in the level, was eagerly wishing to seize the apparent advantage, and indeed so happy and easy an expedient rarely occurs in subjects of drainage. The late Mr. Yeoman in his report, speaks of having had experience of other rivers under the same circumstances; for my part I have seen many rivers; but the Stour in this respect stands alone. Could Sandwich have been as easily moved to Stonar, as the Stour carried through this cut, every thing would have been well on all sides; but notwithstanding the assurances of Mr. Yeoman, it certainly was contrary to the experience of us all, to suppose that the running of the height of the flood waters by another channel should tend to improve the harbour of Sandwich; unless he meant that part of the Stour that lies below the new cut at Stonar, to the sea, as being the harbour of Sandwich, of so much importance to that town. The thing is however now fixed; and under certain regulations by an Act of Parliament, which cannot be deviated from by either interest without the consent of the other. But it appears to me that there is nothing to hinder a different mode of operation by consent of parties; if it can be made appear, that they both will be bettered by the change; and this I will now endeavour to point out.

It is the great land floods proceeding from great downfalls of rain, or dissolution of snow, or both conjointly, that is the great operative cause in clearing the channels of rivers to sea; and this chiefly when the tide is out; and most of all when they happen at the low ebb of a spring tide: at such a time more work is done in a few hours than could be done in months by a leisurely or moderate reflow; this may, as the engineers affect to term it, grind away a little of the mud, sand, and silt; but the former tears it away with violence. A gentle reflow, experience evidently shews, disturbs not a particle of sand or mud; on the contrary, if the preceding tide of flood has had power to bring in a quantity of silt along with it, such a reflow will give it leave quietly to subside; and where nothing is done towards its removal in one tide, no number of repetitions will amount to any thing. It seems therefore to little purpose for the Harbour of Sandwich, I mean that part of it that is contiguous to the town, when the top waters are drawn off through the flood gates of Stonar, to leave the dribblings of the floods and ordinary current of the river, to go quietly round by the way of Sandwich.

The

The work being now executed is not, as before the paffing of the Act, a matter of fpeculation; the effects may be evidently feen and with more certainty judged of. I muft take for granted that the ftandard mark was fixed as the law directs, at the medium of nine feet above the bed of the river at the head of the cut: let any one now found the depth of it for fome fpace below the head of the cut and he will perceive how miferably it is here diminifhed, both in depth and width. This evidently fhews how much the water has loft of its cleanfing and fcouring power, fince the paffage of the top waters have been diverted. On the other hand let the channel of the Stour be viewed above the head of the new cut, he will foon perceive how greatly it is augmented; how ftriking the contraft! all this in favour of the drainage.

For my own part I am profeffionally as great a friend to drainage as to navigation; and therefore if I can fhew, in confequence of what has occurred to me from the view I have taken, how the drainage may be very materially improved, even beyond the advantages already gained, by means that will improve the navigation alfo; and that without any new charge, I fhall be doing an evident fervice to both.

I have always underftood, and believe it to be true, that the flooding of low grounds in the vallies of rivers is an advantage to the grounds, provided it is done or happens at fuitable feafons of the year; for the frefh waters that come down from the high grounds in floods, conftantly bring with them a quantity of foil enriched with the manure that wafhes down from the higher country, which being depofited upon the furface of the low lands, very greatly improves their fertility. It is the misfortune of low lands which are not in a ftate of perfect and ready drainage, that the water when brought upon them by the winter rains is apt to lie upon them fo long in the cold feafon of the year, that it greatly damages the roots of the ufeful vegetables (which in general are not aquatics), fo that the occupiers of thefe lands, lofing much more by the continuance of the water upon them than they gain by the depofition, generally reckon and imagine they never can get them dry too foon. Now, fuppofe at the beginning of a flood, if the waters were kept upon the furface of the vallies for a fortnight, they would be fertilized thereby, provided they were run off more fpeedily afterwards than they ufed to be. The height or depth of the water upon the land is not what does the damage, but the contrary, the injury arifes from its long continuance at a fmall depth; and in cafes of difficult drainage the difficulty has not been to get rid of the top waters, which by their height have a greater declivity to the fea; but the bottom waters, by the riddance of which the ground is made competently dry and fitted for the purpofes of agriculture.

Imagine

Imagine now in the winter months ; fuppofe, November, December, January, February, and a part of March, that inftead of drawing the gates when the water was juft above the mark after a dry feafon, they were to remain fhut for a certain number, fuppofe for three or four days, that the water might completely overflow the meadows and take its ancient courfe ; that then the gates were drawn, and without being reftricted to wait for an hour's ebb on the haven fide, and then to run only for five hours each tide ; they were to run them as foon and as long as the land flood water would override the tide, for the fpace of four tides running, then let them remain fhut for four tides, and four tides run as before ; and fo alternately, four tides and four tides till the flood was run off, and the water never rofe in the time fhut to the ftandard mark.

In this way it is very certain that not only more water would run through the flood gates in a given number of days, than can do now in five hours each tide ; but by draw-ing lower in confequence of a continued draft, the defcent of the water from the meadows would be more quick and effectual, and particularly fo in the decline of floods. On the other hand the greater abundance of top waters that would this way be forced to pafs Sandwich Bridge, would much more effectually tend to keep the whole channel of the Stour open, in its long meandring courfe from the head to the tail of the new cut, than it can now do, when it is never fuffered to have its full effect, owing to the conftant diverfion of a great part of it by the new cut. By this means a fcouring power would act at intervals through Sandwich Harbour, of much more confequence for the reafons given, than the weakened one it can now have ; and which, combined with the advan-tage that the lower part of the haven, from the tail of the cut to the fea, muft neceffarily receive from the flood gates, appears to me fufficient to keep the harbour of Sandwich in as good order, as it has been known in the memory of man ; provided a little help from the hedge-hog and fpade be given in fome particular places.

The harbour of Sandwich, properly fo called, I find is complained of rather on account of its growing narrower than fhallower. The contraction of its fection is a natural confe-quence, of the lofs of the force of the top waters of the land floods ; but I think it is particularly fortunate that the contraction has rather been in width than in depth. The expedient therefore that I have propofed will, together with the means already ufed, in all likelihood recover the former width, or at leaft prevent a further contraction.

When a frefh water river makes its way to fea through the loofe fand and filt that the fea has originally depofited, its courfe is continually varying, as there is nothing ftable to fix it ; for when by any accident it gets into a curve out of a ftraight line, the water by

**its**

its superior action on the concave side of the curve, tends to make it more curved. To attempt to cut off the points by the spade is an endless and therefore an useless work. Nor can the matter be mended by jettees, because these will make curves where there were none before. All that can be done to the purpose in this case, is to alter the perches and buoys conformably; so as to point out the channel.

Upon this view I found not only the outward haven, but the harbour  of Sandwich in reasonable good order; indeed in better than I could have expected, considering the natural difficulties to its subsistence; and especially since the total subtraction of the  top floods from the harbour; which affords an instance what a little help will do when properly and seasonably applied, as I understand the hedge-hog has been made a considerable use of; for when the water runs with a brisk current, a great deal of sand and silt will be carried by it down stream to seaward.

The river Stour above the limits of Sandwich Harbour grows sensibly more and more contracted and obstructed with weeds, till we arrive at the head of the new cut. This district in several places wants the spade as well as the hedge hog, which I also understand to be here used by the Commissioners of sewers; and it appears to me that every year this evil will increase, unless prevented by the different mode of running the flood gates I have recommended, and which will be by far the cheapest expedient.

The Commissioners of sewers in this way, will also be relieved from the necessity of maintaining the new cut at the width of 40 feet; the fulfilling of which condition I expect will otherwise occasion a very considerable expense; as the operation of the flood gates seems likely to widen it apace.

I shall conclude with observing, that what I have recommended being a matter of stipulation, may be tried for a year or two certain, till the effect of it is seen; and then a more solemn engagement entered into, as the effect of such experience shall evince.

*Grays Inn,*
15th *August* 1789.

---

* This distinction seems necessary to avoid ambiguity; the haven being supposed that part between the tail of the new cut and the sea.

## DOVER HARBOUR.

(See Plan, Plate 8.)

The REPORT of JOHN SMEATON, Engineer, upon the Harbour of Dover.

THIS harbour appears from old accounts to have been a national object for ages paft, as being the neareft port, and confequently from its fituation the key between England and France; on this account great fums of money have been from time to time expended in keeping it open and rendering it as commodious as the nature of its fituation will admit; yet notwithstanding every endeavour for this purpofe it ftill labours under fome natural inconveniencies, which it is greatly to be wifhed were removed. For this purpofe, at the defire of the Right Honourable the Earl of Holderneffe Lord Warden of the Cinque Ports, I went down to view and examine the harbour of Dover in February laft, where I had every affiftance the place would afford.

The port of Dover has in length of time gone through many changes; the mouth or entry thereof being at prefent in a very different place from what it was within the compafs of record, as appears by accounts thereof collected by Mr. Hammond of Dover, with the perufal of which he was fo obliging as to favor me. This great change has been evidently brought about by the fame caufe that has at all times been, and ftill continues to be its greateft annoyance; viz. the conftant motion of the beach or fhingle which by the action of the feas is driven coaft ways from weft to eaft; for as the Britifh channel opens to the weft and contracts to the eaftward, the feas are much more violent and heavy from the fouth weftern than from the fouth-eaftern quarter; and in confequence, though it may be apprehended that on violent ftorms at S. E. the fhingle may in fome degree be moved weftward, yet the general prevalence being the contrary way, the moft apparent and obfervable motion is coaftways from weft to eaft.

This fhingle or beach (as it is called) confifts chiefly of flints that feem to have originally proceeded from the chalk cliffs that inveft a confiderable part of the fouth coaft of England; which cliffs being gradually undermined by the action of the feas at the foot thereof, tumble down, and often in very large quantities; where by degrees the chalk diffolves by the action of the fun, the fea, and the frofts, and the flints being broken and inceffantly

rubbed

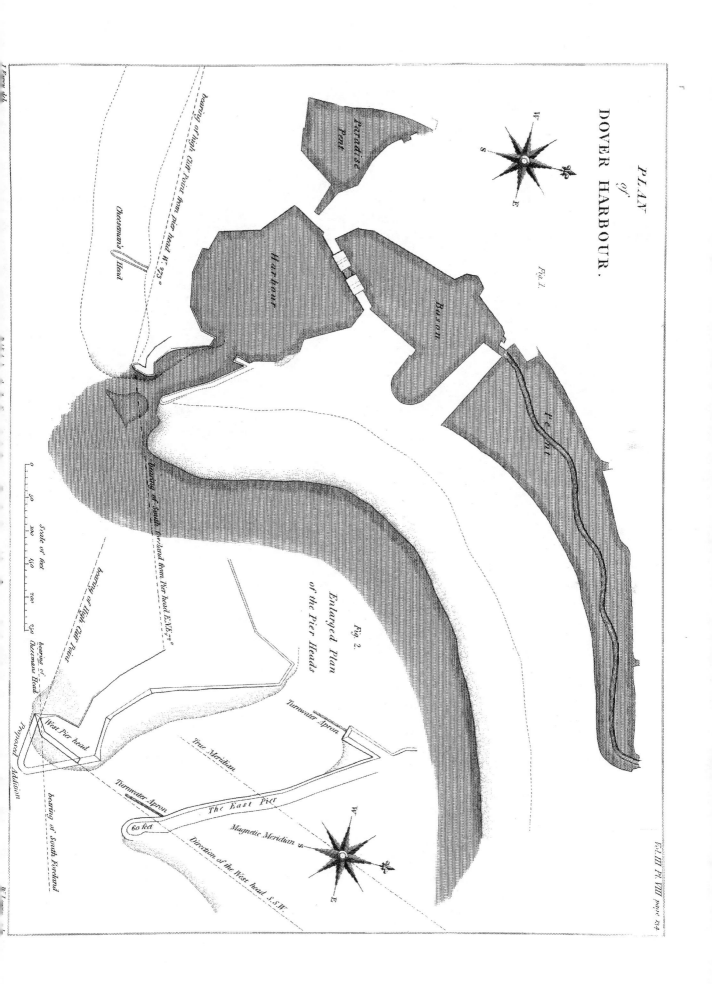

PLAN
of
DOVER HARBOUR.

Fig. 1.

Paradise Pent

Harbour

Basin

Pent

Chesman's Head

bearing of high Cliff Point from pier head N. 27½°

bearing of South Foreland from the head E.N.E. 7°

Fig. 2.
Enlarged Plan
of the Pier Heads

Scale of feet
0    50    100    150    200    250

bearing of High Cliff Point

bearing of Chesman Head

West Pier head

Proposed Addition

bearing of South Foreland

Turnwater Apron

True Meridian

Turnwater Apron

The East Pier

60 feet

Magnetic Meridian

Direction of the West head S.S.W.

Vol. III Pl. VIII page 254

rubbed againſt each other, form a conſtant ſucceſſion of beach. This however is obſerv-
able, whether this ſucceſſion is to be attributed to the above or any other cauſe, that an
immenſe quantity of beach is in a ſtate of continual motion along the coaſt from weſt to
eaſt, part of which lodges and fills up every receſs where it can be depoſited and lie in
quiet.

This beach has formerly been the deſtruction of the old harbour, and it appears from
the above accounts that the mouth has been more than once entirely ſhut up, and has
remained ſo for years ; and that the mouth of the preſent harbour was originally a cut
through the beach to let off the land waters pent up in the inſide of the harbour, in order
more effectually to view and examine the ſtate thereof, and to enable the engineers to
conſtruct ſuch freſh works as might appear neceſſary for its re-eſtabliſhment.

From this ſtate the preſent harbour has been gradually improved ; the entry whereof is
defended by two piers compoſed chiefly of wooden piles, the inſide filled with rough
heavy ſtones. After paſſing the throat or entry, the veſſels arrive in a capacious
outward harbour, where they may lie defended from all winds ; but having an open
communication with the ſea the water flows and ebbs therewith, and at low water ſpring
tides the whole is left dry. Above this the natural capacity of the harbour (as it ſeems)
is divided by a dam, or as it is called the *croſs wall*, in which is an opening of 38 feet
wide at top, and about 36 at bottom ; and in this is placed a large pair of gates pointing
to landward, through which veſſels at high water may paſs out of the exterior harbour
into the interior harbour or baſon (as it is called) where occaſionally they are kept
afloat.

This croſs wall beſides the great gates, has two other openings of about twelve feet wide,
in each of which is placed a pair of draw gates.

The interior harbour or baſon is again divided by a ſecond dam or croſs wall having
an opening of more than 20 feet for the paſſage of ſmaller veſſels, which is alſo furniſhed
with a pair of gates pointing to landward ; this dam has likewiſe another opening furniſhed with three draw gates, by which the water can be occaſionally let off ſo as to
ſcour the baſon. Into this upper reſervoir, which is called the *Pent*, the freſh water river
which ſprings from the chalk hills north of Dover empties itſelf, and makes its way
through both ſets of gates, through all the three harbours, and laſtly betwixt the pier
heads to the ſea.

This

This general difpofition of the harbour appears to me as judicious as can be contrived, and is upon the fame general idea as the port of *Cherbourg* upon which the French had expended an immenfe fum of money, in order to compleat every thing in the moft fub-ftantial manner, before it was deftroyed by the Englifh in the laft war.

According to this difpofition, when by hard gales of wind and feas from the fouth-weftern quarter, a quantity of beach is brought round the weftern head and lodges itfelf between the heads; the bafon and pent are then filled partly by taking in the fea water and partly by frefh water afforded by the river, and there retained till it be low water. The draw gates of the fluices in the crofs wall are then opened with all poffible expedition, and the body of water contained in the bafon and pent, by making its way between the pier heads cuts down and removes the bar of the beach, which at the time of fpring tides is done with fo great effect, that at one fingle operation, as I am informed, a good paffage is opened for veffels; and at two tides the whole mouth of the harbour can be cleared; and could this be done with equal eafe and expedition at all times when wanted, then would the evils that are now complained of not fubfift, and this port would be in nearly the beft condition its fituation is capable of; and which indeed is very refpectable as a tide harbour, having a good capacity, with from 16 to 18 feet water at common fpring and from 11 to 13 feet water at common neap tides: but it fo happens when there are ftorms or hard gales of wind from the fouth-weftern quarter and at the fame time fhort or low neap tides, that fuch a quantity of beach will be lodged between the pier heads, and to fo great a height, that according to my information a veffel drawing but four feet water can hardly get into or out of the port, at a time when if the mouth were clear as ufual, there would be good ten or eleven feet water into and out of the outward harbour.

At thofe times there remains at low water fo great a depth without the heads, that the water from the fluices has not a fufficient fall and power to drive out the beach from between the heads, but it is obliged to lie till the fpring tides come on; which as it may fometimes happen to be an interval of a week, produces great obftruction to the paquets eftablifhed between Dover and Calais, as well as to the mercantile trade of the place; and yet more to general trade, as veffels may want the port for fafety during thefe intervals but cannot enter it.

The remedy for this evil, or as far as it is capable of remedy, is as I apprehend the object of the prefent enquiry; and towards this end two general methods prefent them-felves; viz.

1ft. The

1ft, The prevention of the beach from getting into the harbour's mouth, and

2dly. A more effectual way of clearing it out when it happens to get in at the times above specified.

It has long been observed that when, by the washing of the sea at the foot of the chalk cliffs, any considerable fall of the cliffs happens to the westward of the port, this ground so fallen making a projecting point or promontory further out than usual stops the course of the beach coastwise, so that when the quantity which happens to be laid eastward of the falls (as they are called) has got beyond the port, the quantity passing the pier heads is so small that the port is very little annoyed therewith; but as those falls are chiefly composed of chalk, and much broken by the shock in falling, the sea in the course of a few years washes them away, and then not only the beach is let to pass in its ordinary quantity, but also the quantity before retained by the falls, which gradually escapes as the fall (or artificial point) washes away. This has given occasion to a supposition, that if instead of these temporary promontories (which are often so considerable as to cover some acres of ground), fixed heads or jettees were run out into the sea, which were not capable of being washed away, they would for ever prevent the beach from getting eastward into the harbour; for, say they, so long as these falls or natural jettees laft, so long is the harbour free from beach; especially in such quantity as to prevent its being easily kept clear by running the sluices, and therefore is no annoyance thereto. This matter, as it depends upon facts that are in themselves at first sight striking, and on that account strongly insisted upon by many, I shall endeavour to set in a clear light.

When I was at Dover in the month of February last, a very large fall, about three miles to the westward of that port, had happened but a little while before; as well as another large one nearer to Dover, which had happened some time previously; both these falls I went to view and considered very attentively. I judged that the great fall (as had been represented) covered six or eight acres of ground, and will undoubtedly take a considerable time to wash it away. I observed that the quantity of beach lining the shore gradually diminished, as we approached both these falls from the eastward, so that near thereto the shore was in a manner clear of beach on the east side, while there appeared to be a quantity gathering on the west sides, and which was in a state of increase, as must necessarily happen by the gradual approach of the beach from the westward.

Undoubtedly till these promontories get charged to the full with beach, the greatest part will be there retained, and in consequence the constant supply being cut off, the quantity eaftward of the falls will gradually diminifh by removing ftill further eaft. That thefe falls if rendered permanent would permanently retain a quantity of beach fufficient to charge them, and in confequence make an addition to the coaft for fome fpace weftward thereof, I can readily admit; but that after they are full charged with beach they will continue to ftop the conftant fupply from getting round thefe heads, and again driving along the coaft eaftward as it had done before, is what is by no means clear to me. The matter rather prefents itfelf to me in this light; that in fact by fuch time as thefe falls get fully charged with beach, or perhaps fooner, they get fo far wafhed away as to begin to loofe it again; and as by this means their power of retention, *after they are full,* never comes to the proof, this makes it to be imagined that the beach begins to move eaftward merely in confequence of the falls wafhing away; and therefore had the point been per manent, the beach would always have been ftopped.

That the time taken up in the wafhing away of thefe falls may in fome meafure cor-refpond with the time they take to fill, appears hence, that the largeft falls are the longeft in wafhing away; but then they will confine and lodge more beach before they are full. I can therefore readily admit that if jettees were run out to the fame length as thofe falls, they would if properly maintained permanently retain a *certain quantity* without after-wards letting that quantity go again, as the temporary falls or jettees now do; but as I am no ways convinced that they would tend to ftop the conftant fucceffion of beach from getting round their heads after they are full, and then driving along the coaft as before; I am of opinion that the good that is to be got to Dover harbour by raifing arti-ficial jettees in order to lock up a certain quantity of beach as in a cheft, will be no ways adequate to the expenfe of raifing them, but that the fucceffive quantity will in fact fill them as faft as they can be carried out by men's hands; fo that a remedy this way muft confift in an eternal work of building jettees, which as they will require maintaining as well as building, will, together with the common repairs of the harbour works, induce a very great expenfe.

It is faid indeed, that though the heads may not retain the beach beyond a certain quan-tity, yet that if it be forced to go out into deep water, it will be loft in the fea and never return upon the coaft. But in anfwer to this, I fear it will not be poffible by the hands of man to carry out thofe jettees into fuch deep water upon this floping coaft, as to prevent their return; nor indeed, when I obferve how oddly this beach gets along the coaft through paffes where it muft go through deep water, and afterwards appears again, I am not inclined to truft altogether to the fhoving it out into deep water, even if it could be done. I am therefore more ftrongly induced (without neglecting any advantages that

may

may be drawn from thofe cafual and temporary reliefs of nature, the falls) to confine my views, operations, and expenfe to fuch purpofes as have a determinate end, and which fuppofe the beft to be made of it that can be, under the fuppofition of a conftant fucceffion of beach from weft to eaftward upon the coaft as heretofore.

Conformably to this doctrine of the movement of the beach, may be reckoned the real benefit found by the jettee that has been erected at the Caftle Point. It feems that formerly the breadth of ground between the pent and the fea was fo narrow, that there was great danger of the feas making a thorough breach into it; but that upon a large fall happening at the Caftle Point, a quantity of beach was lodged and the partition between the pent and the fea was greatly ftrengthened as long as the fall lafted; but upon the wafhing away of this fall the barrier was again greatly weakened; which being obferved, an artificial pier or jettee was erected at the fame place, and ever fince the beach has been fo far retained as to lie in a confiderable breadth and ftrength between the pent and the fea, and fo as to put that matter out of all danger. This was certainly a very judicious piece of work, and the effect was fully accomplifhed, that is, of retaining a quantity of beach to the weftward of it, fo as to make an *addition to the coaft*; but this being once made to as great a degree as the projection of this head is capable of retaining it, the further quantity coming from the weft and paffing by the mouth of Dover harbour, is not retained by this head but gets round it, is again gradually wafhed up upon the fhore, and purfues its former courfe, and probably gets round all the heads and forms the beach in the Downs.

To the weftward of the harbour's mouth is erected a pier, jettee, or breakwater, called Cheefe-Man's Head, whofe effect has likewife been to lock up a quantity of beach, and thereby to make an addition to the coaft. While this was doing, the good effect thereof to the harbour was experienced, but being now in a ftate of decay, the beach it formerly retained is coming down; however as the natural fupply will undoubtedly be cut off for fome time by the falls to the weft, it feems to me, on mature confideration, to be more advifeable to take the benefit of this intermiffion, and to employ the prefent powers in raifing a work that will more permanently and directly tend to the relief of the harbour than the repairing of Cheefe-Man's Head.

The natural direction of the entry of the harbour is S. E. by the prefent magnetic meridian, or about E. S. E. by the true meridian; but to avoid confufion I fhall confine myfelf in the mention of the points of the compafs to the magnetic bearings only. The fhape of the weftern head is not only very uncommon but to me very extraordinary; for

T 2

after

after it has been carried out in the natural direction of the harbour's entry, for about 30 feet, in a line at or about S. S. E. it fuddenly turns away to S. S. W. in which direction being carried on between 60 and 70 feet, it is terminated with a falient angle pointing to the fame quarter. The line of direction of this flank of the pier being continued in an op-pofite direction, cuts within the eaftern pier head about 60 feet, fo that with all winds betwixt S. S. W. and E. S. E. this flank is ftruck obliquely by the feas, and acts in the manner of a tunnel, towards bringing the feas (with wind from S. to S. S. W ) and con-fequently the beach that happens to be lodged before the mouth of the harbour, directly into the throat thereof. The fouth-eaftern feas indeed are fo fhort that they do not much affect the mouth of this harbour any way ; but by the pier turning fo much to the weft it greatly facilitates the beach after it has got round its falient point to get along this flank, whofe line of direction being overlapped as already pointed out by the eaftern head, is thereby equally caught and retained, when the wind is more to the weft than the S. S. W. direction of this flank; for it is very obfervable that the feas will wrap them-felves round a head, and act with great power feveral points of the compafs from the wind that caufes them. Nothing therefore, as it feems to me, could have been formed more im-properly with refpect to the bringing of beach into the throat of the harbour, nor indeed, in all foutherly winds, with refpect to bringing in the feas into the harbour, had not this laft effect been in fome meafure prevented by the jettee or tongue projected from the eaftern pier at the inner entry into the harbour which catches them as they run along-fide of the eaftern pier.

I would therefore advife, by way of leffening as much as poffible the quantity of beach that can get round and lodge between the pier heads, and as the firft and moft important work that can be done, to prolong or carry out the firft mentioned line of the head in its direction S. S. E. and that far enough to come into a S. S. W. direction from the ex-tremity of the eaft head, which will be done by extending this face about 90 feet, as fhewn in the plan hereto annexed, and then returning the outfide fo as to fall in with the falient points of the prefent pier. The additional work will form a fort of triangle, whofe bafe will be principally formed by the prefent S. W. flank, and whofe projection forward towards the S. E. in a line perpendicular to the bafe, will be but little above 60 feet further out than at prefent. By this conftruction all the good that arifes from the fhape of the prefent pier head, in regard to locking in the beach, will be retained ; and as all winds that are farther out than S. S. W. (that is than fouth of the true meridian,) which include all thofe which are the moft prejudicial, will meet the outface of the new work obliquely, they will rather tend to fend the feas and beach to feaward, than to bring them round the head into the throat of the harbour ; and as to all thofe winds that are eaftward of the S. S. W. they

are

are found to be productive of no ill effect upon this harbour. The intent then of this projection is not by way of lengthening the head, so as to make it lock in or retain a greater quantity of beach, but by making it to overlap the eastern head sufficiently, to cause the beach brought coastwise by the great seas at W. S. W. to shoot beyond the eastern head, before it is brought up again upon the shore, and by giving it such a shape as shall also tend in the most effectual manner possible to make the beach drive out to sea till it has passed the harbour's mouth.

I am sensible at the same time, that neither this nor any other shape or prolongation of the piers will totally prevent the beach from coming into the harbour's mouth; for though this reasoning would hold good in case the waves of the sea were reflected from fixed objects, like light from polished surfaces; yet the seas, as already observed, so wrap round the surfaces of bodies that oppose them, that they will in some degree go round even while they come into an opposite direction. They may therefore more aptly be compared to the nature of sound than light; yet as the gross and violence of the action is spent according to the angles and directions wherewith they first strike, the main force or stress of the action will conform to those laws; and hence (as it may be expected) if two thirds of the whole quantity of the beach that now would lodge itself in the harbour's mouth is diverted so as to pass by it without entering, it will follow that the third part will not be of one sixth of the evil consequence and inconvenience to the harbour; and that rendered more easily and readily to be removed by the action of the sluices.

It is perhaps impossible to make a compleat artificial harbour, for what improves it in one sense will often be of detriment in another; for this reason the main drift and purpose thereof is to be principally attended to, and when that is done to as great an advantage as may be, that harbour may be said to be in the most perfect state its situation is capable of. I am therefore aware that this addition to the head here proposed may meet with some objection from seamen; for it may be argued that the present westerly turn of the head admits ships coming from the westward with a scant wind at N. N. W. more easily and readily to shoot up into the wind and get between the heads, (from whence they warp into the harbour) with less risque of over shooting their port, than if this prolongation were to take place. To this I answer, that as there is frequently a bank of beach lodged high against the pier in the very place where the new work is proposed (ready to be driven in between the heads by the first shift of wind more southerly), this will either by the reality or the fear thereof prevent vessels from making the turn of the west head sooner, especially if they have no pilot on board, than if this vacancy were occupied with a solid pier, which they will always have the advantage of seeing above water, and against which, from its

shape

fhape and pofition, no material quantity of beach can ever lodge. I muft alfo obferve that if a fhip is coming up the channel intending for this port, fhe always has either the wind at large, fo that fhe can keep a proper offing and alter her courfe proper to run right into port, or elfe it is an off fhore wind, wherein if fhe pleafes fhe may drop an anchor before the harbour's mouth and afterwards warp in.

2dly. It may alfo be alleged that the prefent face of the pier head is fometimes ufeful for fhips to lie along fide; in order to caft off when the wind is fair to go up or down channel, and with fuch a' wind as does not enable her to *fail* out of port: but it may be obferved in anfwer that befides the objeƈtion, that when a bank of beach is lodged here no ufe can be made of this fide of the pier for this purpofe; it may be further obferved that for all fhips going eafterly, the propofed eaft face will be better adapted in foutherly winds than the prefent; and for fhips going wefterly, a tranfport buoy fixed at a proper diftance to fouth-eaftward of the harbcur's mouth, would anfwer far better for this pur-pofe than the prefent pier, and that, even independently of fuch a buoy, there are fome winds which the propofed face would anfwer to better for fhips to caft off from to proceed wefterly than the prefent. I am told that fuch a buoy has been formerly fixed and main-tained, and it feems very advifeable that it fhould be replaced *.

This in my opinion is the whole and moft effeƈtual means that can be ufed to prevent as much as poffible the beach from getting into the harbour's mcuth. I come now to confider the moft effeƈtual method of removing whƒat does come in.

In the time of fpring tides, as has been already obferved, there is no difficulty; and in regard to neap tides as the quantity that can get in will after the execution of the propofed head be far lefs than at prefent, it may with more eafe be removed, even with the prefent power; but as that does not feem fufficient to be abfolutely depended upon, I fhall now fhew what is in my opinion the moft effeƈtual way to improve it.

On Thurfday morning the 23d of February, being the third day after the full moon, a full head of water was penned in and the turn-water next the eaft pier head fet, I obferved the flate of the beach about and between the heads, (which indeed were tolerably clear) before the fluices were played at low water, in order that I might fee the operation thereof. I obferved that the gates of the fluices were five minutes in drawing, and that the water

* The prolongation of the weft head and fixing a tranfport buoy, I find was adopted by Captain Perry in his report on this harbour 1718.

took

took up the fame time in getting down from the fluices to the pier heads, that it was five minutes more in getting to its full ftrength, which continued for about half an hour ; but after that, though it continued a quarter of an hour longer, the effect was inconfiderable. After the operation I obferved that wherever the beach had laid in the way of the water, it was greatly reduced and carried fo far without the heads that the wefterly feas would infallibly carry it beyond the eaft head, and prevent its returning into the harbour.

I obferved that before the fluices were drawn there did not appear to be above fix or eight inches of fall from the apron of the outmoft turnwater to the fea, and about an equal declivity from the ftone apron of the great gates to the faid turnwater apron, fo that from the ftone apron of the great gates to fea, the fall could not much exceed a foot, or at moft eighteen inches of declivity, but this will be variable according as the tide ebbs more out ; I only beg leave to obferve that this will be the ordinary ftate of it in fpring tides, as the prefent ones were faid to be of a middling kind. I obferved further that the water from the fluices when in their full power fcarcely overtopped the turnwaters, which appeared to be about three feet and a half high, fo that the fall of the water's furface from the turnwater to the fea was fcarcely more than four feet ; and yet in this ftate the fluices are capable of keeping the harbour clear. I was informed that at neap tides fometimes the water will ebb down to the ftone apron of the great gates, but ordinarily fo as to leave about twenty inches upon it, fcarcely ever more than two feet, if it be not penned by beach caft into the harbour's mouth. Hence it appears that at neap tides there is not above three feet more depth of water at the harbour's mouth than at fpring tides, and confequently by a power of water that will overtop the fea water at neap tides as much as it now does at fpring tides, the harbour's mouth might be cleared as effectually at neap as it is now at fpring tides, and this would be effected if the capacities of the fluices were doubled ; for the fame defcent of the furface would produce the fame velocity and effect upon the bottom, and being confined to the fame breadth a double quantity of water would produce a double height, which would then have as good or a better fall into the fea at neap tides, than it now has at fpring tides.

It is true that being difcharged in double quantity it would be fpent in half the time, fo that inftead of lafting half an hour in full vigour, it would laft only a quarter ; but as a much greater body and weight of water will act at once a confiderable effect muft be produced, fo as greatly to relieve the harbour's mouth, which though not made perfectly clear till the approach of fpring tides, would, in conjunction with the relief that is to be expected from the propofed addition to the weft head, prevent its ever being barred up at neap tides or prevent its ufe to all middling veffels ; for even a fingle difcharge of the

fluices

fluices with the power I have mentioned, would in a manner remove the grofs of any obftruction that then could happen.

I was informed that in fummer they can generally gather a full bafon and pent of water
in four days, and one in a week in the very drieft feafons and fhorteft tides; if fo the
pent might always be kept full againft neap tides, fo as to fill the bafon or nearly fo on
fhutting in the firft tide that fhall happen after the approach of any emergence; I fay the
approach of any emergence, for whenever it comes to blow at fuch points as are found
by experience to bring in the beach, it is not neceffary to wait the event but to prepare
immediately to get a full head of water.

There are two ways by which the capacity of the fluices may be augmented. One is by
building a new tunnel at each end of the prefent ftone wall. The width of the prefent
tunnels or archways for the draw gates is twelve feet, but as I am perfectly clear that
the advantage of the fluices does not fo much depend upon the length of time they
play, as upon the great body that can be at once difcharged; I would advife that the new
tunnels would be fifteen feet wide each, and not be fhut by draw gates but by turning
gates, by which the water can be inftantly difcharged without lofs of time or addition of
hands. The execution of this method will require the moft time and expenfe, but when
done will be the moft durable and require the leaft repairs.

The fecond method is by placing turning gates in the great gates, which can be difcharged and will operate like the former, and anfwer the end in all refpects the fame:
the only objection to this method, is, that folding gates made with turning gates encafed
are not only more expenfive to conftruct, but lefs durable than folding gates made plain
and whole; otherwife the bufinefs will be done in this way at far lefs expenfe and in
much lefs time than with ftone tunnels, efpecially as the great gates having been lately
renewed may be made ftill to ferve by having turning gates adapted to them.

As I do not know of any turning gates encafed in folding gates in England, it perhaps
may be doubted whether the thing be practicable, but for their information I beg leave
to mention that this very thing is done and practifed with fuccefs, not only at the Briel
but in the great gates of the fluice of Helveot in Holland, and made ufe of for fcouring
the outward harbour and pier heads, and that it was formerly done in the great fluice
of Mardick near Dunkirk, before it was demolifhed by treaty.

As

As therefore this method can be put in practice in a fhorter time, and at a lefs expenfe, I would the rather advife it at prefent to be adopted, that when the good effects thereof are feen, the larger and more durable work in ftone may be doing while the firft pair of gates are in wear, which may afterwards be changed for thofe of the prefent conftruction when the others are executed.

The fluice of Helveot is forty-eight feet wide : the gates, which I faw the year 1755, by a date upon them, appeared to have been conftructed in the year 1722, fo that they had been in thirty-three years, and were then in perfect good repair ; they had however originally been exceedingly well conftructed and very ftrong. I happened to have an opportunity of feeing the operation of this fluice, which I well remember was far beyond the prefent ones at Dover, though the turning gates encafed in the great gates were the only apertures they had for difcharging the water from the bafon, which, according to my idea, is much lefs than the Dover bafon and pent together ; the whole operation was in a manner over in a quarter of an hour.

The addition to the weft head I would advife to be of ftone, which will not only be more durable and in the end cheaper than wood, but will probably fhew the way of re-building the prefent piers with ftone as they may occafionally want it, which it is to be wifhed had been done at firft.

The outfide of the new work I would propofe to be conftructed of Portland blocks, but the infide may be done with the large rough ftones that are brought from Folkftone, fuch as are ufed for filling the prefent piers. I would however advife, that to the height of fix or eight feet, the fouth face of the propofed addition may be done with Cornifh moor ftone, as the expenfe will not create any material difference upon the whole, and its hardnefs will prevent its being affected by the beach rubbing againft it.

It would be befide the prefent purpofe to make out defigns for the execution of the propofed works, till the execution fhall be refolved upon ; and yet without this it is not poffible to come to any complete eftimate : however by way of giving fome general idea of the expenfe, I fuppofe the ftone head may be done for the fum of £7,500. and the gates made upon the plan I have mentioned, for the fum of £1,500, including an addition to the apron to prevent the action of the water from gulling the bottom.

Either the increafing the water way or the addition to the pier, I expect will be of great ervice to the port, and as one is chiefly mafonry and the other carpentry, if proper funds can be raifed they may be both carried on together; but if they are to be carried on feparately, I look upon the alteration of the head as of the moft immediate confequence and importance.

It may be proper to fuggeft that when the port of Dover is by the means above fpecified put upon the beft footing its fituation is capable of (that is, according to the beft of my judgement) I apprehend it will be very eligible to be more frequently ufed by the king's floops and leffer frigates, and therefore that for fitting them out it would be very practicable to build a dock in the place now called the Paradife Pent, which is in a manner a wafte piece of ground, and where the excavation for the purpofe is in a great meafure ready; this having been in ancient times, as I underftand, a very material part of the harbour.

*Aufthorpe,*
*17th June 1769.*
                                                    J. SMEATON.

---

## OPERATION of the Turning Gate Sluices for Dover Harbour.

When the gate is to be fhut for penning in a head of water, the rope from the crab is to be hooked to the downftream fide of the gate, and by that means brought clofe to its bearings. When the water comes againft it the other half of the gate will be fupported by its threfhold and cheek, in the fame manner that common fluice gates generally are, but as the part of the gate that lies downftream when open, receives no fupport from either, in order to give it that fupport, a triangular frame called a Valet, turning upon two gudgeons is brought round, and acting as a lever, its turning poft is made to bear hard againft the clapping poft of the gate, and thereby firmly fupports it, and preffes it upward againft its cheek, which it would otherwife be inclined to depart from; and in order to make it the more tight, an eye upon the end of the diagonal lever of the valet is brought towards an eye upon the fixed beam by a luff tackle, and in that fituation is lafhed by feveral turns of a cord round the pummels; this done, the crab rope is to be hooked, and then the gate is fet for penning in a head of water.

In

In order to difcharge it, the lafhing is to be cut, and a part of the gate being then unfupported, will throw open the valet, and in part open itfelf, but to fet it right fquare the crab is to be made ufe of, and will be ftopped in its proper pofition by the counter threfholds. The water then runs through in a full bore, and by its fuperior accumulation in the exterior harbours, will act in proportion more ftrongly upon the beach at the entry of the pier heads.

N. B. By giving the greater fide of the gate ftill more advantage, it might be made to open wholly of itfelf, but as its operation would then proceed with violence, it is judged more fafe to have recourfe to two or three turns of the crab, at which a fmall power will ferve, the gate being nearly in equilibrio.

## REMARKS.

According to my notes, the overfalls of the draw gates are about eighteen feet above the floor of the great gates. The floors of the turning gates are propofed to be one foot higher than that of the great gates, and being fifteen feet above the floor, they will pen in fixteen feet head of water, which will be fully fufficient for the neap tides, at which time thefe gates are principally fuppofed to be of ufe. When the full head of eighteen feet is to be penned; the water is to be ftopped from going over betwixt the top of the turning and the crofs beam, by flafh boards, and as on all occafions the draw gates are to be firft opened, by fuch time as that is done the head of water will be run off to the level of the turning gates which may then be difcharged alfo : this being judged more fafe than to raife the turning gates to the full height of eighteen feet.

*Aufthorpe,*
18th *June* 1771.

J. SMEATON.

ESTIMATE

ESTIMATE for erecting one of the Turning Gate Sluices, propofed at Dover Harbour, of fifteen feet water way, by John Smeaton, Engineer.

|  | £ | s. | d. |
|---|---|---|---|
| To erecting a dam of rabbetted piles fo as to pen the water in the bafon to its proper height, and defend the foundation of the fluice from the rife of the tides, the timber work being fuppofed to be completed in place at 5s. per foot   - | 400 | 0 | 0 |
| To a fimilar dam towards the outward harbour to pen out the neap tides from the foundation   -   -   -   -   - | 400 | 0 | 0 |
| To a horfe engine for draining the water, complete   -   - | 150 | 0 | 0 |
| To piling machines and other utenfils   -   -   - | 200 | 0 | 0 |
| Preparation   - £ | 1,150 | 0 | 0 |

N. B. The materials of the dams together with the engines and utenfils will be equally ferviceable in the fecond erection as the firft, and at laft fell as old ftores.

### The FLOOR.

|  | £ | s. | d. |
|---|---|---|---|
| To 1008 feet of fir timber in bearing piles, fuppofed ten inches fquare, and eight feet long, preparing, fhoeing, and driving included, at 4s. per cube foot   - | 201 | 12 | 0 |
| To the main fell of fir timber, framing, jogling, bolting, and laying down, containing 175 feet, at 5s.   -   -   - | 43 | 15 | 0 |
| To 648 feet of fir timber in the common fells, preparing, levelling the pile heads, and trenailing down, at 3s.   -   -   -   - | 97 | 4 | 0 |
| To 180 trenails of $1\frac{3}{8}$ in., at 6d.   -   -   - | 4 | 10 | 0 |
| To 972 feet fuperficial of four inch plank piling of beach, elm, or fir, at 3s.   - | 145 | 16 | 0 |
| To 150 cube yards of chalk rubble, to be rammed under and between the fell timbers, at 3s.   -   -   .   -   - | 22 | 10 | 0 |
| To planking the floor, containing 1060 feet fuperficial of three inch fir plank to be lathed and put together with tar and hair, at 1s. 6d.   -   - | 79 | 10 | 0 |
| To forty-two feet of oak timber in threfhold framing, laying down and bolting, at 6s.   -   -   -   -   - | 12 | 12 | 0 |

### IRON WORK.

|  | £ | s. | d. |
|---|---|---|---|
| To 8 cwt. of iron work in bolts, plates, &c. for the main fell and threfholds, at 6d. per lb.   -   -   -   -   - | 22 | 8 | 0 |
| To 12 cwt. of fpikes for the pile planking and floor, at 36s.   - | 21 | 12 | 0 |
| Total of the floor   - £ | 651 | 9 | 0 |

The

## The SUPERSTRUCTURE.

### MASONRY.

|  | £ | s. | d. |
|---|---|---|---|
| To 10,638 cube feet of Portland ftone to be fet in pozzelana * or terras mortar, at 3s. per cube foot, all work included | 1,595 | 14 | 0 |
| To 9,751 cube feet of rubble backing, the mortar to have a mixture of pozzelana equal to one-third of its proper quantity of fand, at 1s. | 487 | 15 | 0 |
| To an allowance of 3d. per foot, upon the top area for dreffing ftones, for paving, letting in cramps, &c. containing 1,908 feet | 23 | 17 | 0 |

### IRON WORK, &c.

|  | £ | s. | d. |
|---|---|---|---|
| To iron work in cramps 3 cwt., at 4d. per lb. | 5 | 12 | 0 |
| To lead for fixing them, 3 cwt. at 2d. per lb. | 2 | 16 | 0 |
| To oil, &c. | 0 | 10 | 0 |
| Total of the fuperftructure | £ 2,116 | 4 | 0 |

## The GATE, VALET, and CROSS BEAM.

### CARPENTRY.

|  | £ | s. | d. |
|---|---|---|---|
| To 138 feet of oak timber in the turning poft and crofs beam meafured in fquare in place at 6s. | 41 | 8 | 0 |
| To 248 feet of oak timber in the ribs, clapping pofts and valet meafured in fquare, at 5s. | 62 | 0 | 0 |
| To 220 feet of two inch fir plank for planking the gates on both fides, at 6d. | 5 | 10 | 0 |
| To framing and fixing the above, and letting in the iron work, containing 423 feet folid, at 2s. 6d. | 52 | 17 | 6 |

### IRON WORK.

|  | £ | s. | d. |
|---|---|---|---|
| To one ton of caft iron in gudgeons, pots, and boxes, at 22s. per cwt. | 22 | 0 | 0 |
| To 28 cwt. of forged iron work, in plates, bolts, &c. at 6d. per lb. | 78 | 8 | 0 |
| To 2 cwt. of fpikes for the planking, at 36s. | 3 | 12 | 0 |
| Total of the gate, valet, and crofs beam | £ 265 | 15 | 6 |

* N. B. Pozzelana has been imported from Italy to different parts of Great Britain, at the price of 40s. per ton, containing about twenty-four bufhels each, and preferable to the fame quantity of terras.

The

|  | £ | s. | d. |
|---|---|---|---|
| The crab with ropes, blocks, and iron work complete  -  - | 50 | 0 | 0 |
| To drainage of the fluice while building, fuppofing the labour of four horfes for fix months, or 180 days, at 3s. per day each, with two drivers at 2s. per day each, in the whole 16s. per day  -  -  -  - | 144 | 0 | 0 |
| To 600 tons of Folkeftone rubble ftones of promifcuous fizes, for fecuring the apron of the fluice, at 6s. 6d.  -  -  -  - | 195 | 0 | 0 |

## ABSTRACT.

|  | £ | s. | d. |
|---|---|---|---|
| The preparation  -  -  -  -  - | 1,150 | 0 | 0 |
| The floor  -  -  -  - | 651 | 9 | 0 |
| The fuperftructure  -  - | 2,116 | 4 | 0 |
| The gate, valet, and crofs beam  -  -  - | 265 | 15 | 6 |
| The crab, ropes, &c.  -  -  - | 50 | 0 | 0 |
| Drainage of the fluice  -  -  - | 144 | 0 | 0 |
| Folkftone rubble for fecuring the apron  -  - | 195 | 0 | 0 |
| Neat eftimate  -  £ | 4,572 | 8 | 6 |
| Allow for unforefeen accidents and expenfes, 10 per cent. upon the above  - | 457 | 4 | 10 |
|  | £ 5,029 | 13 | 4 |

N. B. No falaries or gratuities to engineers or furveyors are fuppofed to be included in the above eftimate.

*Aufthorpe,*
18th *June* 1771:

J. SMEATON.

P. S. I cannot be anfwerable for prices, for want of a fufficient knowledge of particulars at Dover, but the prices herein fet down are fuch as I think fhould rather exceed than fall fhort. But the quantities, being correctly eftimated, the whole is capable of being corrected by any perfon well verfed in the prices of work and materials at Dover.

PLAN
of
AYR HARBOUR,
1772.

The Iron Point

South Point of Arran

ROCKS

The UCHER

Low Water Mark

Low Ground
Overflowed at Half Flood

Mr Montgomery's
Engine

Road to the Coal Works

High Water Mark

Newtown Loch
Overflowed at
High Water

Ropery

High Water Mark

New
House

N.B. At Present there is said
to be 8 F. 6 In. Water at Neap
Tide, and n F. 6 In. at Spring
Tide over the Bar.

Section of the South Pier

Inside

Section of the North Dyke

Inside

Section of the North additional Pier

Inside

0    5    10    15    20    25    30    35    40    45

Low Water Mark

Nicholas Rock

Magnetic Meridian obs.d by Mr Smeaton

Point of Ayr

Isle of Arran

Meridian of Mr Greggs Plan

N

J. Ferry Junt delin.

## AYR HARBOUR.

(See Plan, Plate 9.)

### The REPORT of John Smeaton, upon the Harbour of Ayr.

FROM a view taken the 2d September 1771, it appears to me that the harbour of Ayr which is near the mouth of the river Ayr, and which falling into the wide and open part of the Firth of Clyde, upon a flat coaft with a drift fand, is by nature badly circumftanced for keeping open, having conftantly a bar before its mouth, as almoft all others in like circumftances are more or lefs fubject to ; and were it not for the river Ayr, which being a river of middle fize, and fubject to great fpeats in rainy feafons, which drive the fands out to fea, that the furge with wefterly and north wefterly winds bring in and heap upon the entry of the harbour, it would be very fpeedily choaked up and deftroyed. Seeing therefore, that the fpeats of the river Ayr are in a manner the only means by which the harbour is kept open, it follows, that to improve the fame in depth of water, which is the prefent object, it is neceffary to attend to thefe circumftances, by which the force of the land waters can act with the utmoft advantage in driving off the fands, and to the forming and improving fuch artificial barriers as fhall have the moft direct tendency to prevent the effect of the winds in bringing it in.

Something for this purpofe has been done many years fince, in the erection of the north and fouth dykes or walls, to confine the channel of the river in paffing the flat fands, but not having been effectually executed, the depth of water in the harbour is not fo good as it might have been ; however, though I have reafon to fuppofe when what I have to propofe is executed, that there will be a gaining of full three feet additional depth of water at a medium, yet as the action of the winds to bring in fands, and the floods to drive them out, are the two powers that are oppofed to one another, but as at different times and in different degrees, it follows that from the nature and operation of fuch inconftant forces there will always be times when the depth of water will be alternately greater and lefs.

The principal fault of the prefent dykes is their lownefs, which being in a great meafure covered at half flood, the fpeats are left unconfined from half flood to half ebb, whereby a material part of the power of the fpeats to fcour is loft. Another material thing, is, that

the

the south dyke being so low, that the seas find a passage over a great part of it, even at one quarter of flood; and the sands appearing to drive coastways from the S. W. to the N. E. by the westerly winds, the south dyke does not prevent the sands from being washed over it into the channel of the harbour, and which must then remain till the speats come to drive it out again. Lastly as the north-west and northerly winds act very powerfully on the mouth of the harbour, and the seas having a long fetch from the north down the firth of Clyde, they have power enough to prevent the larger gravel that comes down the river from getting out to sea, and thereby form the north bank extending from the north pier to the north-west, so as to bar up the harbour entirely to the north, the bank being dry at low water, and forcing the land waters to take their course to the westward: when a number of large speats happen, this bank almost directly opposing the land current, the bank is driven out more to the northward, so as to admit the vessels to steer in from the N. W.; but by a series of northerly winds without freshes, this bank is brought closer in, so as to oblige the vessels to steer in from the W. by N. as it was when I saw it, and sometimes from the west, and as this is the direction nearly of the two present heads, vessels then going in with a scant wind, are apt to be driven upon the north pier.

The most obvious remedy for those evils in part is to build up both the south and north dykes up to high water mark, for by this alone a considerable part of the obstructions will be removed; the sands will be prevented from driving into the harbour from the south west and the speats and land floods being confined, they will act more powerfully upon the bottom, and therefore being preserved in one column till they pass the heads they will act more forcibly upon the north bank to drive it off, and by that means not only keep the Channel more to the north but also make deeper water over the Bar; it is not a matter of calculation to determine what additional depth will be hereby gained, but I should expect it to be not less than eighteen inches over the Bar.

The shape and position of the north bank shews how the powers of nature are balanced; and though it may be very desirable to force a passage right through the bank directly out to sea, yet this seems too great a thing to be attempted. I apprehend it better to conform to nature than attempt to force her, as I take it for granted it will be more desirable to have a good deep passage out at N. West than an indifferent and obstructed one at N. or N. N. West.

It was with a view doubtless of keeping the channel to the North, that the south dyke was carried out so much further than the north, but yet it has not this effect; for though the seas have power enough to keep up the north bank of such a height as to prevent

vessels

veffels going over it, yet being covered at young flood, it is fo low that fpeats find their way over it in fo great a breadth, as not to be able effectually to remove it ; by this means a great part of the force of the land waters is fpent in a fruitlefs effort to do what they can never fufficiently accomplifh to anfwer any good purpofe ; I therefore propofe to turn their whole effect to fuch a quarter as by co-operating with nature they may act with full advantage, and thereby maintain the greateft depth of water poffible in fuch a fituation.

This will be done by lengthening out the north dyke or pier towards the weft, as fhewn in the plan, by which means the whole force of the back-water being fpent in a N. W. direction, and the north bank prevented from being driven in by the feas into the harbour, I expect there will be more water over the bar than at prefent by at leaft three feet, and the direction in and out permanent to the beft fituation it is at prefent capable of.

The only objection I fee to this prolongation of the north head, is that the N. Weft feas will fall right into the mouth, and carry a fwell into the harbour ; but as this harbour by the fouth point of the mull of Cantire and the north of Ireland is in fact land locked from every quarter, there never can be any very violent furges fuch as coafts open to the ocean are fubject to ; and the quay is fo far upwards from the mouth, that what comes in will be in a manner fpent before it comes to the place where the fhips lie. However, I apprehend this will be the leaft of thofe evils the harbour is fubject to ; and in all artificial harbours, we have only the choice of the leaft evil.

The prefent irregularity of the north and fouth dykes makes it very difficult to form an eftimate of the expenfe of compleating them ; but taking it for granted that ftone can be procured at the fame price as that employed for the new coal quay, the beft I can make will be as follows.

ESTIMATE for the Harbour of Ayr.

|  | £ | s. | d. |
|---|---|---|---|
| The North Pier being fuppofed 16½ chains long = 363 yards; and fuppofing half of this to want raifing at a medium 4 feet high and 5 feet thick, will contain 403 cube yards at 7s. | 141 | 1 | 0 |
| To rubble for footing the prefent dyke, viz. 2 yards per yard running for the harbour fide, and 1 yard per yard running for the outfide, viz. for the whole length 1089 yards at 6s. | 326 | 14 | 0 |

To making up the fouth dyke, being fuppofed 19 chains or 428 yards long, being fuppofed 12 feet thick at bafe, 5 feet at top and 10 feet high, will contain 4042 yards at 7*s.* - - - - - 1,414 7 0

To lengthening the north dyke by an additional pier of 5 chains long or 110 yards, to be at a medium 15 feet bafe, 5 feet top, and 15 feet high, will contain 1833 yards at 7*s.* - - - - - 641 11 0

Neat eftimate - £ 2,523 13 0

Allow for contingencies at 10 per cent - 252 7 4

Total - £ 2,776 0 4

N. B. The prices of the above eftimate are formed upon the expenfe to me delivered of building the new coal quay on the north fide of the river, after deducting the expenfe of timber and lime, which will here be unneceffary. The quantities in refpect to length are taken from the plans of Mr. Gregg to me delivered, upon a fuppofition of its being laid down by a fcale of 3 chains to 2 inches,* there being no fcale annexed to the plan; but if I happen to have taken my meafures from a wrong fcale, the expenfe muft be increafed or diminifhed in proportion as the real meafures turn out greater or lefs than I have fuppofed them.

*Aufthorpe,*
22d *January* 1772.

J. SMEATON.

* This is commonly marked fifteen among the plotting fcales.

# HULL QUAYS.

### REPORT on projected Quays at Hull, 17th April 1773.

Mr. Smeaton is defired to give his opinion upon the following queftion, viz.

SUPPOSING a platform or quay be advanced to the diftance of fifteen feet beyond the fronts of the ftaiths or warehoufes on the weft fide of the river Hull, within the town of Kingfton upon Hull, from Hickfon's ftaith fouth of Walton's fhip yard north, to be fupported upon piles, leaving the courfe of the river open underneath the faid advanced wharf, or otherwife making the fame up folid within the faid advanced fpace :— Quære, whether this will have any, or what effect, upon the drainage of the carfe lands and low grounds that difcharge their water into the river Hull, above the north bridge of the town ; and alfo whether it will have any, and what effect, upon the wharfing or jetties on the eaft fide of the river oppofite to the faid advanced wharfs ; on fuppofition alfo that docks for the reception of veffels are to be conftructed in the town's ditches on the north weft fide ?

### The Anfwer of John Smeaton, Engineer.

Having carefully viewed the river Hull from the Holdernefs fluice, a little above the north bridge to the garrifon jetty ; and having perufed the feveral reports of Meffrs. Mylne and Robfon, of Mr. Grundy and of Mr. Wooler, refpectively ; having alfo made diligent enquiries of thofe whom I efteemed beft able to inform me ; I beg leave firft to lay down the following facts :—

1ft. That when it is low water in the river Humber, the river Hull in the greateft land floods does not fill the channel thereof (within the fpace above mentioned) fo high as half-tide mark, and then goes off with great rapidity.

2dly. That when it is high water in the Humber, the greateft land floods make a current of but very moderate velocity.

3dly. That the fea doors of Holdernefs fluice continue fhut feveral hours during the time of high water, even in wet feafons.

4thly. That

4thly. That the frontages of matter formed before the ſtaiths and warehouſes ſpecified in the above queſtion, are at a medium uncovered to the breadth of fifteen feet and upwards at half tide.

> N. B. As there was a conſiderable freſh in the river, from preceding rains, on my view thereof, as well as a conſiderable diſcharge from the Holderneſs drain, the information whereon the three firſt articles were eſtabliſhed was in a great meaſure confirmed by my own obſervation, and the laſt occurred upon my own view.

From the above I infer, that by the great declivity in the river in the above ſpace, (which from former obſervations I make to be above four feet), notwithſtanding the many obſtructions therein, as all the natural flood waters are carried off in the bed of the river, without ſwelling it ſo high as the half or middle tide mark, the greateſt obſtruction to their getting away when above that level is the oppoſition of the tide of flood from the Humber, which alone cauſes them to riſe above that level ; and though the ſeveral impediments mentioned in the ſaid ſeveral reports, undoubtedly contribute ſomething ; yet theſe impediments altogether, when the ſurface of the river is above half tide mark, are ſmall in proportion to the oppoſition of the Humber's tide.

2dly. That the drainage of the Holderneſs and other carrs muſt neceſſarily be carried on principally while the river Hull is under the half tide mark, that is from half ebb to half flood ; and conſequently, as during this time a projection of fifteen feet from the fronts of the ſtaiths above mentioned, would reſt on ground that in general would be uncovered by the water ; the drainages above mentioned would ſuffer no ſenſible change, whether the ſaid projected ſtaiths are built up hollow on piles, or ſolid with walls.

3dly. Since the velocity of the water of the river Hull, within the ſpace above mentioned, is very much ſlower when above the level of half tide than when below it ; and ſince it is a known fact, that the ſcouring of rivers by land floods chiefly takes place when they run over the beds of the rivers at low water, it follows that the projection aforeſaid, whether ſolid or otherwiſe, will make no alteration in the action of the river upon the wharfing or jetees on the eaſt ſide.

I am therefore of opinion, that the propoſed projection of 15 feet cannot be of detriment to either the drainage of the low grounds into the river Hull, nor to the wharfing or jetees on the eaſt ſide of the river, even if made ſolid, but if built on piles, this appears to me

to

to take away all colour of objection; and in cafe a wet dock for the reception of fhips and veffels be made in the town's ditches on the north weft fide of the town, as this will proportionably diminifh the number of veffels that now lie as obftructions in the ftream of the river at low water, the obftructions upon the whole muft be leffened, efpecially at the time of low water, when every impediment to the free courfe of the river is of the moft confequence to the drainage and fecurity of the faid wharfs and jetees.

*Hull,*                                            J. SMEATON.
17th *April* 1773.

---

**An APPENDIX** to the REPORT of JOHN SMEATON Engineer, upon the propofed Quays and Docks at Hull, 17th April 1773.

BEING called upon this bufinefs without previous notice, and being through want of time obliged to deliver my opinion with more brevity and lefs explanation than I otherwife fhould have done; being alfo lately called to meet Mr. Grundy at Hull, who was invited to give his opinion upon a propofition of the fame nature, as that which formed the fubject of my report above mentioned; I have on this occafion had a further opportunity of extending my obfervations, and converfing with Mr. Grundy upon the nature of the cafe; and having alfo the pleafure to find, that when the fame propofition is ftated to each of us, we do not materially differ in our conclufions; yet as each of us has proceeded in his feparate report, to form his opinion by a mode of reafoning fomewhat different; we find it will be more eafy to ourfelves, and more fatisfactory to our employers and all concerned, to deliver what we have further to fay, according to our own words and conceptions: And I find no reafon now to deviate from any material fact or opinion contained in my faid report of the 17th April laft; what I now offer I fhall confider as an appendix to that report, and in further explanation and confirmation of the conclufions therein contained.

In Mr. Grundy's report of the 14th October 1773, he has given the admeafurements of Hull harbour or haven from the outlet thereof at the garrifon jetee to the north bridge, at thirteen different places; which dimenfions, as I doubt not of their having been taken with all due care and fidelity, I beg leave to refer thereto, and found myfelf thereupon. The firft and the two laft of thefe admeafurements lying without the limits of the propofed undertaking, I put on that account out of the inveftigation of a mean fection of the harbour.

I think

I think it right, however, to take the feveral capacities within the limits above mentioned altogether, and draw the medium from the whole; and not from the moft contracted places only; becaufe the feveral parts can obftruct more or lefs, only in proportion to the contraction along each fpace refpectively, the fum of all the obftructions being made up of the fum of all the parts taken feparately.

The meafures oppofite the three interior jetees, contained in numbers 3, 5, & 7, being contractions of a peculiar kind, without appearance of real utility, I clafs them by them-felves, and then taking a mean of all the remainder, viz. 2, 4, 6, 8, 9, 10, & 11, I find that by proceeding to draw a mean fection therefrom, in a manner fimilar to that whereby Mr. Grundy has obtained a mean fection of the narroweft places in his report.

|  |  | feet. | inches. |
|---|---|---|---|
| The mean high water line will be | | 195 | 6 |
| The low water | | 112 | 6 |
| The depth at low water | | 5 | 9 |
| And the total area of the fection | | 4100 fq. ft. | |

Proceeding in like manner with the three fets of dimenfions at the faid three jetees, the mean dimenfions will be, viz.

| | | feet. | inches. |
|---|---|---|---|
| Mean width at high water | | 168 | 6 |
| Ditto at low water | | 102 | 9 |
| Mean depth | | 5 | 3 |
| Mean area of the fection | | 3573 | |

Now, as thefe things at prefent ftand, thofe jetees have their fhare in obftructing the current of the water; and fuppofing their influence equally extenfive as the former, (which is more than there is need to allow), the mean fection of the harbour as it now ftands will be the mean of the areas of thofe two mean fections.

The whole will ftand thus :—

| | feet. |
|---|---|
| The mean or average fection of the harbour, exclufive of the jetees | 4100 |
| The mean or average fection of ditto at the jetees | 3573 |
| The difference | 527 |
| The mean of the average | 3836 |

So

So that 3836 feet I confider as the mean fection of the harbour, as it now ftands, when clear of fhipping; and if from this we deduct what Mr. Grundy has deducted for the mean or average fections of four fhips lying abreaft in this harbour, viz. 448 feet, we fhall then have a neat mean fection, with the fhipping, viz.

| | | | feet. |
|---|---|---|---|
| Neat fection of the haven, clear of fhips | - | - | 3836 |
| Deduct at a medium for fhipping | - | - | 448 |
| Remains clear water way, with the fhipping | - | - | 3388 |

The fection of the propofed projection of fifteen feet for a fufferance quay, if made folid, will, as I compute it, occupy $152\frac{1}{2}$ feet of the area above mentioned, which is fcarce $\frac{1}{23}$d part of the whole fection, with the fhipping therein, fo that

| | | | feet. | inches. |
|---|---|---|---|---|
| Deducting from the above area | - | - | 3388 | |
| The whole fection of the quay | - | - | 152 | 6 |
| There will ftill remain after the quay is made, clear | - | - | 3235 | 6 |

Mr. Grundy has in his report alfo given the admeafurement of the river Hull in ten different places above Hull bridge, from whence and from the before cited admeafurements it appears, that the faid clear fection of $3235\frac{1}{2}$ feet, mean fection of the river before the propofed quay, fuppofing all impediments to remain, is yet larger than any clear fection of which the admeafurement is given above the faid propofed quay; the largeft of them all, viz. below Holdernefs new clough, which is below the outfall of all the drainages, Mr. Grundy has now accurately computed to be 2969 feet, that is $266\frac{1}{4}$ feet lefs than the mean fection oppofite the projected quays. If therefore the faid largeft fection of 2969 feet taken below all the drainages, is efteemed fufficient, no detriment can be apprehended from the propofed quay, as it leaves a clear water way of $266\frac{1}{4}$ feet more than fufficient; but as I apprehend this argument, though juft, is inconclufive, upon the principle of my own reafoning, viz. that every fection of a river is more or lefs an impediment to the currency of water; fo that as impediments they only differ in quantity, as the fection is relatively greater or lefs, the fum of all the impediments being the fum of all the particular parts as above ftated; we will therefore now examine the matter as to the queftion.

At the time of high water of a high fpring tide with a great land flood in the river, when it is fuppofed the river rifes to the height of twenty-two feet above the ordinary low

water

water line, in which state the sections above referred to are stated by Mr. Grundy and myself; it is allowed that in this state the quay will occupy $152\frac{1}{2}$ feet of the section, being at a medium $\frac{1}{13}$d part of the whole nearly; but at the time of high water even in the greatest land floods, the water being then in Hull haven nearly stagnant, it cannot in that state be any impediment at all; for were there a sheet of piling across the river, while the water continued stagnant, it could have no effect. At low water also, as the greatest land flood would be below the face of the quay, as the ground now lies, it then can be no impediment; the impediment, if any, must therefore be the greatest at some point between high water and the time that the tide leaves the face of the quay, which in the ordinary state of the river will be at or very soon after half tide; and from a computation that I have made for this purpose, in which every circumstance is exaggerated in disfavour of the quay, it can never have the effect of causing the water to swell at the head of the quays above $\frac{3}{4}$ an inch; and as this will be when the water has subsided below high water about 7 feet, so that the swell of $\frac{3}{4}$ an inch at that time cannot be supposed to have any effect upon the running off of the water in the superior parts of the river or so minute a one that it escapes calculation to say what it is.

With respect to the discharge of the land flood waters, as affecting the barrier banks of the adjacent lands, it is evident that the pinch of the whole affair is at high water, when it has been shewn that the water being stagnant, the impediment is reduced to nothing: again, as the internal drainage sluices must do the greatest part of their business after the water has left the face of the quay, its impediment to them is reduced to nothing also; or to so very minute a quantity as to justify the opinion I have already given in my report of the 17th of April last, that the proposition of 15 feet cannot be of detriment to the drainage of the low grounds into the river Hull, or to the wharfings or jetees on the east side of the river, even if made solid.

However, as it may be expected by the proprietors of land, and desired by the promoters of this undertaking, that every cause of apprehension, however minute, should be intirely removed or fully compensated, it will follow, that if an addition of water may be made equivalent to $152\frac{1}{2}$ feet, that this will at all events counterbalance the supposed obstructions; and if the said addition of water-way is of such a nature as to operate during the whole time of tide, or when the water is at a lower level than the said quays can impede, instead of a detriment this must be considered as a benefit.

Now, if but a single row of vessels that now lie in the harbour are supposed to be moved into the wet dock, and supposing them at a medium to occupy an area equal to 20 feet wide,

and

and 8 feet deep, this produces 160 feet, which will equally operate at low water as at high ; and as it is probable that the dock will entertain as many ſhips as will amount to a double and even a treble tier in the harbour, it will follow that the circumſtance alone of tranſ-ferring a part of the ſhipping to the dock, will be a moſt full and ample compenſation for any impediment that can be ſuppoſed to ariſe from the projeсtion of the quays, even if they were made greater than now intended.

A further advantage will probably ariſe from the removal of part of the matter before the face of the quays, that according to the preſent ſeсtion now lies there ; for as the ground is too ſteep for ſea veſſels to lie upon, it will certainly be the intereſt of the proprietors to remove it, in order that large veſſels may unload at the quay ſide ; this will undoubtedly be gradually done, and will add to the ſeсtion of the River towards the bottom, where one foot will be of more ſervice than two near the top.

It is alſo to be remarked, that the difference of the mean ſeсtion of the haven at the jetees and clear of them is no leſs than 527 feet, which is a quantity above three times greater than the ſuppoſed addition by the quays. It appears to me that thoſe jetees are of no uſe, but on the contrary an impediment to the harbour, and that they may ſafely be taken away, or at leaſt in the greateſt part ; and in caſe that can be effeсted without objeсtion, I recom-mend it to be done, not only as an improvement to the harbour, but as of itſelf much more than an equivalent to any impediment that can be ſuppoſed to ariſe to the drainage from the quay.

The North Bridge is alſo another ſubjeсt of improvement to the paſſage of the water ; as upon tide of ebb from the multiplicity of timbers and obſtruсtions there is a ſenſible fall here, and as the ſpaces between the croſs rows of piles for ſupporting the dormant part of the bridge ſeem filled with ſtones or ſome kind of hard matter above the bottom of the river, thoſe being removed will increaſe the ſeсtion in this, now the moſt obſtruсted part.

As it ſeems neceſſary to make a communication with the Humber by draw gates from the wet docks, it will be praсticable on extreme of floods to give paſſage to part of the waters of the river Hull that way ; and upon emergencies there is no doubt but that this will be done, as it may be without detriment to the works ; but as the Humber's tide at high water will equally oppoſe the paſſage of the waters by way of the dock as by way of the river, its motion will become ſo ſlow till it is got down to ſuch level, as the gates can be no longer permitted to ſtand open, that in general its benefit to drainage can be ſo ſmall as not to demand, what on every account feems ineligible, viz. for the drainage to have command over the gates of the dock.

Upon the whole matter, having with Mr. Grundy carefully viewed and examined the ftate of the channel of the river between Fore Dyke and Hull Bridge at low water, I am of opinion that the obftructions in the channel of the river, particularly from Scullcoats Church to Hull Bridge, are fo many and fo great, that even if the river Hull from Hull Bridge to the Humber were dug as wide and as deep as the Humber itfelf, yet the drainage above Scullcoats would remain in nearly the fame ftate as at prefent, till thofe impediments between Hull Bridge and Scullcoats were removed.

On examining alfo with Mr. Grundy the ftate of the bottom of the river between Hull Bridge and the South End, I find that the bottom of the river in every place, where the crowded fhipping would permit a trial to be made, is compofed of ftones, and fo hard compacted together, even in the middle of the river, that I did not find it practicable to pierce through it with a fpit; this continued nearly the fame till we got down oppofite to the loweft of the three internal jetees, viz. Church Lane jetee, where in the middle of the river we firft began to lofe the ftone, and below this (it being here the wideft part of the harbour) we found the bottom a watry fand or filt, but yet ftones on the wharfing on each fide; from whence I conclude as before; that as the principal rapidity of tide of ebb will be when the water is below the face of the quays, that the eaftern wharfs and jetees will be unaffected thereby; and that were the current much ftronger than it is, it could make no impreffion upon the bottom of the river above Church Lane jetee, and below from the increafe of width the current has a lefs tendency to make an impreffion, which tendency will be ftill prevented at the fides by the ftones, in like manner as now that the three internal jetees by difturbing the regular fet of the current are of differvice to the wharfing taken collectively; and if thofe jetees were removed, the ftones will remain a better defence to the wharfing than in the prefent ftate.

J. SMEATON.

*Aufthorpe,*
18th *December* 1773.

# HULL DOCK.

The REPORT of John Smeaton and John Wooler, Engineers, upon the feveral Matters referred to their Confideration by William Walker, Efq. Chairman to the Commiffioners of the Dock Company at Hull, by his Letter dated the 14th Auguft inftant.

HAVING carefully viewed the feveral matters referred to our confideration, fo far as the prefent ftate of the works would admit, affifted by the information of Meffrs. Holt, Turner, and Wefterdale, we are of opinion as follows :

1ft. Refpecting thofe parts of the dock wall that are bilged out of their line, containing a length of 187 yards, or thereabouts, we are informed that the ground upon which they ftand was not only more foft and fpungy than any other parts of the dock, but that part being raifed the latter end of a feafon which turned out rainy and wet, the mafonry had never an opportunity of drying or fetting to that degree of firmnefs that otherwife it would have done ; and further, it appears to us by the report of Mr. Holt, as to the depth at which the whole was founded, videlicet, nearly even with the excavated bottom or bafon of the dock, that the greateft caufe of failure in this part was the not going to a fufficient depth below the general bottom in this foft ftratum, fo that the weight or re-fiftance of matter before the foot of the wall might be a fufficient counterpoife to prevent the bottom from preffing forwards, and at the fame time raifing the ground before it. And as this appears to us to be a radical defect that cannot now admit of an adequate remedy, we fee no certain method of preventing its going further, but that of taking down the whole length fo bilged, and new founding it upon fafer principles ; but as this will be attended with a confiderable lofs of time, as well as of expenfe, we fubmit the following confiderations to the Board :—

In the firft place, we are willing to think that this derangement would not have happened, if the bottom and fides of the bafon had been charged at that time with the weight of water intended to be pent therein ; that is to fay, had the dock been full of water. Secondly, we are of opinion, that the letting in the water, and keeping it pent up to its proper height, may afford the means of preventing this bilging from increafing, and in all probability would have the defired effect, were the dock always to remain fo filled, or always with a confiderable depth of water therein. Thirdly, that the greateft danger to be apprehended will be, when the water on account of any neceffity or other-

wife

wife is let out again, and as we cannot take upon us to fay that this part of the wall would come down, or fail upon fuch a retreat of the water, we therefore cannot help recommending it to the Company, as every thing feems nearly ready for the trial, and nothing appears to us to forbid its being done, to put this matter to the teft; for if any part of the wall fhould fail thereupon, it can be nearly as well rebuilt as now, and if it fhould thus bear the teft, much time and expenfe will be faved. On this head therefore we beg leave to remark, that it would be proper to introduce the water fo gradually, that it may for example be a week in filling, which may be done by plug holes cut through the frame dam. Afterwards to let the water remain at its full height for a fortnight at leaft, fo as to give time to the earth that backs the walls to fettle and confolidate, and then to let it out, not all in one tide, or fo faft as the tide ebbs, but to bring the ftrain or ftrefs upon the walls by degrees, allowing, for example, to ebb out five or fix feet the firft day; the fecond to let it ebb out as low as the former; and after this, in the courfe of the recefs of the tide to let off by the draw fhuttles in the lock gates about three feet more; and thus day by day making the reduction by two and a half or two feet, till either the whole is let off that will go out, or that fomething fhall appear to prohibit it; fo managing the water difcharged by the fhuttles, as to take all the time in each tide poffible, fo as only to be drawn down to its intended gage by the time the tide has re-flowed nearly to the fame height.

It is to be remarked, that in one place in this damaged wall, for about twenty feet in length, near the greateft fwell of the weftern bilge (where the top of the wall is got out of the line two feet eight inches and a half), that the bottom part below the off-fets is further bilged and feparated from the internal part; which part being of confequence weaker than the reft, we wifh to recommend, by way of giving it an equal chance with the reft, that after the water is let in, and the frame dam removed, to form an artificial buttrefs againft it, by dropping down Hazlecliff ftones, brought into the dock by lighters, fo as to form a flope againft the wall higher than the fet-off, this buttrefs by its weight will prefs equally againft the wall, and prevent the bottom from rifing; and fhould the whole by this means fucceed, and thefe ftones be found materially in the way, they may be afterwards removed, and the place otherwife fecured in fuch a way as to be lefs liable to objection on that account; at any rate this place may be avoided without lofing more than one birth by the quay fide.

In regard to the reft of the wall round the dock, having the advantage of better ground, and being got up in better feafons, it appears to us to ftand as it was built, with-out crack, or any apparent failure; we therefore apprehend and expect that it will prove fufficient, obferving at the fame time, that the whole is in our opinion defective in

not

not having the buttreffes or counter-forts much nearer together; in confequence hereof, fo great a ftrefs lies upon the prefent ones, as in the bilging parts of the walls to have produced a thorough feparation between the wall and buttrefs. There alfo appears a deficiency in the walls, in the omiffion of a proper quantity of throughs or large bond ftones which ought to have been diftributed throughout at due diftances, in order to tie the brick-work properly together. We alfo beg leave to mention another difadvantage attending them, and that is, in not feating their bafes or foundations a couple of feet or there-abouts, deeper in the ground, or in other words, below the general bottom or floor of the bafon, in order to afford them the proper abutment or balance, when their weight joined to that of the earth they retain, comes to act with full force thereupon.

The fecond point referred to in the above letter is the rifing of the floor of the lock in the middle, which we beg leave to obferve arifes from a natural caufe, and will neceffarily do fo, more or lefs, according to the firmnefs of the ground upon which it is laid, and the ftrength of the timbers which compofe it, joined with the perfect ftability of the wing walls which form the lock. From the information of Mr. Holt, as to the fcantling and diftribution of the timbers employed therein, we are of opinion, that the floor is too flight for fo maffive a piece of work, which occafions it to fpring more than it otherwife would have done. This fpring or rifing is in general about two inches at or near the threfhold of the gate where the floor is double, but in the fingle floor of the chamber it is about three inches, but in the aftermoft or ftern-gate receffes, where the breadth or width of the floor is fix feet greater, the fpring or rifing is four inches. We are therefore willing to hope the whole floor may remain or continue as it now is, and that no further inconvenience is to be apprehended therefrom, than what regards the facility of opening and fhutting the gates. We had them tried, and found the north of the head gates, and fouth of the ftern gates, move fomewhat ftiffly, yet as the letting in of the water will be likely to produce an eafement, both by fetting down the middle of the floor, and buoying up the gates, we cannot recommend an alteration till the whole is tried, and then it is poffible feveral things may be better remedied than they can fpecu-latively be forefeen. Refpecting the walls of the lock they have the appearance of being well built; we however obferve fome fmall fets therein, which we impute to the want of ftrength in the foundation timbers.

Thirdly, to make the entry into the harbour, as eafy and convenient as poffible.

We have made a rough fketch of the ground adjacent to the entry, and apprehend, that to make all poffible convenience the Dock Company fhould be poffeffed of Mr. Walton's ways or frontage before his wharfs into the river.

In

In anfwer to the fourth, regarding what is further to be done for the completion of the plan, we apprehend that the dock in its prefent ftate is wanting in fome effential matters, the principal of which is the means of fcouring it from the filt that will undoubtedly gather therein, by taking in the muddy tides from the Humber. Had the lock been made with gates pointed both ways, that is, towards the river as well as to the land, fo as to fhut out all fuch fpring tides as would rife above the gage height of the conftant water in the dock, much of this would have been excluded ; alfo the depth of the dock being fuch as to retain four feet at low water, and being in breadth 255 feet, we are of opinion that no fcowers from any practical refervoirs can take place, fo as to clear the filt into the Hull, it therefore remains, that by introducing the current of the river Hull at low water, (and particularly when there happens a frefh in the river at fpring tides in the fummer), to pafs through the dock by means of a canal down towards the Humber, this clearance with the aid of men may be effected, which canal muft be dug to a depth at leaft equal, or indeed rather of fuperior depth to that of the dock, with fuitable fluices or locks thereon.

The four draw-gates now formed in the crofs wall at Beverly Gates, amounting together to a breadth of fifteen feet only, we look upon as infufficient to let all the current from the river Hull pafs through, that may be introduced to advantage, unlefs fuch a pen and fall be brought on at that place as would materially injure the declivity of the current in its paffage through the length of the dock, and in confequence the ftrength of the fcower. There being therefore a neceffity for the removal of this wall, we would advife a fluice to be erected of equal depth with the prefent great lock, or indeed fomewhat deeper, with a fingle pair of gates pointed towards the dock, and a draw bridge over it, for the Beverly roads.

At Myton Gate it will be fufficient to conftruct a bridge only of fufficient paffage for the water, and at Hazel Gate a fluice of fome kind, alfo of fufficient water paffage, with a bridge over it. If this be made a navigable paffage even for fmall veffels, fome of them, fuch as the large keels from Thorn, &c. that cannot ftrike their mafts, will render it proper that the bridges at Hazel and Myton Gates, fhould alfo be draw or moveable bridges, and we would recommend that the fluice at Hazel Gates, if made a navigable fluice or lock, fhould be conftructed with double pairs of gates, pointed both to landwards and feawards ; and we look upon it that a navigable fluice at Hazel Gates would be found very convenient for the following reafons :—

When a confiderable trade comes to be eftablifhed upon the bafon, we apprehend it will be found very troublefome to open the great lock for the paffage of keels, lighters,

and

and fmall veffels, as well as reduce the water of the bafon in neap tides. This defect may in fome meafure be fupplied by conftructing a proper lock for fuch veffels to pafs in and out at the Hazel Gate, which if laid one foot deeper than low water at the Humber, will afford a fufficient paffage for the fcower water, if made eighteen feet wide, but may be made as much wider as fhall be thought neceffary for the convenience of trade.

Now, as there will be but about five feet of defcent of the furface of the water from the entry of the dock at the river Hull, to its paffage out into the Humber oppofite Hazel Gate, we look upon it that its velocity through the dock will of itfelf be too feeble to drive out the filt, it will therefore require to be aided by fetting in a large number of men with col-rakes and other inftruments to raife the mud and filt, as many tides, once or more in a year, as fhall be found fufficient for the purpofe of getting rid of the intermediate accumulation; beginning in the middle, and going as near the fides as fhall be judged convenient to the fecurity of the walls, for the middle being cleared to a fufficient extent, the grounding of veffels near the fides will continually be preffing down the mud from the fides towards the middle.

It is to be noted, that nothing but the prefent dock bafon will require the aid of men to raife the mud; the external paffage from the lock to the river Hull, as alfo from the propofed fluice at Beverly Gates to the Humber, may be kept clean by fcowers from the great bafon.

We muft conclude with remarking, that, with regard to the keeping the great bafon clear of mud, there is doubtlefs a poffibility of doing this by machines worked by men and horfes, properly contrived for the purpofe; and though doubtlefs this labour will be very great, and at the fame time an almoft continual incumbrance to the bafon, yet as on the other hand the charge will be very great for the additional works above propofed, it might be wifhed to bring the matter to a computation; but as it would take up a confiderable length of time to enter into a detail of this bufinefs, which would be neceffary for the purpofe of an eftimate, we muft beg leave to poftpone the confideration of this matter to fome future opportunity; and indeed till the trial that we have already mentioned to be made of the dock fhall fhew for a certainty what works it will be moft needful for the company to take in hand.

*Hull,*
17th *Auguft* 1778.

J. SMEATON.
JOHN WOOLER.

To the Honourable the Commiſſioners of the Cuſtoms, and the Chairman of the Dock Company of Kingſton-upon-Hull for the time being.

Hull, 24th November 1779.

WE, the underwritten perſons being ſeverally appointed by the Honourable Commiſſioners of His Majeſty's Cuſtoms, and the Chairman of the Dock Company of Kingſton-upon-Hull, in purſuance and by virtue of an Act of Parliament made in the 14th year of the reign of his preſent Majeſty George the Third, chap. 56, for making and eſtabliſhing public quays and wharfs at Kingſton-upon-Hull, having carefully ſurveyed and examined the ſundry works executed by the ſaid company in purſuance of the above mentioned Act, humbly beg leave to report as follows :

1ſt. That in conformity to the clauſe, page 9 of the printed Act, directing the works to be done, the ſaid Company have made the baſon or dock with a large ſea lock, or double ſluice thereupon, for the commodious paſſage of veſſels, in the whole extending from the river Hull to within twenty yards of the road, formerly paſſing through Beverly Gate, being as near to the ancient ſcite thereof as in our opinion it conveniently could be.

2dly. They have dug, and made the ſame in all neceſſary parts within fifteen inches of the depth of the bed of the river, as we found it at, and near to the entry into the baſon from the ſaid river, and that it now continues of the ſame depth except in ſome places near the entrance of the ſluice and baſon, where the mud has gathered, but which can be cleanſed away whenever the ſame ſhall become a real inconvenience or impediment, and which as we apprehend comes under the clauſe for cleanſing, page 13 of the ſaid Act to be quoted bye and bye.

3dly. That the ſame is capable of admitting loaded ſhips of great burthens, as it will receive upwards of twenty feet depth of water in ſpring tides, and as it ſtands at preſent, will retain by means of the lock or ſluice eighteen feet or thereabouts.

4thly. And that the width or breadth of the baſon or dock (excluſive of the ſluice and entry which are of neceſſity contracted) is of a medium of above ſixty feet greater than it could have been if the Company had not purchaſed ground beyond the limits which the ground granted by this Act from his Majeſty would have admitted.

5thly. That

5thly. That the faid Company have built a commodious quay or wharf, ranging along the fide of the faid bafon or dock next the town, which as it appears to us is of a fufficient and convenient length for the trade and bufinefs of the faid town and port, and which we can the more readily judge fo to be, becaufe it is not only of the full length required by the faid Act, but of a greater length by above thirty feet than it could have been made had not the Company purchafed the additional ground aforefaid on the oppofite fide, being now of the width or breadth of feventy feet.

6thly. That refpecting the conftruction of fuch " refervoirs, fluices, bridges, roads, and works, requifite matters and things as they (the faid Company) fhall from time to time judge neceffary for the more convenient ufe of the faid bafon and dock," as directed in the 9th page of the faid Act, it appears to us that the Company have done fuch of the above articles as are at prefent neceffary for eftablifhing the trade and commerce intended by the faid Act, upon the faid bafon or dock, and the faid quay or wharf adjoining, and that the faid trade now is, and has been for fome months paft, carried on with convenience.

7thly. With refpect to fuch of the above difcretionary articles, as further time and experience fhall fuggeft to the faid Company to be neceffary, for cleanfing or repairs of the faid bafon, quay and works, as it is directed by an exprefs claufe, page 13 of the faid Act, as follows: " That the faid Company fhall from time to time and at all times hereafter well and fufficiently repair, maintain, fupport and cleanfe the bafon or dock, and the quay or wharf, and all other the works, matters, and things by them to be made, built, and provided by virtue of this Act;" it feems to us that the manner of performing the fame is left to the difcretion of the faid Company, whenever it may become neceffary.

Laftly. That the public being now in poffeffion of the object required by the act to be done, and within the limited fpace of feven years,—we do therefore report and adjudge that the works of the faid bafon and dock, and quay or wharf, are completed according to the intent and meaning of the faid Act.

JOHN WOOLER.
J. SMEATON.

# WORKINGTON HARBOUR.

The REPORT of JOHN SMEATON, Engineer, upon the Improvement of the Harbour of Workington in the County of Cumberland.

IN confequence of a view and neceffary admeafurements taken of the harbour of Workington upon the 25th and 26th days of November 1776, at the inftance of Sir James Lowther Bart. I have confidered the fame, together with the fix firft articles of a paper delivered by the gentlemen of Workington, to the following Purport, viz.

The gentlemen of Workington having taken their harbour into confideration, think that it ought to be enlarged as well as repaired, and for that purpofe wifh to have an Act of Parliament.

What they chiefly want, is,

1. To extend the frame work down the north fide of the river Derwent.

2. To repair, join, and preferve the quays on the fame fide of the river, and to give powers to fix pofts for mooring veffels, &c.

3. To fecure the river Derwent in its prefent courfe fo far as may concern the harbour.

4. To repair and extend the middle or merchants quay, and fix pofts thereon for mooring veffels, &c.

5. To cut open, widen, and cleanfe the fouth gutt, or mill race, as far up as to a point twenty yards northwards from the north corner of John Smith's houfe at the low end of the town; and between the faid gutt and a breaft-work or quay, which is intended to be made between the eaft end of the prefent fouth quay and the faid houfe; and to take off the marfh twenty yards broad along the breaft-work or quay already made and to be made between the faid houfe and the low gutter, for ways, moorings, lading and unlading fhips, &c.

6. To

6. To make a mole on the fouth fide of the river, to break the force of the fea, and to extend the quay down the Rectory ground to the faid mole.

Having therefore confidered how far Sir James Lowther's intereft and property may be affected by the above propofition, as well as what may be proper to be done for the good of the whole, I am of opinion as follows:

1ft. To the firft propofition there appears this objection; that by extending the frame work down the north fide of the River Derwent, being nearly in a direction N. W. by the compafs, and the great Seas that affect this harbour the moft being thofe from N. W. to S. W. all thofe feas will be catched by the faid extended frame work, and will range along-fide the north quay in a more violent manner than hitherto they have done, there being no beach on that fide for them to break and fpend themfelves upon, fo that the births of the ships lying to load coals, &c. at the north quay and hurries thereupon, will be ren-dered more unquiet than at prefent, and which are now the moft uneafy fituations of any part of the harbour; and not only this, but at the fame time Sir James Lowther will be precluded from making a very capital improvement, by an additional harbour on the north fide.

2d. To the fecond propofition I muft obferve, that this being the property of Sir James Lowther, it is his bufinefs to execute: nor can I apprehend that Sir James can reafonably object to fuch pofts being fixed upon the quays on the north fide of the river, as may be proper and neceffary for occafionally hauling ships in and out of the harbour; and from the quays on the fouth fide, provided the places of thofe pofts be previoufly afcer-tained, and limited in fuch manner as not to interfere with fuch improvements as Sir James Lowther may think proper to make upon the faid north quays and his property there.

3d. To the third I do not fee any reafonable objection, provided it be done fo as not to interrupt the free courfe of the river Derwent.

4th. To the fourth the fame obfervations occur as to the fecond.

5th. To the fifth I do not fee that Sir James Lowther can have any reafonable objection, except any ſhould arife from matters of property that do not pertain to confiderations of engineery.

6th. I alfo

6th. I alfo fee no objection on the part of Sir James Lowther to the works propofed in this article, provided the head of the faid mole be terminated at the place marked out in the plan for the termination of the faid mole or pier on the fouth fide, which place is nearly the fame as where a heap of ftones were placed, which were fhewn to me as put there by the direction of the gentlemen of Workington for marking the place of the head or termination of the faid intended mole ; and provided nothing be contained in the act to prohibit Sir James Lowther from building  mole on the north fide, whofe  head or termination fhall be refpecting that on the fouth fide, the fame as is fhewn in the plan, that is, in a direction nearly N. E. and S. W. from each other by the compafs, and at a diftance not exceeding 200 feet.

I come now to defcribe the improvements that I would propofe to be made by Sir James Lowther on the north fide ; and look upon it that the fpace propofed to be inclofed by the new projected pier on the north fide  will be the moft proper ; and which appears to me by far  the moft improvable  and eligible fituation ; for, from the foundings taken, there will be here nearly three feet more water than can be eafily obtained in any fituation further up the river, which is indeed a very material confideration, becaufe at common fpring tide there may be made here by deepening the ground as per eftimate full fixteen feet water, and as the neaps are faid to run about four feet lower than the fprings, there will then be full twelve feet water, equal to the draught of the largeft veffels that now frequent Workington harbour as I am informed ; an advantage either to the movement of the prefent veffels, or to the reception of larger that I need not infift upon.

As the bringing out a new mole from the land at high water mark will of itfelf be a work of confiderable expenfe, it feems proper, in order to reap all the  advantage that may be from it, to give the area inclofed  a competent breadth.  The breadth inclofed here is propofed of about 200 yards, which feems fufficient to contain every improvement that can at prefent or in future be requifite at this place.

This difpofition will alfo be a very confiderable improvement to the ftillnefs of the water at the prefent quays ; for inftead of the feas after they have  entered the outward heads ranging along the face of the north mole, as would be the cafe if the frame work were continued to the fame point ; by the fpace being  left open between the propofed north head, and the prefent termination of the north frame work, the  weight of the fea will in part fall in there, and ultimately meeting a floping beach, will there be broken and fpent, inftead of being reflected from one object to another, as is in a great meafure at prefent the cafe.

Furthermore,

Furthermore, the feas after paffing this fecond opening, finding fo very confiderable a fpace to be expended in, the veffels arranged alongfide the interior face of the pier will lie in a quiet harbour, and in failing in they will conveniently bring up by coming upon the floping ground of this new harbour, and this I apprehend is the whole of what may be originally done: But if the fpace comprehended between an interior wharf wall, and the N. E. fide of the pier, be dug out to the greateft depth propofed, and thus feparated from the reft of the floping bay, the feas that will be propagated through this laft contracted entry, will be fo broken and difperfed, that the veffels will lie comparatively fpeaking as ftill as in a pond; while the outer bay or harbour will remain fufceptible of any other improvement that may hereafter be fuggefted; and in the mean-time be a receptacle for fmaller veffels occafionally coming in to take refuge.

It is a poffible thing, though not certain, that the fands that drive coaftways from S. W. to N. E. may in part get between the outward heads, and meeting ftill water within the opening between the north pier head and the termination of the frame work may fubfide and lodge there; and being out of the courfe of the Derwent at low water, cannot be driven out thereby: for remedy whereof, if this fhould prove to be the cafe, I would propofe to conftruct an under-ground tunnel of five feet wide, to communicate the head of the interior harbour with the river Derwent above the iron works quay, in which fpace there being a confiderable fall at low water, fuch fands would be driven out thereby again into the courfe of the Derwent, which will take them beyond the outmoft heads.

J. SMEATON.

*Aufthorpe,*
8th *January* 1777.

---

ESTIMATE for building a new North Pier at the Harbour of Workington.

|  | £ | s. | d. |
|---|---|---|---|
| To the building of 50 rods of the pier on the north fide, beginning from the pier head, to be of the mean height of 21 feet, 36 feet bafe, and 30 feet top, with a parapet 6 feet bafe, and 9 feet high, at the price of £114 per rod - - | 5,700 | 0 | 0 |
| To building 38 rods of the pier, to be the fame infide as the former, and of conformable dimenfions without, with parapet mean height 5 feet by 4 feet thick, at £79 per rod - - - - | 3,002 | 0 | 0 |
|  | To |  |  |

To building 8 Rods of pier conformable at top to the former, and same parapet, at £32 per rod - - - - - 256 0 0

To digging out the harbour so as to form a clear area next the north pier 85 rods in length and 12 rods width, so as to make 16 feet water at common spring tides, besides what will be used and paid for as fillings between the pier walls including foundations, viz. 40 rods, at £6 per rod - - - 240 0 0

45 do at £12 do - - - - - - 540 0 0

| | | |
|---|---|---|
| Neat estimate - - | £ 9,738 | 0 0 |
| Add 10 per cent for contingencies - - - - | 974 | 0 0 |
| Total of the exterior pier and clearance - - - | £ 10,712 | 0 0 |

*Austhorpe,*
8th *January* 1777.

J. SMEATON.

## ESTIMATE of interior Works, viz.

| | £ | s. | d. |
|---|---|---|---|

To 65 rods of interior wharf wall 7 feet base and 5 feet top and 20 feet high, to be faced with squared stones, at £41 per rod - - - 2665 0 0

To 56 rods of tunnel 5 feet wide, for opening an occasional passage of water from the river Derwent, if found necessary for scouring out the north harbour, at £15 per rod, including extra work about a draw gate for opening or shutting the same at pleasure - - - - - 840 0 0

| | | | |
|---|---|---|---|
| | £ 3505 | 0 | 0 |
| Add 10 per cent for contingencies - - - - | 350 | 0 | 0 |
| | £ 3855 | 0 | 0 |

*Austhorpe,*
8th *January* 1777.

J. SMEATON.

## Second REPORT of JOHN SMEATON, Engineer, upon the Harbour of Workington.

PURSUANT to the requeſt of Sir James Lowther, Baronet, ſignified to me the beginning of September laſt by Martin Dunn, Eſquire, I took the opportunity upon the 20th December laſt to reinſpeót the harbour of Workington, attended by Mr. James Spedding and Mr. Thompſon Ship maſter of the ſaid place; and the ſubjeót of my preſent view being, as ſtated to me by Mr. Dunn, to conſider by what means the entry into the preſent harbour might be beſt facilitated, 1 accordingly made ſuch obſervations and enquiries of the gentlemen who attended me, as tended to make me clear in this matter; and alſo took ſuch meaſures as appeared neceſſary for the founding an eſtimate; from all which I am of opinion that if the frame work were extended in ſuch manner as was agreed upon in London, when the clauſes relative to the engineery part of the propoſed aót were ſettled; and which was particularly deſcribed in the beginning of the clauſe marked D. that this would be the moſt likely means of facilitating the entry of the harbour; the deſcription whereof in the ſaid clauſe is as follows : " And whereas it is apprehended it will be uſeful for the navigation " of the ſaid river, that the ſaid frame work of the north ſide the ſaid river ſhould be ex- " tended and carried downwards from the north-weſt end of the ſaid work, one hundred " yards, according to the preſent direótion thereof, the top of which to be not leſs than " one foot higher than the higheſt ſpring tide that has been known at Workington. To " be filled ſolid to not higher than within four feet of the height of what is ſolid in the " preſent frame work; and that the pile work from the ſaid frame work ſhould be ex- " tended downwards to where the low perch now ſtands."

The above extraót was the colleótive ſenſe of the different parties concerned at the time, and which if enaóted into a law muſt have been executed accordingly; but this not having been the caſe, and being as I now apprehend, called upon to give my opinion what it may be proper for Sir James Lowther of his own free will to execute; I look upon myſelf at liberty to give my particular opinion, which is, that I do not apprehend it neceſſary to conſtruót the frame work ſo high as ſpecified in the above extraót, for a very extraordinary tide having happened at Workington in September 1776, that was near upon five feet above the common ſpring tides; and this frame if to be one foot higher than that muſt be ſix feet above the ſame; it appears to me quite ſufficient if the new frame work be raiſed three feet above ordinary ſpring tides, which will be ſo much higher than the top of the upper rail of the preſent frame work, as this is barely even with the ordinary ſpring tides; alſo the pile work propoſed to extend from the termination of the intended extenſion to the low perch, being deſigned ſolely for the keeping the water in its preſent channel; it appears to me that this work may be deferred till the river has ſhewn ſome ſymptoms of its being likely to change it; for as, according to the accounts ſet forth in Mr. Jeſſop's report, it has continued in this channel ſixteen years; and the channel itſelf is now formed in a very compact bed of gravel to a depth from five to ſix feet below the general ſurface of the flat gravel beach to the north, with very gradual ſlopes on the ſides, it does not at preſent appear to me very likely to change its courſe; and unleſs poſitively neceſſary, the pile heads propoſed to be driven may occaſion damage to ſhips if they ſhould happen to ſtick upon them, if they ſhould be driven out of the channel, when otherwiſe they might not touch, or elſe ſtick upon a gravel ſurface much nearer to a level.

It alſo appears to me that the propoſed extenſion of the frame work will add to the ſecurity of the channel's remaining where it is, without the uſe of pile work; and that in caſe the funds will not properly allow the execution of the whole of the propoſed one hundred yards at once, that fifty yards may be firſt done, which will alſo aſcertain the utility that may be expected when the remainder is executed at a future time; for this reaſon, and as the propriety of the execution of almoſt every thing depends upon what it will coſt, I have very carefully conſidered the probable expenſe of extending the frame work per yard ruuning. The top to be three feet higher than the preſent frame work, and the ſolid to be four feet lower than the ſolid of the preſent frame work, which will obviate the objection to the extenſion of the frame work (according to its preſent dimenſions) contained in my firſt report of the 8th January 1777, and ſuppoſing the extenſion to be carried on in a firm and ſubſtantial manner, with Riga or Memmel fir balks of twelve inches ſquare for the main timbers firmly bolted together; and the frames at no more than five feet diſtances middle and middle; this I make to amount to the coſt of £13 4s. per yard running, and therefore for 50 yards £660, and for the 100 yards, to £1320.

This work may be done at a leſs expenſe if done in a ſlighter manner, that is, with timbers of leſs ſcantlings, the frames at greater diſtances, and leſs iron work; but the ſavings that will thence ariſe, I cannot recommend.

*Auſthorpe,*
3d *March* 1778.

J. SMEATON.

## PLYMOUTH YARD.

To the Commissioners and principal Officers of His Majesty's Navy.

The REPORT of JOHN SMEATON, civil Engineer, upon the defective Works in Plymouth Yard.

HONOURABLE SIRS,

CONFORMABLY to the matters fuggefted in your letter to the Right Honourable the Lords Commiffioners of the Admiralty of the 5th April laft, a copy of which you were pleafed to put into my hands, and agreeably to your requeft to me for that purpofe, I have viewed the defective works in Plymouth Yard, and having been conducted in my view thereof by the officers of the yard, I beg leave to report to your Honors fuch opinion thereon as by their appearance in their prefent ftate, affifted by the faid officers and arti-ficers, I am enabled to do; and having viewed the whole range of wharfs and frontage works from Mutton Cove at the fouth part of the yard to and inclufive of the Graving Place, my obfervations were as follows:

1ft. That the cove, formed at the fouth part of the yard by a circular pier or wall, is carried up upon a rock foundation to the height of about eight feet with Portland blocks, in front, of a breadth from three feet to three feet fix inches in bed; and which being flat jointed and fet in terras mortar, appears to ftand firm and unmoved. Above the height aforefaid, the external of the wall both infide and out are built with limeftone or marble afhlers, of above one foot height of courfe or thereabouts; and faid to be in bed from fourteen to twenty inches, which were alfo fet in terras mortar; but on the outfide prin-cipally and chiefly, for about four courfes above the Portland blocks, the marble afhlers are bulged outwards; and that in fo regular a manner as can only be accounted for by fome regular caufe which has taken place at the fame height, and not from accidental violence or derangement: and though this bulging is moft confpicuous in the circular part of the wall above mentioned, yet the fame kind of appearance extends in a greater or lefs degree from the circular pier above mentioned, through the whole range of wall that fences in the ponds before the maft houfes; and as the greater length of this range is built upon rock and the other upon a timber foundation; and as that on the timber is in general as free from the derangement of bulging as that built upon rock, I muft conclude that the bulging of the wall throughout this diftrict has not been occafioned by any failure in the foundations.

2dly. On this head I am informed, that the backing, or what may more properly be called the fitting of the wall betwixt the two outfides, (both of which were fquared ftones fet in terras,) was wholly done with rubble mafonry, built with common mortar. Now it appears to me from the whole courfe of my experience, that no compofition of mortar made with lime burnt from Plymouth marble with common fand will ever concrete into a hard ftony fubftance if conftantly under water, or fubject to be frequently wetted by the return of the tides ; or even remain hard after getting fome degree of induration in firft fetting, unlefs it had the opportunity of becoming dry, which never can happen to works thus expofed to the frequent influx of the tides. The confequence that naturally follows, is, that the weights that are upon the interior parts of the works refting upon matter that is foft and yielding, its endeavour is to burft the external coat ; the Portland blocks being large, well fquared, and weighty, have been able to refift the action of the femi fluid contents ; but the marble afhler being lighter and lefs in fize, with a greater number of joints (and which in the ufual way of working, are lefs clofe behind than before) though they have proved firm enough to prevent the joints from being actually opened ; have not been fufficient to prevent that kind of fwelling which conftitutes the bulging of the walls now to be feen. Had the backing of thefe walls been performed with the fame kind of rubble ftone that was really ufed, laid in a mortar that would grow hard under water, then thofe derangements could not have happened from the caufes above-mentioned. If alfo the walls had been backed with brick and common mortar, as the joints would have been comparatively fmall, and all the matter reduced to horizontal bearings, the action of that infinite fyftem of wedges, compofed by rubble, and laying in every direction, and aided by common mortar, as in its original ftate like greafe, to make them flide, would have been avoided, as well as the lateral tendency to fpread thence arifing.

I took notice of one place in the wall before the maft pond, where the marbie afhler had not only bulged, but the joints had opened on the backfide next the pond, where it was tolerably fair without, which was generally the cafe ; but this was accounted for by a heavy Dutch veffel driving againft it. On this occafion I muft obferve, that thofe walls, having no fupport by a weight or backing of earth lying againft them (which together with land tyes, &c. receive the greateft part of the fhock) appear to me too flight in point of thicknefs (being only about fix feet at the top) to fuftain the fhocks of veffels in hard weather ; if they had been built for veffels to lie againft them. This obfervation will appear confirmed by this remark, that the wharfing fronting the flips, which are filled up

with

with earth folid behind the walls, are in better condition than thofe behind the maft pond which ftand fingle; though thofe walls, like a part of thofe againft the maft pond, have been built on timber, the front of which has been deftroyed by the worms.

3dly. Refpecting the article of the deftruction of the foundation timbers by the worm, I muft remark, that the proper and effectual remedy has already been taken by the officers and artificers of the yard, viz. That of laying a footing of rubble before them, fo as to engage the mud, warp, and fullage to cover them, for being fo covered, all experience fhews that the worm cannot fubfift to hurt them.

4thly. No part of the works appears to have fuffered fo much derangement as the fouth pier leading into the fouth channel, though this has been built with Portland blocks from the bottom to the top, it does not want for backing of earth, is of a circular figure, and the Portland blocks have not only like the reft been walled with terras mortar, but a certain portion of thicknefs of the rubble on the backfide of the wall has been walled in terras alfo. This part is built upon plank and pile, which in like manner as the other has been deftroyed by the worm, and which in like manner as the other has been underpinned and footed with rubble, fo that in that refpect it may be faid to be fecured. The middle part however of the backing of this pier, was like the reft done with rubble work in common mortar, and which probably aided in a degree by fome-kind of fettlement from the foftnefs of the foundation, as well as from the deftruction of the timber foundation by the worms, has had fo great an effect by endeavouring to burft the outward cafe, which though of Portland blocks it has by fpreading opened the upright joints in feveral places to the extent of two inches, though from the largenefs of the beds of Portland, it does not appear to have bulged in any confiderable degree. This work having been underpinned and footed as already mentioned, and many of the joints filled up with thin ftones in terras mortar by the artificers of the yard, it would appear from thefe joints not opening afrefh, that the work is now come to fuch a fettlement and bearing, that being made good as it now ftands, it may be likely to prove lafting.

6thly. The north pier of the fouth channel, together with both the entering heads of the graving place, have been all built in a fimilar manner to that of the fouth pier of the fouth channel; they are underpinned and footed with rubble on account of the fame effects of the worm, and have fhewn themfelves fubject to the fame kind of deficiencies, though not in an equal degree; and which have been alfo in part repaired. There is however one open upright joint in the weft face of the fouth head of the graving place that remains open nearly upon three inches; and one of the terminating ftones of the fouth altar of the graving

place

place has its joint open to the fame extent, but this has been apparently deranged by external violence.

7thly. The fteps in the fouth channel are greatly deranged below the high water mark, (though very entire above it) notwithftanding they have been twice repaired by the artificers of the yard ; which fhews that as the fteps at the head of the graving place have been very effectually repaired by the faid artificers, there is fomething wrong in the original defign thereof ; and this appears to me to have been a want of folidity of thicknefs in the frontage wall, againft which the fteps have abutted; and a want of cramping the fteps to the faid frontage wall; befides all which, the interior part of the folid or backing work upon which they fubfift having been built with rubble in common mortar (except that part behind each ftep that fupports the tread of the ftep next above is fet with brick in terras mortar) and there being in this place an outburft of falt water from the main wall, this wafhes out and paps the common mortar, fo that the leaft degree of froft will rend a work of this kind, however well it might be done with the materials under the circumftances mentioned.

8thly. The laft thing viewed was the clerk of the rope yard's office, which as I am inform-ed was built about eighteen years ago. A fettlement has taken place in the whole except at the north end ; which had it fettled alfo, the whole front would have ftood fair. The north end it feems was built upon a wall of a refervoir which is upon a rock ; the reft is fuppofed to have been built upon planking, which having decayed about eight years ago, the above fettlement took place, amounting to about two inches and a half perpendicular ; and as this fubfidence is nearly regular, it is moft naturally accounted for as above. On this head I muft obferve, that it is generally accounted the moft difficult of all foundations, that of being partially upon a rock, and partly upon a fofter matter, becaufe here being a manifeft inequality, it is one of the moft difficult problems in architecture, to form a judge-ment of what may be fufficient as an artificial ftrengthening, to make it equal with that of a rock which can fuffer no compreffure. Planking is the common expedient, and where it lies under water and fo buried in the ground as not to be fubject to drying, it appears from the works of former ages to be fufficiently durable ; but where laid in loofe or made ground, fo as to get a partial drynefs as may be prefumed in the prefent fituation) it appears fubject to the rot in a moderate courfe of years, and therefore to compreffure. A much better expedient in fuch a cafe is to pave the foundation and build upon the pavement (laying a courfe of flat ftones upon it) rather than upon planking. The whole is only a fmall building ; and it is faid not to have fettled further of late years, the rebuilding of the whole front is no great matter, but it appears to me that the taking it down to the chamber and raifing that floor may be fufficient, or even drawing and rectifying the facings, putting on the additional height under the roof may anfwer the end.

9thly. Having

9th. Having faid as much upon this laft fubject as feems neceffary, and as it is a matter of a very different kind from all the reft, I fhall now lay it out of the account, and proceed to further obfervations upon the reft. It appears to me that the general caufes of all the failures in the works before fpecified, have been a want of perfection in the art of making mortar, and a knowledge of the proper materials from whence it is beft to compound it, for works that are either conftantly under water, or in fuch a fituation where they never can become *dry* ; nor can this be properly afcribed as a fault to any one, as the firft inveftigation of this bufinefs was brought forth by neceffity thirty years ago, in the building of the Edyftone lighthoufe, and though the fame compofitions and methods have been applied and improved in my fubfequent works ever fince, and confequently known not only to the refpective artificers employed, but always freely communicated by me to all fuch as defired the knowledge thereof, yet not having been publifhed, I find even at this day they are very far from being generally known.

10thly. As I have already intimated, I lay it down as a fundamental pofition, that no compofition can be made with lime burnt from Plymouth marble, nor any limeftone of the nature thereof, or of common white chalk mixed with fand, which compofes the common mortar, that will ever acquire a ftony hardnefs under water, or where it can be perpetually fupplied with moifture from the daily flow of the tides. But lime from the fame kind of ftone or chalk when reduced to a dry powder (as ufual by quenching) and mixed with half its quantity (by meafure) of Dutch terras ; and well beaten together for many days fucceffively, without fand, will acquire a ftony hardnefs under water, after it is firft fet in building, which it generally fpeedily does. This compofition, commonly called terras mortar, has been in ufe time immemorial, and if well performed and diligently applied, very found and good work has been performed therewith, but even this mortar will never acquire the hardnefs of good Portland ftone, which I have by trials and experiments been able to attain : but though this lime ufed for terras mortar is generally of the cheapeft kind, yet the terras is fo dear a material, having been for many years from three to four fhillings per bufhel in the light dry powder, and the quantity of mortar made therefrom is fo fmall, and that with fo great an expenfe in the labour of beating it (in general nearly half the value of the material) that it has in general been very fparingly ufed : in terras brick-work, one brick in breadth on the outfide is generally deemed fufficient, very rarely more than a brick length ; and in aifler work, fometimes four inches and a half of the joint within the face, fometimes nine inches, and at moft as in the works of his Majefty's dock yards, the whole aifler or Portland blocks are fet in terras mortar.

11thly. This

11thly. This dearnefs of the material, and conceiving the above as fufficient to defend the infide works from the immediate action of the water, has induced the backing as it has been called, to be generally performed with common mortar: but brick has for this purpofe been generally preferred to rubble or rough ftone, becaufe the thicknefs of the mortar joints will be lefs, and the furfaces lying level one upon another, they fupport one another like pillars, with very little tendency to fpread; and this kind of work when built within dams, which often for months together give time to the mortar to acquire a degree of hardnefs and confiftency, as is the cafe in building the King's docks; I fay this kind of work feldom fhews any deficiency. But in the wharf walls which are generally done by tide work, where the water flows over the whole every tide, the common mortar work never acquires any competent hardnefs, but when covered up by the heaping on of frefh materials of the fame kind, though thereby defended from the immediate action of the water, it has never in any length of time where I have had the opportunity of feeing old works taken to pieces, acquired more than the confiftence of compreffed *curd*, which with frefh beating will become mortar again fit for the trowel. This kind of work therefore (tides work) coming to the moft fevere trial of any, we muft not wonder, if while it is taking time for compreffure, it fhews fymptoms of internal weaknefs: the want of certainty therefore of anfwering its end, with credit to the artift and artificer, has led me further to ftudy how to get a compofition of mortar for fuch kind of works, that will internally as well as externally acquire the hardnefs of ftone, and that without greatly exceeding the price of mafonry in common mortar.

12thly. The inveftigation of a compofition for the Edyftone has in its iffue led me to this point. To recount the fteps taken, would be tedious as well as not to my prefent purpofe; fuffice it to fay, that as all the work there was clofe jointed, the *expenfe* of terras mortar was no object; but the length of time it required to beat it to the beft confiftence, was likely to prove a very material impediment. This induced me to make trial of lime burnt from every fpecies of lime ftone I could then hear of or procure, and I found that as much depended upon the quality of the ftone from whence the lime was burnt, as upon the indurating ingredient; and that the defired effect was produced by the ufe of a lime from *Aberthaw* in Glamorganfhire; but the lime actually ufed, was of exactly the fame nature, and came from Watchett in Somerfetfhire, on the oppofite fide of the Britifh Channel, where the fame ftratum runs, and is in both places called *Blue Lyas*.

13thly. In thefe original experiments, I had the fatisfaction to find not only that terras mortar made up with this kind of lime, in lieu of the common fort, greatly

exceeded

exceeded in hardnefs common terras mortar, but needed no more beating than common mortar *ought* to have; and alfo, that this kind of lime beaten up with common fand, (*without terras*) was almoft as ftrong as the common terras mortar could be rendered by any degree of beating, and fuperior to it, if the latter was but indifferently beaten; and that falt water to this kind of lime for under water work, was as good, if not better than frefh, which was a circumftance very material to my views at that time.

14thly. It was not till fome time after this building was erected, that I found this fpecies of lime ftone was not confined to the neighbourhood of the two places mentioned, but it was a frefh difcovery of the moft agreeable kind to find it in feveral and very diftant parts of this kingdom. This put it in my power to try further experiments, and to ufe it in works where a much greater degree of œconomy was required, than in the mortar of the Edyftone, and from various trials I found that two meafures of this kind of lime to one of terras, or Pozzelana from Italy, beaten up with three meafures of fand, coarfe and fine mixed, made a compofition of equal goodnefs with that of the Edyftone mortar, but doubled the quantity, or that could be made with the fame quantity of terras, in the common mixture; and that eight meafures of this lime to one of terras or (Pozzelana) and fixteen meafures of fand (fine and coarfe) made a mortar for backing very nearly equal to the common terras mortar; and as the greateft part of the bulk of this compofition is common fand, its price per hod in moft fituations will not much exceed one-fourth of that of common terras, nor much more than double the price of common mortar. In many places where ftone naturally rifes in flat beds, as the blue lyas itfelf does (and is ufed for building-ftone in the countries where it is found) as alfo the Purbeck aifler, &c. and the ell and edge ftone, brought as flat paving from Yorkfhire, all of which being ufed as backing, where from fituation it can be procured at a moderate price, this greatly leffening the quantity and vacuity of joints even compared with grey ftock brick work, and thereby leffening the quantity of mortar in a rod of work, thereby reduces the expenfe of mortar to be no greater than that of the common fort, and at the fame time making a ftronger bond throughout, produces firmer mafonry; but in all cafes as the bond made in converting the *whole* into one *mafs of ftone*, takes away the neceffity of ufing fuch maffive ftones for the front (which is generally the moft coftly part of the work) it follows, that in all cafes of water buildings, by the ufe of proper water lime, inftead of the common fort, they may be done at an equal, and in many cafes at a lefs expenfe, as well as a much greater degree of folidity and certainty, than by the common conftructions, with the common materials.

15thly. I call

15thly. I call the lime for thefe purpofes by the general name of *water lime*; for after I found its properties not confined to the blue lyas ftone, but even refiding in a certain fpecies of chalk which is to be met with near Petersfield in Hampfhire, as pointed out in the faid Committee of engineers' report, and which at Portfmouth is commonly known by the name of grey lime; but I had alfo found it near Lewes in Suffex, and there called clunch lime, as alfo at Guildford and Dorking in Surrey; but what is of moft confe-quence to the prefent bufinefs, as I am informed by fcientific friends whom I can depend upon, that the true blue lyas is in quantity at Lyme in Dorfetfhire, and I have myfelf obferved it not very far from thence at Axminfter in this county.

16thly. I am fenfible I might have fpared myfelf the greateft part of this difquifition, but as I humbly conceive it may tend to lead your Honours into the true nature of this bufinefs, it will in effect tend to fhorten what I fhould otherwife have to obferve upon particulars; to return therefore to my immediate fubject.

17thly. As from information I do not find that the defective works (the ftairs excepted) have lately given way, I would fuppofe they have fettled till the parts are come into contact; and therefore may probably remain fo for many years, before they need rebuild-ing, as before noticed; and as the effect of the worm has been remedied in the moft judicious way of which, as far as I know, the fubject is capable; I do not fee that any thing can be done more effectual, than carefully to point and make good the joints wherever they are found defective, with the beft compofition above mentioned, and when the upright joints will admit thereof, to cut Purbeck flat paving ftones, a very fmall matter wedgewife; and drive them in with a foft wooden maul, and before clofure with the laft pierce, to cram in as much mortar as poffible, of the fame kind mixed with pebbles, where the joints will take them, with the end of a board or battan; the mortar being rendered a little more fluid, by a fmall addition of water. This will not indeed take away the *appearance* of the bulging courfes, but for this I fee no remedy but to take down the walls to the top of the Portland blocks, and then if the rubble contents in common mortar are taken out, and the whole built in with the different kind of mortar fpecified (though of the fame ftones) I fhall not be apprehenfive of any further bulging or fettlement, except it fhould arife from the foundation; and of that where there is nothing but foft matter, into which the piles can be driven, I do not look upon in the nature of the thing there can be any abfolute certainty.

18thly. Refpecting the fteps in the fouth channel, I would advife their being taken down till there appears fomething folid to build upon, and to wall the infide core or backing

with

with the compofition given, called Portfmouth terras, made with the water lime from the lyas ftratum at *Lyme:* to build the front wall with a ftring courfe of at leaft eighteen inches thick, cutting the outward ftones into fteps out of the folid; and cramping every other ftep to the front wall, which would be ftill more fatisfactory if thefe front ftones of the ftring courfe were formed from blocks of moor ftones cut into three or four fteps each: the fteps themfelves to be carefully bedded in the beft compofition, and carefully pointed, fo as entirely to exclude the water; and proper conduits to be carried through the work, to take off the outburfts of falt water from the main wall: the work will become ftill more folid, if there be no occafion to continue the fteps fo low, as fug-gefted by the officers.

19thly. It now only remains that I advert to the queftion contained in your letter of the 5th April, before referred to:—that is, " whether rough mafonry at the back of " the aiflers would not (on account of the irregular figure of the latter on the outfide) " if carefully performed, anfwer as well as brick, which being of a regular form " cannot be fo well worked into the back of the aiflers, nor bonded with the rough " mafonry in the interior parts of the work, in which cafe the whole would not be fo " well connected together?"

From what I have already faid, your Honours will perceive, that the refolution of the queftion depends upon that of another; with what fort of cement is the work to be per-formed? If it is to be with the beft compofition of water lime, as I have mentioned, it will together conftitute one folid ftone, and then it will be of little confequence what may be the relative figure of the parts: and that this will be the cafe in a certain degree, if done with well-beaten terras mortar, your own obfervations have proved; for on examination of the back fide of walls fo built (being thofe on the fouth fide of the graving place, and at the north fide of the fouth channel) though you found them " not done with fquare mafonry as the contract directed," yet the rubble mafonry in terras wherewith they were done " appeared *very found and firm.*"

20thly. The rough mafonry in terras has been reported by the mafter bricklayer to be more expenfive to the contractor, than if done with fquare mafonry in terras, and this is confirmed by my own calculation, and will readily appear to your Honours when it is adverted to, that in rubble mafonry there is at an average nearly double the quantity of vacuity that there is in fquare mafonry, even if but rudely fquared, and which muft be filled with the moft coftly of all building materials, videlicet, terras mortar. Yet I cannot agree with the mafter bricklayer in fuppofing it *better*, becaufe as the getting of mortar of any kind perfectly well beaten, may be esteemed the most difficult part of the whole; I should always prefer that construction which at the same expense had the most stone in it, and the least quantity of vacuity to be filled with mortar of any kind.

21st. If the question regarded dry work, as common mortar of common lime and sand will, if well beaten and proportioned, come by length of time to the hardness of stone, then I should judge the rubble work so performed to be equal to brick; because it would then be in the same predicament as water work performed with proper water cement. But if the work is supposed to be under water, as the common mortar as already observed will never come to a stony hardness, I judge that kind of rubble work to be greatly inferior to brick for backing, as pressure will make it tend to spread, which it has been observed from their figure is not equally the case with brick. Besides, as brick is a more dry and absorbent body than marble and most kind of stones, it drinks up the moisture, and causes the mortar to set more speedily, and by that means often gives a greater opportunity of its acquiring some degree of firmness before the water comes upon it, which in this case is a considerable advantage. I therefore cannot recommend rubble backing in works under water, where better materials can be had, and more especially in tide works, where there is the least opportunity of the mortar's setting by the drying quality of the air.

J. SMEATON.

## BRIDLINGTON PIERS.

The REPORT of JOHN SMEATON, Engineer, upon the ſtate of the Bridlington Piers, with the moſt probable means of preſerving the ſame from the deſtruction of the worm.

THIS Harbour for ſhipping, formed artificially by the piers at Bridlington quay, being at the bottom of the great bay conſtituted by the great projecting point of Flamborough Head, with the coaſt tending from thence ſoutherly, is not only ſituated commodiouſly for the reception of the coaſting traders, when they cannot make the Humber to the ſouth, or Whitby on the north, but the piers are ſo diſpoſed as to afford all poſſible ſhelter for veſſels within them. A harbour therefore ſo ſituated, and ſo conſtructed, being found to be of great utility, its piers have been erected and upheld at a great expenſe, but they have at firſt been built with timber probably with a view to the ſaving in the charge that would have accrued by conſtructing them with ſtone, it has unfortunately happened, not only that theſe piers are ſubject to the gradual decay that neceſſarily muſt attend all works of wood, when expoſed to the action of the ſea, but alſo to a particular kind of decay ariſing from the continual eating of a certain ſpecies of worm, that infects the timber work of this harbour, greatly differing from the common worm whereby ſhips are deſtroyed, and which is ſaid to have been originally brought from the Weſt Indies.

The object of my view of theſe piers made at the deſire of the Commiſſioners for the preſervation thereof, the 29th, 30th, and 31ſt of December laſt, was not with any intent of improvement of the diſpoſition of the piers and works, but altogether to conſider and give my opinion upon the beſt method to be taken in future repairs, and for the preſervation of what is now ſtanding.

I have now very ſeriouſly conſidered theſe propoſitions, and however difficult it might at firſt ſeem, I now hope to be able to lay down a very practicable method whereby a great deal of expenſe ariſing from the decay of the worm may be avoided; but that what I have to offer may be clearly underſtood, I beg leave to ſtate what was pointed out to me by the pier maſter, and which appeared to me upon inſpection to be the caſe reſpecting the nature of this worm, and its operation.

This

This worm appears as a fmall white foft fubftance, much like a fmall maggot, fo fmall as not to be feen diftinctly without a magnifying glafs, and even then a diftinction of parts is not eafily made out; it does not attempt to make its way through the wood longitudinally, or along with the grain, as is the cafe with the common fhips' worm, but directly, or rather a little obliquely inward; the holes made by each worm are fmall proportioned to the fize of the worm, but as they are fo many in number as to be but barely clear of each other, as they do not appear to make their way by means of any hard tools or inftruments, but rather by fome fpecies of a diffolvent liquor, furnifhed by the juices of the animal itfelf; it follows that as the animals which overfpread the whole furface of the timber expofed to their action, proceed progreffively forwards into the body of the wood, the outward cruft becomes macerated and rotten, and gradually wafhes away by the beating of the fea, fo that in fact the timbers, planks, &c. gradually wafte in fize and thicknefs, till at laft becoming too weak to fupport the ftrain upon them, they are obliged to be replaced and new done many years fooner than would happen by the natural decay of timber in fuch circumftances, if unaffected by the worm.

The worm is found lodged in a cruft of wood, generally from a quarter to half an inch deep, that part of the wood under this cruft remaining perfectly found. The rate of progreffion, as I am told, is, that a three inch oak plank will be deftroyed in eight years by action from the outfide only.

It is furthermore obferved, that thofe animals do not live except where they have the action of the water almoft every tide, for they are not found in the timbers above the level of common neap tides, high water, or indeed fcarcely fo high; fo that it is to be inferred, that if any happen to fix fo high as the common neap tide mark, if a few low tides fall out together with ftill water, as frequently happens in fummer, the worm thus unwafhed dies, and a ftop is put to its further progrefs higher.

Again, it is very obvious that fo high up as the piles and work are covered with fand, or as foil lies againft it, the wood is perfectly free from the worm, fo that the parts affected are what are alfo expofed to the air, that is from the furface of the ground or fand to high water neap tides. Whether a deprivation of the action of the fea water, accompanied by a continual change every tide, or a deprivation of the benefit of the free circulating air each tide, occafions the lofs of what is neceffary to their fubfiftence, may be a a queftion, but which indeed it does not feem neceffary now to refolve; this however is the fact; for on the infide of the planking againft which the ballaft, fand, or gravelly matter which is heaped by way of filling and giving folidity to the piers' bafe, and which in general is filled above the high water neap tides, is alfo found to be a prefervation to the

infide

infide of the planking, though from the outfide only they will wafte, as has been faid, three inches in eight years.

The rebuilding thofe piers with ftone would doubtlefs be an abfolute cure of the evil complained of; and if they had fortunately been done fo at the firft it would have been well : but as it appears to me that it would be impracticable now to conftruct thofe piers of ftone, without a much larger fund than there is any likelihood of being raifed for the pur-pofe, it remains only to make the beft of them as they now ftand; I fay impracticable without a much larger fund, becaufe if attempted to be proceeded with gradually with ftone upon the prefent funds, the progrefs would be fo flow, that a great part of the pre-fent piers would be down, and the harbour lie open, before it would come in turn to re-place them with ftone, fo that the repairs of the timber work muft ftill go on, and if it exhaufts the prefent funds to keep the prefent works in repair, a double outlay would in confequence be incurred to which the prefent funds would be inadequate; and though the timber repairs would gradually leffen as the ftone advanced; yet it would be feveral years before the outlay on the current repairs of the timber piers would become a very material and fenfible eafement; for thefe reafons, defpairing of feeing any thing very extenfive done here with ftone, I have very ferioufly applied my mind to the propofitions laid before me, which in fubftance are, however, to make the beft of what now is.

It will alfo be neceffary, for the clear apprehenfion of what I have to deliver, that it be underftood how the outfide of the timber work of the piers, that is, what is fubject to the worm, is formed. It is obfervable that the outfide of the piers is formed by a ftrong row of fquared piles of oak, in general about a foot fquare, and near the pier heads the fpaces between them not much more than the breadth of the piles. Infide of thofe they are planked with three-inch plank, in the general old fhip plank, but of late years there being a fcarcity of this, fir plank has been in fome places tried, which is found ftill more fubject to the worm than oak; this planking is to keep in the ballaft, wherewith, as already mentioned, the piers are filled; and which preferves the infide of the plank from the worms, while the outfide being expofed to the free action of the fea, is fubject to their incroachments in the manner that has been mentioned, and not only the fpaces be-tween the piles, but alfo behind them; becaufe as there is feldom a water-tight joint between the plank and the back fide of the piles, the fea drives in, and nourifhes the worms; as alfo in the joints between plank and plank, and thereby widens them, fo as in time to let the fand and fmall parts of the ballaft wafh out through them, which on this accoun needs replenifhing from time to time : it is alfo obvious that the piles themfelves will be expofed to the worm on all fides, for by the fame rule that the worm gets into the plank behind them, they will affect the back fide of the piles, the other three fides being open to them.

It

It is doubtlefs on account of the fpeedy decay and confequent weakening of the piles, in the very part of them where the main ftrain lies, that it becomes neceffary to plant the piles fo near together, and hereby a much greater expenfe is incurred than would be neceffary if they were fubject to no other decay than what length of time would produce, that is in cafe there were no fuch worm, or what would amount to the fame thing, in cafe the worm could be effectually fhut out.

This latter, viz. the fhutting out the worm, appears to me very practicable, the method whereof for new outfide work I will now defcribe. Suppofe the piles driven and lined with planking in the ufual way, only a little more pains taken than heretofore in fquaring the fides of the piles that look one towards another; when the work therefore is completed, I would begin by amending the fquaring the fides of the piles, fo as to reduce them to ftraight-fided figures; then I would adapt chocks of fir to thefe fpaces between pile and pile to be driven lengthways, and which being feparately fitted and driven in may be very nearly adapted to the place that each is to occupy; and between every layer there is to be a bed of tar and oakum, or tar and hair, which is to be crammed into every little vacancy that appears after the driving in of each chock, and fo to go on ftratum fuper ftratum between every vacancy between pile and pile, till you come juft above the height where by experience you find the worm to ceafe; in fhort you may then confider all the fpaces between pile and pile completely walled up with fhort blocks of fir, bedded in tar and oakum, and all the vacuities completely filled with the fame fubftance. This done, the chocks being left a little fwelling before the face of the piles, the whole is to be dubbed off with an adze to one general furface, upon which a fheathing is to be put on in the manner of the fheathing of fhips, the whole furface of which to be filled with fheathing nails, which, affected by the falt water, will throw a coat of iron ruft over the whole furface, and preferve the work from the worm, in the fame manner as in the fheathing of fhips. It is not improbable that this cruft of iron may be wholly impervious to the worm for many years; but fuppofe that in the compafs of ten or twelve years it fhould be found that the worm had got through the fheathing, new fheathing may then be put on at a very light expenfe, and the work guarded for as many years more, and fo on as long as the main timbers will laft from the abfolute decay of time.

In the execution of this method I would obferve, that I would begin the loweft tier, layer, or ftratum of chocks, as low as I could well get for water, or as low as there is any probability of the fand or foil being taken away, and continue them up to where the worm ceafes, and thereby putting in kant pieces, the fheathing will be reconciled with the back planking, and fo continue from thence to the top of the pier.

Again,

Again, I have mentioned that the fides of the piles ought for the fake of the more regular fitting in the chocks to be a little better fquared than common, yet I do not mean by this that the timber ought to be wafted or weakened by being hewn dye fquare; it is fufficient to bring the prominent places to a fquare, leaving the wainy parts to be brought nearer by being notched in with a chiffel after the main work is fixed ready for the chocks; for as I fuppofe every chock particularly adapted, it is no matter that they fhould be all exactly of a length, or their ends be a perfect fquare; and I would alfo further remark, that if the chocks are made out of ftuff about fix inches fquare, fo as to take two breadths in the thicknefs of the piles, they may be both made much more readily, and more completely fitted than if attempted to be put in from whole balks. I would alfo add, that in cafe in time the foil fhould happen to wafh below the ground tier of chocks, fo as to leave the under work expofed to the worm, on fuch accident it will be better to drive a row of ground piles about fix feet from the pier, and fill the intermediate fpace with chalk ftones, &c. fo as to retain foil againft the chock work, than attempt to put in chocks under the former, to which in fuch circumftance it may be difficult to make a good joint.

This kind of work will doubtlefs require a little care and attention, but will be fubject to no real difficulty or hazard, nor be attended with any confiderable expenfe; and indeed were this expenditure altogether additional, the quantum would be no ways proportionable to the duration it will communicate to the work; for if it be confidered how much folidity this method will give and the whole being preferved of its original ftrength, there will be no need of allowance for wafting on account of the worm, fo that it will rather appear in the light of a faving; for I am fully of opinion that if the piles are put at two feet fpaces, where they are now put at one, or at three feet fpaces where they are now put at two, they will in reality be in proportion more ftrong and valid than at prefent; and that the faving of fo much prime and valuable oak timber will nearly pay all the other expenfes.

I come now to lay down what is to be performed to preferve the work already done, and not as yet fo far decayed as to require new; and I fuppofe the fides of the piles not having been driven with any view to this kind of treatment, will be found to be irregular and require too much cutting to bring them to plain furfaces every where, but in many places this may be done, and indeed every where a chock may be fixed in front by going in bevil, and fixing by a couple of fpikes; by this means at leaft the front may be made good; and as the decay of the outface of the planking may be attended with fuch irregularities as cannot be fitted with chocks, I would recommend thefe fpaces between the front chocks and back planks to be made up courfe by courfe with brick and terras mortar, which will grow hard in the water, and enable the whole of thefe irregular cavities to be filled completely full of matter, fo that the worm will be thereby as effectually killed, and

a frefh

a frefh growth prevented as in the new work. The faces of thefe chocks being then neatly dubbed off along with the faces of the piles, and a fheathing put on as before directed, I have not the leaft doubt but that fuch parts of the work as, though confiderably decayed, ftill remains ftrong enough to refift the violence of the fea, will be preferved for a confiderable term of years, that otherwife would be in a few years in a ftate of ruin.

Befides the above defcribed works that compofe the outfide of the body of the piers, there are, at certain intervals, ftill larger detached piles driven as fender piles, which are equally eaten away by the worms as the others; to thofe I would only advife that their whole furface on every fide be driven or filled with fheathing nails from a little below ground to neap tide mark as before recommended.

As it appeared to me the whole of the north pier head lies dry to the ground at fpring tides in moderate weather, but if it fhould happen that any part cannot be eafily come at to fix the low tier of chocks within the ground, I would begin by laying a tier of the large block ftones clofe againft the foot of the piles, and in fome degree adapted thereto, kanting off the outward angle, and fixing each ftone by two ftrong dogs or iron cramps one near each end of the block, the dogs to be firmly fixed at one end to the piles or timber works, and the other leaded into the upper furface of the ftone, then ramming full every cavity between the blocks and the timber work with bricks and terras mortar, the chocks may be founded on this, a little below the upper furface of the ftone blocks.

A further requifition from me, was, in what manner the block ftones already provided may be moft effectually applied?

As I have already fhewn how the worm may be prevented, and even ftopped where it has already entered, without much ufe from ftone work, it will follow that much the greater quantity, faid in the whole to amount to 750 tons, will remain on hand.

I obferve that the cliffs to the eaftward of the north pier are greatly wafted by the continual beating of the fea, and in a way of wafting ftill more; fo that in procefs of time, it may be juftly apprehended, that unlefs fomething be done to controul it, the fea will make its way into the harbour; and indeed had it not been for a great number of the block ftones that have been promifcuoufly tumbled down into the external angle, between the north end of the north pier and the main land, that has prevented its direct progrefs that way, it is not unlikely but that this would have already happened, a circumftance that would be fatal to the harbour, as the cafe would be very different here from what it is from the want of junction from the weft end of the fouth pier with the main land, as the fea only comes through the beach by recoil, after having in a great meafure fpent its fury upon the beach; whereas here, it would come in by the moft direct paffage into the very place where the principal fhips lie.

It

It appears therefore to me that thofe ftones being already upon the place, cannot be better difpofed of than by fecuring this neck of land, in lieu of ufing a quantity of timber work in this place alfo, that otherwife might be neceffary. The moft eafy and fimple way of conftructing fuch a bulwark, and what feems to me beft calculated for receiving and diverting the fhocks of the fea, is for the face of the bulwark to be difpofed in a circular form, and the ground courfe let down its whole thicknefs, or thereupon into the folid clay. I do not mean that there fhould be any work upon the blocks, except what is done to bring them to a ftea ly bearing in place ; and except what may be neceffary for the ftairs in cafe a ftair is thought indifpenfable, and which I would otherwife advife to be avoided, to fhun the making any break into the regular face of the works.

As the furface of the clay bottom appears to be in a ftate of wear by the action of the fea though feemingly flow, yet it may be expected that in procefs of time it will fo wafh away as to expofe the matter upon which the ground courfe is founded to the action of the fea, and therefore it would be the regular courfe of bufinefs to found the ground tier of blocks upon a fheet of plank piling ; but if it is confidered that this fheet of plank piling is equally liable to be laid bare, and when fo, muft for the fame reafon as the piers, be defended by fomething elfe ; and as it is probable that it will be feveral years before fuch defence will become neceffary ; and when neceffary may be defended by letting down another row of blocks at the foot of the former ; it feems to me, that the original expenfe of pile work may for the prefent be difpenfed with ; and as befides doing the work fpecified a confiderable quantity of the block will ftill remain, they may not only be applied to the occafional purpofes before mentioned, but alfo in making good the infertion of the bulwark into the main land, as the cliffs gradually wear away.

I do not mean that any mortar or cement fhould be ufed in the conftruction of this bulwark ; but to prevent the gravelly matter from wafhing through the joints of the blocks, I would advife a ftratum of heather to be laid upon the chalk ftone fittings ; and alfo the joints of the back fides of the blocks to be well crammed up with the fame ; or in lieu thereof hay may be ufed for the fame purpofe.

*Aufthorpe,*
15th *May* 1778.

J. SMEATON.

## SUNDERLAND PIER.

The REPORT of JOHN SMEATON, Engineer, upon Mr. SHOUT's plan for rebuilding and extending the old pier of the Harbour of Sunderland.

HAVING carefully viewed the harbour and piers of Sunderland, and received the information of feveral eminent mafters of fhips belonging to that port, having alfo confidered the plan produced to me by Mr. Shout their Engineer for the improvement of the faid harbour, I am of opinion, as follows :

1ft. That nothing appears to me fo likely to improve the depth of water over the bar (which is unanimoufly reprefented as the moft defirable circumftance wanting towards its improvement) as the prolongation of the old pier to the low water mark at the leaft ; and as the direction of the laft ftretch of the old pier has been now proved by many years experience to afford a fufficiency of fhelter to the fhipping from out winds and feas, and alfo to be in a commodious direction for going in and out ; I am of opinion, that what has been fhewn by experience to be right, ought not to be effentially altered ; that is, that the head of the advanced pier fhould be in the line of direction as the pre-fent pier.

2dly. That as a confiderable part of the old pier from the head inward appears fhaken, and its foundation fapped, I fee no more eafy or proper mode of repair, than that of re-erecting it on a new foundation ; and as by doing this, according to the plan exhibited by Mr. Shout, a projecting part of the pier will be taken off, that now appears to difturb the current of the river, and break its force in time of great land floods, which are the principal agents towards fcouring the channel and keeping down the bar ; the current of the river will thereby have a more direct courfe to fea ; and being preferved in one di-rection by the pier from fpreading, will not only make a deeper channel alongfide of the pier, but, acting more directly and immediately upon the bar, deepen the channel over it in like proportion.

3dly. I therefore approve of the plan of Mr. Shout, but look upon it that it will be an improvement to the fafety and fhelter of veffels within the head, that inftead of carrying it out in a direction parallel to the terminating direction of the old pier, gradually to bring the new pier round, fo that its terminating head may be in the fame individual place as if the old pier were extended to the fame point.

4thly. As

4thly. As the new pier is carrying out, the effect of it will be seen, and the degree to which it ought to be extended more certainly judged of; but I am of opinion it cannot be too far extended, even beyond the bar, if circumstances of time and expense will admit.

5thly. I apprehend it will be proper to begin the extension of the pier from somewhere near upon abreast of the present old pier head, and to carry it on to some length, suppose about two years work, before any of the old pier is removed; then to work backward removing the old pier, as the new one advances towards the point of junction, by which means the shelter to the harbour will be preserved.

6thly. As I am of opinion, that as the extension is carried on, the north sand will follow it nearly at the same distance as at present, whereon the seas in like manner breaking and spending themselves, will keep the harbour in the same degree of quietude, with the advantages mentioned of a deeper channel over the bar.

*Gateshead,*
16th *January* 1780.

J. SMEATON.

N. B. The old pier as it is removed must be as entirely eradicated as possible, that nothing of wood or stones may be left to prevent the land floods from deepening the channel in the ground upon which the old pier now stands.

## TINMOUTH HARBOUR.

### REPORT on Tinmouth Harbour's Mouth.

June 21ſt, 1769.

HAVING this day viewed the harbour of Tinmouth, and particularly that part of the entry thereof lying between the Low Lights and the Black Middens, I am of opinion, that it will be a great addition to the ſecurity of veſſels entering and going out of the ſaid harbour, if the looſe ſtones that are ſcattered upon the north ſhore, between the Muſcle Scalp and the Prior Stones, were removed. Theſe ſtones in ſome places extend themſelves into the channel, ſo that ſhips that happen by accident to take the ground, often receive damage that would be prevented by their removal, and alſo by affording a greater width and capacity, would greatly add to the conveniency of veſſels working to windward either into or out of the harbour.

The rocks called the Prior Stones, I am of opinion for the ſame reaſon ſhould be removed and made level, together with ſuch others as obſtruct the channel as far as they can be got at.

The ſtones that lie near the Black Middens may be ſtowed away in the hollows of the ledge of rocks, ſo as to ſtrengthen and increaſe the utility thereof as a break-water, and which if carefully depoſited in the method of a pile of cannon ſhot, I am of opinion will lie where they ſhall be placed. Thoſe at a greater diſtance may be laid in the ſame method on a ridge extending over the higheſt part of the ground between the Black Middens and the ſhore, which will alſo add to the break-water, and in ſome meaſure prevent the current from dividing near high water into two channels; and as to thoſe ſtones that lie up nearer the Muſcle Scalp, they may be depoſited under the high cliffs, and will tend to the defence and ſecurity thereof.

Thoſe ſtones may be removed in different ways, each of which according to circum-ſtances may have its advantages; and indeed I would adviſe them all to be tried, from whence it may be with certainty determined, which is executable at the leaſt charge.

1ſt. The ſmaller kind of ſtones may be moved by hand-barrows.

2dly. The larger kind may be ſplit with gunpowder and mauls, and then removed by hand-barrows.

3dly. They

3dly. They may in fome places both large and fmall be removed by Carts.

4thly. They may be removed by lighters laid on fhore, and loaded at low water, and removed on tide of flood.

5thly. They may be drawn over the rough ground by windlaffes and tackles, to be moved from place to place, as need fhall require.

The ftones may be laid hold of by eye-bolts, fixed in holes bored by a jumper, or they may be harneffed with chains.

I would advife that the workmen at all times employ themfelves upon thofe ftones that lie loweft and neareft the channel, and as the flow of the tides obliges them to retreat, to work upon thofe that lie higher.

I obferve that a great many ftones to be removed are very good lime ftones, which, if broken up, I apprehend many would be glad to remove for the fake of the lime.

J. SMEATON.

P. S. In coming down by way of South Shields, I obferved a place called the Mill Dam, which, if the trade requires it, may be very properly converted into a wet dock, that would hold a great number of fhips.

## SCARBROUGH PIER.

### To the Commiffioners of Scarbrough Pier.

GENTLEMEN,

I WAS duly favoured with your order of the 15th of May laft, from the hands of Mr. M'George, referring to a cafe to be ftated by Mr. Gilbert, for my opinion thereon, and alfo refpecting the carrying on the works of the pier in future, which cafe was alfo inclofed, defcribing the method by which the works were propofed to be carried on. All which having carefully confidered, and having referred to my former plans and eftimates, I find, that in the courfe of twenty-five years, there have been executed from my defigns and directions upon principles fimilar to the mode of operation defcribed in the cafe inclofed piers for the following harbours, videlicet,

Aberdeen, Aymouth, Portpatrick, in Scotland; and St. Ives in Cornwall: all of which have anfwered and given entire fatisfaction. The outfide ftones of any of the above were not above three tons weight, nor were they backed with any thing but promifcuous fittings, from a ton to about a cube foot in a ftone; the general weight of the fitting ftones were about half a ton in a piece. It is however to be ob- ferved, that the whole were quarry ftones, rough and angular, and not rounded by the action of the fea. We ufed no gravel for wrecking the work, but in what was above low water, the interftices were filled by hand, after fet with fmall fharp rubble ftones of the fame kind; and it is to be further noted, that the outfide blocks, after being roughly fquared, were not laid upon their beds, but fet with their angles upwards, fo that every ftone was jambed between two, fet in a fimilar manner below. I am therefore of opinion that the method propofed will be fufficient to refift the violence of the fea, when got up to its full height, and covered with an entire platform of large blocks upon the top.

In regard to the economy that may attend this mode of operation, I muft obferve, that in all the piers above mentioned the whole or by far the greateft part of all the ftones employed were not found by the fea fhore, but by neceffity brought from quarries by land carriage, fo that the expenfe upon the whole was not lefs than four fhillings per ton; and would have been far greater, had nothing but large blocks been employed. It may therefore in fome fituations be better economy to make ufe of the ftones that are

there,

there, if large, though rounding, and to fill all folid, than either to break them to render them angular, or fetch this kind of ftones from quarries within land. It feems therefore probable that fomething may be done by way of eafing the expenfe, and increafing the expedition; but the proper mode of operation depending altogether upon local circum-ftances, I am unable to form my opinion fully thereon without a view of the premifes.

I am with great refpeft,

Gentlemen,

Your moft obedient humble fervant,

J. SMEATON.

*Aufthorpe,*
9th *Auguft* 1781.

---

ESTIMATE for improving and enlarging the Harbour of Scarborough, in the County of York.

|  | £ | s. | d. |
|---|---|---|---|
| To extending the prefent pier 150 feet, containing 115 ton per foot running, at 6s. per ton | 5,175 | 0 | 0 |
| To building a new inner pier, containing 320 feet in length, and 31 ton per foot running, 6s. | 2,976 | 0 | 0 |
| To making a wharfing to confine the fands, extending 1,300 feet, at £2. per foot running | 2,600 | 0 | 0 |
| To clearing the prefent area of fand, fo as to make 11 feet water at a fpring tide by the fide of the wharf, containing 31,300 yards, at 6d. per yard | 783 | 0 | 0 |
| To making a groyne or jetee near the fpace, extending 900 feet, for preventing the fands from circulating into the harbour | 900 | 0 | 0 |
| To making a bafon 100 feet fquare, and laying pipes for keeping the fand from lying before the wharf | 1,000 | 0 | 0 |
| To erecting 5 new dolphins, and other pofts for mooring veffels; and other contingencies | 500 | 0 | 0 |
| To contingencies upon the whole | 2,600 | 0 | 0 |
| | £15,934 | 0 | 0 |

# SHIELDS DOCK.

ESTIMATE for lengthening the Dock at North Shields, fo as to make it in the whole 237 Feet Length between the heel of the prefent Gates and the crown of the Arch at the upper end of the Dock ; and alfo for the erection of a pair of interior Gates, as per Defign of John Smeaton.

£ s. d.

TO digging out the upper ftratum fuppofed pan rubbifh to the mean of 9 feet deep, the ground being opened 63 feet wide at top, and to be 60 feet wide at bottom of faid depth, will contain     cube yards   2133

The lower ftratum fuppofed a clayey fand   2246

Total excavation digging 6d. loading at 3d.   4379 at 9d.  - -  164 4 3

Drainage of the water, fuppofe 4 men employed at a time, that is 8 men for 24 hours, and this continued for three months or 90 days at 18d. per man per day  -  54 0 0

To leave for laying the ftuff  - - - -  25 0 0

Clearing the foundations  - -  £ 243 4 3

## CARPENTRY IN THE FLOOR.

To 1 pile per yard running under the ftring piece in the front of the wall, being 9 inches diameter, and 9 feet long, of beach, elm, alder, or fir timber, will be in number  - - - - - 81

To 2 piles under the middle of each beam or groundfell, which, if at 3 feet diftance, will be  - - - - 70

To piles, timber, making and driving  -  151 at 10s. 75 10 0

To 35 beams or groundfells at 3 feet diftances middle and middle, 33 feet long 12×12 containing   cube feet 1155

To 242 running of ftring piece 12×6   121

To timber and workmanfhip  - - 1276 at 1s 9d.  - -  111 13 0

To 2 piling machines at £3 each  - - - -  6 0 0

To tranails of 18 inches long for the ftring pieces 81 at  - -

To do   24   groundfells, 70 at  - -

To 3332 feet fuperficial of 3 inch fir plank for the floors, timber and workmanfhip at 6d. per foot  - - - - - 83 6 0

To caulking ditto at per foot fuperficial  - - - -

To 13 cwt. of fpikes at 36s.  - - - - 23 8 0

Carpentry in the foundations, exclufive of trenails and caulking  - - £ 299 17 0

|                                                                                                                   | £ | s. | d. |
|-------------------------------------------------------------------------------------------------------------------|---|----|----|
| To ramming in between the walls fells, with a mixture of clean London ballaſt and quick lime, containing 83 cube yards, at 1s.    -     - | 4 | 3 | 0 |
| To fluſhing the fame with a layer of pozzelana mortar, to lay the planks upon, containing 259 ſquare yards, at 7d.     -     -     - | 7 | 11 | 1 |
| To filling in the foundations     -     -     - | £ 11 | 14 | 1 |

## MASONRY.

|                                                                                                                   | £ | s. | d. |
|-------------------------------------------------------------------------------------------------------------------|---|----|----|
| To aiſler in the ſide walls, which at the mean thicknefs of $2\frac{1}{2}$ feet, will contain 12,352 cube feet, which winning, leading, and hewing, at $7\frac{1}{2}$d. per foot   - | 386 | 0 | 0 |
| To pozzelana mortar for ſetting the fame $\frac{3}{4}$ per foot, including extra labour   - | 38 | 12 | 0 |
| N. B. 14 tons of pozzelana mortar will be wanted at £2 2s. per ton. | | | |
| To rubble ſtone for backing 1148 yards of 2 feet thick, which will take 1435 fodders of ſtone, which winning at 3d. leading at 1s. is 15d. per fodder   -   - | 89 | 13 | 9 |

Mortar for do.

|                                   | £ | s. | d. |   |   |   |
|-----------------------------------|---|----|----|---|---|---|
| To 162 fodders of lime at 6s.   -   -   - | 48 | 12 | 0 | | | |
| To do. of pan rubbiſh ſifted, at 10d.   - | 6 | 15 | 0 | | | |
| To 100 tons of ſand at 6d.   -   - | 2 | 10 | 0 | | | |
| | | | | 57 | 17 | 0 |
| To ſetting 12,352 cube feet of aiſler, at 1d.   -   -   - | | | | 51 | 9 | 4 |
| To building 1148 cube yards of rubble, at 1s.   -   -   - | | | | 57 | 8 | 0 |
| Maſonry in the whole   -   -   - | | | | £ 681 | 0 | 1 |

This comes to 7s. 5d. per the reduced yard upon the whole ſolid, the maſon ſaid it would be done for 4s.

| To moating the fame with clay, containing at one foot thick, 4838 cube feet at 2d. including the filling in and making good the earth behind the walls   -   - | 40 | 6 | 4 |
|---|---|---|---|

## CARPENTRY in ERECTING the GATES.

### The Threſhold and Barrier at the Gate Heels.

|                                                                                                          | £ | s. | d. |
|----------------------------------------------------------------------------------------------------------|---|----|----|
| To 185 cube feet of oak and elm timber, timber and workmanſhip at 3s. 6d. per cube foot meaſured neat in place   -   -   -   - | 32 | 7 | 6 |
| To fir timber, 133 cube feet, meaſured neat in place, at 2s.   -   - | 13 | 6 | 0 |
| Threſhold and barriers   - | £ 45 | 13 | 6 |

## THE GATES.

|  | £ | s. | d. |
|---|---|---|---|
| To oak for the two heel pofts, and clapping pofts, 78 feet, at 4s. timber and work- manfhip included, neat meafured in place   -    -    - | 15 | 12 | 0 |
| To 52 oak keys for the gate bars at 1s.    -    -    - | 2 | 12 | 0 |
| To 504 cube feet of fir timber, in the gate ribs, timber, workmanfhip, and meafured neat in place at 2s. 6d.   -    -    - | 63 | 0 | 0 |
| To caulking   -    -    -    -    - | | | |
| Carpentry in the gates   -    -    - | £ 81 | 4 | 0 |

## IRON WORK, &c.

### In fixing the threfhold and barriers.

|  | £ | s. | d. |
|---|---|---|---|
| In jagged bolts and anchor at 4d. per lb. 3 cwt. 2 qrs.   -    -    - | 6 | 10 | 8 |
| In fcrewed bolts and ftraps for fixing the main pofts, braces, &c. at 5d. 3 cwt. 1 qr. | 7 | 11 | 8 |
| To fpikes for putting on the ribbands and planking, at 36s per cwt. 1 cwt. 2qrs.   - | 2 | 14 | 0 |
| Fixing the dormant part   -    - | £ 16 | 16 | 4 |

### IN THE GATES.

| | cwt. | qr. |  | £ | s. | d. |
|---|---|---|---|---|---|---|
| To 20 pair of T and L's, weight   - | 7 | 3 | 0 | | | |
| To 2 long fcrewed bolts   -    - | 2 | 2 | 0 | | | |
| | 10 | 1 | 0 at 4½d per lb.   -    - | 21 | 10 | 6 |
| To 2 hoops for the top of clapping pofts  - 1. ½ at 4d.   -    - | | | | 0 | 14 | 0 |
| | | | | £ 22 | 4 | 6 |

### Hinges for the Gates, &c.

| | cwt. | qrs, | | £ | s. | d. |
|---|---|---|---|---|---|---|
| The pair of upper hinges   - | | 2 | 1 | | | |
| The 2 heel hinges   -    - | | 1 | 3 | | | |
| 4 cwt. at 4½   -    -    - | | | | 8 | 8 | 0 |
| To 2 caft iron beds for holding the gudgeon and forming the heel circle, 5 cwt. 1 qr. at 18s. per cwt.   -    -    -    -    -    - | | | | 4 | 14 | 6 |
| To 2 copper plates, each 1f. 6i. long, 1 foot broad, and about ⅛ inch thick 22lb. at 16d. | | | | 1 | 9 | 4 |
| Work for the gates to turn   -    - | | | | £ 14 | 11 | 10 |

ACCOUNT received from Mr. Walton, after making the Eftimate, of the following Articles, not charged for want of knowing the cuftomary price, viz.

|  | £ | s. | d. |
|---|---|---|---|
| 81 trenails, 18 inches long, including labor, at 1¼d.   -    -    - | 0 | 8 | 5¼ |
| 70 do 24 inches long, including labour, at 2d.   -    -    - | 0 | 11 | 8 |
| Caulking the floor of the dock, 3332 feet, including oakum, pitch, and labour at 2d. | 27 | 15 | 4 |
| Do. the dock gates, 576 feet, including do at 2d.   -    -    - | 4 | 16 | 0 |

ABSTRACT

## ABSTRACT of the preceding ESTIMATE.

| | £ | s. | d. |
|---|---|---|---|
| Digging, drainage, and clearing the foundations | 243 | 4 | 3 |
| Carpentry in the foundation, exclufive of trenails and caulking | 299 | 17 | 0 |
| To filling in the foundation | 11 | 14 | 1 |

| | £ | s. | d. |
|---|---|---|---|
| The foundation | £ 554 | 15 | 4 |

| | £. | s. | d. | | | |
|---|---|---|---|---|---|---|
| Mafonry in the walls | 681 | 0 | 1 | | | |
| To moating with clay, and filling in behind the walls | 40 | 6 | 4 | | | |
| | | | | 721 | 6 | 5 |

| | £ | s. | d. |
|---|---|---|---|
| Total expenfe of lengthening the dock | £ 1276 | 1 | 9 |

### ERECTING THE GATES.

| | £ | s. | d. | | | |
|---|---|---|---|---|---|---|
| Carpentry in the threfholds and barriers | 45 | 13 | 6 | | | |
| Do. in the gates | 81 | 4 | 0 | | | |
| | | | | 126 | 17 | 6 |
| Iron work in the threfhold and barriers | 16 | 16 | 4 | | | |
| Do. in the gates | 22 | 4 | 6 | | | |
| Hinges and work for the gates to turn upon | 14 | 11 | 10 | | | |
| | | | | 53 | 12 | 8 |
| Erecting the gates | | | | 180 | 10 | 8 |

| | £ | s. | d. |
|---|---|---|---|
| Net eftimate | £ 1456 | 11 | 11 |

Allow for contingent articles not included in the foregoing eftimate, at the rate of
5 per cent      72   16   7

£ 1529   8   6

*Aufthorpe,*
26th *May* 1775.                     J. SMEATON.

## TREWARDRETH HARBOUR.

The REPORT of John Smeaton, Civil Engineer, upon the Question of a Harbour in the Bay of Trewardreth, Cornwall, exhibited by Brooke Watson, Esq.

THE defign for a bafon and an entering veftibule before me for Trewardreth Bay, appears to me fufficiently well imagined as to its fhape and pofition, refpecting the points of the compafs and the adjacent coaft; but the entering veftibule, which I apprehend is principally intended to fhelter the gates, whereby the water of the bafon is to be held up to its deftined height, appears to me by far too fmall to produce that effect. In this area I have drawn a circle of 100 feet, that is of 33 yards diameter; and the whole area will fcarcely amount to a quarter of an acre. It does not fignify what, or how great the area of the bafon may be within the gates; becaufe this will have no influence on the quieting the outward port or veftibule; and therefore, fuppofe a fea runs in with a ftrong wind at S. E. (I mean right into the opening), and is therein reflected from the different piers and walls, it will manifeftly tend to the deftruction of the gates; as from the meafures they muft neceffarily be placed fo near thefe walls, and receive their ftrokes in one direction or another, fo very frequently, that I do not apprehend any timber work upon that fcale will be found ftout enough to endure the fhocks for a length of time.

If the neceffary occafions for the fize of fhips were lefs in proportion, as it might be more ufeful to lay out a lefs capital upon a project of this kind, then, as the entry of a port might be made lefs in width, in the fame proportion as the fhips themfelves were lefs, it might perhaps be brought about, that a fmall harbour could be rendered as quiet as a large one; but fince the neceffary occafion for width in the entry of the port, is not fmaller in proportion as the area of the harbour is fmaller, but to the width of the largeft fhip intended to enter it, then a very fmall port may have an occafion for as wide an entry as a large one; and confequently the quietude of the fhipping lying therein, will be lefs (perhaps in a greater proportion) than the increafe of width, to that of its area.

I do not know that it has yet ever been afcertained, what is the greateft width to be admitted to the entry of a port, in proportion to the number of acres it contains; or what

amounts

amounts to nearly the fame thing, what will be the proportion between the width of the mouth of the port and the mean diameter thereof. In the prefent cafe the width of the entry is thirty feet, that of the area of the veftibule about 100; that is a little more than one to three; certain it is that I do not recollect ever to have feen the area of a port fo confined, where it has been enclofed with a wall; and where a large pair of gates are to be protected, I fhould imagine the internal fpace fhould be far greater. I fhould think a mean diameter of 300 or rather 400 feet quite little enough, or the veftibule ftill preferving the width of thirty or at moft thirty-five feet to the mouth; and even this mouth will be quite little enough to admit a veffel of twenty-feven or twenty-eight feet to enter with tolerable facility.

It being fuppofed that the figure of the outward harbour or veftibule is near upon right, its dimenfions may be increafed in any proportion; and if there were a want of fpace, a part of the propofed area of the bafon may be taken into the veftibule: True; but here will be a great addition to the walling in proportion to the area of the bafon, and capacity for the entertainment of fhips.

I think it my duty to fuggeft, that every part of the works expofed to the fea in the bottom of this bay, ought to be done in the moft fubftantial manner, otherwife failures may be expected; and how far the occafions of trade in this place will render it a profitable undertaking, to encounter the neceffary expenfes fhould be well confidered before the execution is refolved on.

It is generally allowed that Milford Haven is one of the moft complete natural harbours in this kingdom; but let it be reduced by a fuitable fcale to two or three acres of area, and the width of its mouth reduced from three or four miles, to thirty or forty yards; it will then become a much more unfuitable defign for a harbour than this under confideration.

*Grays Inn,*                                                J. SMEATON.
18th *January* 1792.

# JERSEY HARBOURS.

## (See Plate 9.)

To the Right Honourable his Excellency the Governor, Lieutenant Governor, Lieutenant Baillie, and the Magistrates composing the Committee for the Improvement of the Harbours of Jersey.

The REPORT of JOHN SMEATON, Civil Engineer, upon the Harbours of St. Helliers and St. Aubin.

HONOURABLE SIRS,

HAVING at the requeſt of the honourable the Governor, made a voyage to Jerſey upon the ſubjeċts above ſpecified; and having perſonally reſided in that Iſland from the 5th to the 13th of June laſt, both incluſive; and having during that time diligently obſerved and conſidered whatever appeared material, in conſequence of the matters given me in charge by the honourable Committee above mentioned, I beg leave to report as follows: ſtating firſt, the principal matters reſpeċting the harbour of St. Helliers.

That with reſpeċt to the harbour of St. Helliers, the matters of complaint ſtated were, want of capacity for holding veſſels; and want of quietude when lying therein.

The pier of St. Helliers appears to be of the conſtruċtion that may be properly denominated a ſcreen; being a work of that kind, where, in the firſt formation, the greateſt degree of ſhelter may be procured at the leaſt expenſe; it has therefore the advantage of doing, in its origin, the greateſt quantity of work for the leaſt money; and were this form attended with no diſadvantages, this would in all caſes be the moſt proper way of proceeding. But as this kind of pier naturally turns its back towards the ſea, and its mouth towards the land, upon the ſuppoſition of a ſloping ſhore as commonly happens; if an extenſion of the pier is wanted for an enlargement of the harbour, it naturally brings the entrance into ſhallower water; whereas, when piers are extended from the ſhore towards the ſea, a further extenſion of theſe piers naturally gains deeper water.

The pier of St. Helliers having been fixed and ſupported at a conſiderable expenſe, in proportion to the funds of the iſland, it would be uſeleſs to conſider how it ought to have

been

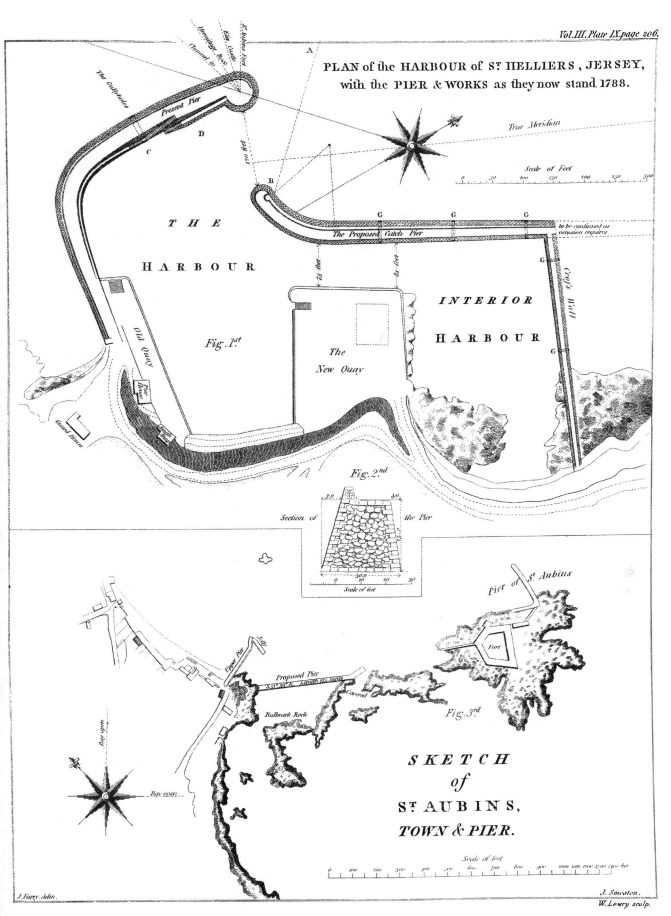

PLAN of the HARBOUR of St HELLIERS, JERSEY,
with the PIER & WORKS as they now stand 1788.

*True Meridian*

Scale of Feet

THE

HARBOUR

Present Pier

The Gullyholes

*The Proposed Catch Pier*

to be continued as
occasion requires

*INTERIOR*

Croy's Wall

*HARBOUR*

Old Quay

Fig. 1st

The
New Quay

Guard House

Fig. 2nd

Section of          the Pier

Scale of feet

Pier of St Aubins

Fort

Upper Pier

Proposed Pier

Covered

Bulwark Rock

Fig. 3rd

Bay open

SKETCH
of
St AUBINS,
TOWN & PIER.

Bay open

Scale of feet

*J. Farey delin.*                    *J. Smeaton.*
*W. Lowry sculp.*

*Published as the Act directs, 1812, by Longman, Hurst, Rees, Orme and Brown, Paternoster Row London.*

been planned, were every thing at this day in a ftate of nature. The propofition is to make the beft ufe of it, as it now ftands; and confidering it as a part of a whole, to procure at the leaft expenfe, more room and better fhelter.

Had it not been for the fhoaling of the water at the entry, the firft thing that would naturally ftrike, would be to extend the prefent pier further to the north, and to meet it with a catchpier from the north-eaft corner of the new quay. This contracting the great width that the mouth bears to the capacity of the harbour, would at once render it more ample and more quiet, and would be the moft eafily done, as the materials for both piers could be carried progreffively from the fhore; but in the opinion of the pilots, as well as the committee and gentlemen confulted upon the fpot, the fhoaling of the water is an invincible objection; in which opinion on full confideration I find reafon to concur.

Of the number of plans exhibited to me, that had been the fruits of the ftudy of different gentlemen of the ifland, the idea moft generally adopted was, to begin a pier, detached from the prefent, at fome point betwixt the prefent pier head and the new quay, leaving about 120 feet or upwards of opening betwixt the prefent and new pier head; then carrying it on in a ftraight direction, fomewhat near parallel to the face of the new quay, to leave a fpace between the new pier and the new quay for veffels to lie; and by extending it northward in the fame direction towards the town, as the funds for carrying it on admit, to give it every fuitable enlargement that the occafions of trade may now or in future require,

This idea in its general outlines and purpofe feemed to me well adapted to the occafions of the Committee, but yet was likely to be fubject to one capital inconvenience. I was told that when the wind was at N. W. which point being clear of Elizabeth Caftle and rocks, the fetch of the fea, then lying open from the whole extent of St. Aubin's Bay, in blowing weather, caufes the veffels to ride unquiet in St. Helliers harbour. Now any pier carried on in any direction parallel to the face of the new quay, or trending more weftward, will have a great effect to bring the waves that will neceffarily range along the face of it with fuch a wind, directly into the mouth of the harbour; and if to avoid this, it has a bearing given it more eaftward, will confine the additional room propofed to be made in the harbour between the new quay and the town, fo near the fhore, that the depth of water as well as room therein will be too much curtailed. A further objection is, that when veffels are coming for the harbour before the wind, through their proper channel, or opening through the rocks, which as I underftand lies at W. S. W. from the pier head; or coming with any fcant wind, from that point to the S. E., after turn-

ing

ing the head to go into the harbour, they will have the wind in their teeth ; and as they may not always be able to shoot into the harbour, there will not be sufficient room to *bring up* before the harbour's mouth, without being in the way of those that may follow them under the same circumstances. It appears to me therefore that every inconvenience may be avoided, and the desired end attained, by forming the pier with a curve towards the head, according to the plan I have now the honour to lay before you ; and conformable to the idea of which I left you a sketch on leaving Jersey, and which being verbally explained to the Committee, seemed to give general satisfaction.

According to this disposition, the north-western seas will meet the curve of the pier towards the head, in such an angle, that they will be turned from the harbour's mouth (as will appear by considering the position of the north-west line A. B. fig. 1.), and not be turned into it, as would be the result, if the pier were to be continued in the straight line, or any one parallel to it, and which in this case would leave too great a width of mouth for a harbour of this size. But this affords ample space to come-to, in the lee of the harbour's mouth, and afterwards warp in at leisure. It will also afford a superior advantage in getting out with north-westerly winds, by getting under sail from the present pier head.

This, according to my judgment, will be the way to make the best of St. Helliers harbour, and furnish the desiderata at the shortest expense. Yet it must be adverted to, that no small harbour can be made a very quiet harbour ; for the magnitude of the waves are supposed the same to all, and the necessary width of the mouth for a ship to enter the same ; seas then that in a degree will inevitably wrap round the heads, will affect a smaller harbour more than a large one, though of similar constructions ; for the effect of the waves in disturbing a harbour is greater, in proportion as the lineal width of the mouth is to the whole area of the harbour. In this respect the harbour of St. Helliers is, and always must be, defective. Another circumstance tends much to render a harbour unquiet, and that is, when they are boundered with walls. The waves of the sea follow the laws of the pendulum, which when once set a vibrating would never cease, if not stopped by friction and the resistance of the air. The same would happen to the libration of the water if there were nothing to stop it ; for meeting with walls and objects comparatively smooth, the waves are not destroyed, but reflected into another direction ; and from that into another, till they are gradually destroyed by friction. The speediest way by which waves are destroyed (that is by friction) is by forming a surf, and breaking upon a sloping beach, sand, or rocks ; in which also the harbour of St. Helliers is defective ; there being no part of its circumference, where the seas have an opportunity of breaking and spending

themselves,

themfelves, the bottom of the harbour excepted, which is a part of the natural fhore, extending on the S. E. fide about thirty fathom, betwixt the old and new quays; and which, if opportunity fhould occur, it would be well to extend. The new quay, which, as I was informed, is principally a cover to rocks, though it feems to afford great utility in other refpects, yet by preventing the feas from breaking upon them, will naturally have the effect of rendering the harbour more unquiet; but which, when the new pier is built, will become extenfively ufeful, efpecially when the crofs wall is formed to the northward, and the north fide of the new quay is wharfed up.

## EXPLANATORY OBSERVATIONS ON THE DESIGN.

THE catch pier as it is here fhewn extends 520 feet: but I expect that it will have a confiderable effect in quieting the harbour, by the time it is carried out abreaft of the north corner of the new quay, that is, from and including the head, to the length of 250 or 260 feet. This will eventually make more room in the prefent harbour, that is, the veffels can lie clofer by being more quiet; and will alfo actually increafe the room by the fpace betwixt the face of the new quay and the new pier; for then I expect two tier of veffels may be laid before the face of the new quay, and be fcreened, not only from the N. W. but all other winds.

When the catch pier is carried out to the length as fhewn in the plan, the crofs wall built, and the area inclofed freed from the encroaching rocks, then the additional room will be fully double of what can now be occupied. It is true, and I apprehend muft remain a truth, that the fands, this way, lie higher than in the prefent harbour; but the interior harbour being confidered as a place of depofition, the very great fpring tides that reign here, (which rofe above low water mark, the fpring tides when I was there, no lefs than forty feet, though that was a time of the year when the tides in general run fhort), will enable large veffels to get into it, and out, for more than a week together during the time of fpring tides; and though the neap tides here in general run fhorter than any place I have before had an opportunity of noticing, in proportion to the fpring tides, being, as I am informed, in general but from nine to ten feet, yet this being con-ftantly the cafe with large veffels in tide harbours, can be no objection here.

Here being no back-water, and the interior harbour a place of great quietude, the fands may be expected to gather; but to prevent this as much as poffible, I would re-commend an expedient, that I obferve has been judicioufly and fuccefsfully practifed; and that is, the making low arches or gully-holes underneath and through the body of

VOL. III.                                E e                                the

the pier; one of which paffes through the fame about 200 feet from the pier head. It is about fix feet wide and four feet high, fo that a perfon cannot walk under it upright. This aperture has the effect of preventing the fand from lying on the infide in quietude; fo that keeping it fomewhat difturbed, by the agitation of the feas, it is ready to be hauled out on the retreat of the tide: and I have no doubt, but that it has greatly contributed to prevent the fands being lodged in the harbour to a greater height than they are, and their gradual increafe; for I obferve the ground of the harbour at and near the interior mouth of the gully, at C. is full eighteen inches lower than the ground at D. being the point equally diftant from the gully and the pier head. Were this ftopped up, I have no doubt but that in a little time the ground there would become higher.

In this view of keeping down the fand, I would recommend gully-holes to be made in the catch pier, at the diftance of 120 feet, and difpofed as pointed out in the plan at **G. G.** This will make four gullies in the main body of the pier, and two in the crofs wall; but being frequent, I would not advife their being above four feet wide, and three feet and a half high in the crown; and the effect of the firft will be fo pointed out, before a fecond is needed to be pitched, that if the dimenfions I have affigned, fhould be too large or too little, the effect will fpeak for itfelf, and be remedied in the fubfequent *.

With refpect to the conftruction of the pier, I would recommend it to be built with rough granite, in the fame manner and ftyle of workmanfhip as the works of this harbour have already been carried on. I would recommend it to be filled with rough quarry ftones to the middle, without any intermixture of gravel or rubbifh. The height of the platform at top of the pier I fuppofe to be the fame as the new quay; and as I do not fuppofe it a place of bufinefs, it is very poffible it may do without a parapet; but if on raifing it to that height, experience fhews one to be wanted, it can be added at pleafure. For that reafon I have not fuppofed any ftairs, as ftrong upright ladders bolted to the walls at proper places, will be a fufficient convenience for the purpofe. The conftruction and dimenfions of the pier will fufficiently appear by the detached plan and fection (fig. 2.); and as there always occurs a difficulty in turning the convex parts of pier heads when compofed of rough ftones, that is, fo to tye in the ftones fo as not to be fetched out by the furge of the fea; in the plan that I have given of the ground courfe of the pier head, I have fhewn the manner in which I directed the catch pier head of Aberdeen † to be formed, which is alfo of granite; that is, with anchor ftones and dove-tails, fomewhat upon the idea of the dove-tail work of Edyftone lighthoufe.

* It is poffible that fome of thefe may be of differvice, while others are of fervice, in which cafe the former can eafily be walled up.

† The reader may refer to this plan in the third plate, page 49 of this volume.

If

If it could be afforded, it would be well to form every courſe in this manner; but I apprehend it may be ſufficient to do this at every five feet in height; ſo that beginning with the firſt courſe that riſes above the ground, there will be ſix courſes bonded in this way; and which will be completed by ſixty anchor, and ſixty-ſix dove-tail ſtones, and which, if not more eaſily had elſewhere, will be undoubtedly afforded by the quarries of Montmado. Though in the plan they are ſomewhat regularly diſpoſed, yet they are no ways required to be of equal ſizes; all that is wanted will be, that each ſtone be adapted to its next neighbour.

A very material article that remains is the expenſe of the work; but this being entirely local it is not poſſible for me to give any ſpecific eſtimate, but what would be likely to do more harm than good, for if the price aſſigned by me be too large, the workmen would endeavour to take advantage of it; and if too ſmall, they would not be bound by it; and the Committee not ſo well ſatisfied in not getting the work done at the rate given out. But it will be very material and uſeful to know, that in the firſt 240 feet, or ſixty fathom of pier, including the head, and ſuppoſing the mean height, including the foundation, to be thirty feet, that this pier will contain 9,500 cube yards of work; and as I eſtimate it, will take 19,000 tons of rough materials, excluſive of the parapet.

## THE HARBOUR OF ST. AUBIN.

RESPECTING this Harbour the following matters having been ſubmitted to me in writing, I cannot acquit myſelf better than in giving diſtinct anſwers thereto. The paper is as follows:

" Experience having ſhewn that the new pier of St. Aubin, called the Upper Pier, intended to bring and lay up ſhips and veſſels cloſe to the town, for the convenience of fitting them out, loading and unloading, is from its ſituation liable to fill, by the great quantity of gravel which the ſea waſhes in from the back of it; and that the depth of water is, from that action of the ſea, diminiſhing more and more; it has therefore been conceived that the building, inſtead of this, another pier further out to reach from the point of land called the Bulwark, to a rock called Cavard, in the direction of the fort, will anſwer every purpoſe of ſecurity, convenience, and greater depth of water, and is not ſo likely to be expoſed to fill as the other.

" It

" It has alfo been found that the want of a proper road for carts to communicate with the aforefaid pier during the hours of high water in fpring tides, is a real inconvenience ; as in that circumftance, carts cannot have accefs to the pier, without going round a part of the town, and up and down again a fteep road ; wherefore it would be defirable, to procure a proper road by taking upon the harbour, fuch an extent as may not be prejudicial, and obtaining by that means the additional advantage of a quay or wharf, againft which boats and other fmall craft could be brought up at high water, to take in and out things upon occafion.

" But as it is found by obfervation that men who want profeffional knowledge and experience, in refpect to the effects of buildings for harbours, and to oppofe the fea, are fubject to be difappointed in the works they project ; it is therefore wifhed Mr. Smeaton would well confider the above ftated work, and give his opinion.

" 1ft. Whether this new projected pier inftead of that now fubfifting, is the moft eligible plan which can be adopted to remedy the inconveniences of the other ? And if not, What other expedient he would advife ?

" 2dly. In cafe Mr. Smeaton approved the faid plan, Whether it is moft advifable to commence the work at the extremity towards the fea, and to continue it towards land to a certain degree of forwardnefs, before the old pier or any part of it be taken down ?

" 3dly. Whether he would advife the taking upon the harbour, and in what extent, in order to acquire the road and convenience above ftated ? And whether this work of the road and quay ought or ought not to be begun, before fome progrefs is made in the building of the intended pier ?

" N. B. The time the want of this road is felt, is for the fpace of about two hours in the mornings and evenings of fpring tides.

" Mr. Smeaton is alfo requefted to impart fuch further advice upon the whole as his fkill and experience may fuggeft."

Anfwer to the 1ft.—As I look upon it the reafon why the upper pier, as it ftands at prefent, becomes liable to fill with gravel, arifes from this circumftance ; that the weft end of the ifland of Jerfey lying open to the fetch of the great feas propagated from the Atlantic Ocean, there being no land to fhelter it in that direction, that is from the W. S. W. to W. N. W. (according to the true meridian), great feas, of courfe, with all wefterly and fouth wefterly winds, will rake the fhore from the Corbiere to Noirmont Point ; and turning that, fouth wefterly winds will drive forward the gravel brought along with it coaftwife, towards the bottom of St. Aubin's Bay ; and in its way meets with the upper pier now erected ; but this pier not extending far enough from the fhore effectually to

interfect

interfect and intercept the courfe of the gravel, thus continually and fucceffively brought, it is driven round the pier head; and there finding an eddy (which will conftitute a place of repofe on the north fide of the pier), will naturally produce the annoyance complained of. Nor does the effort made ufe of, appear likely to ftop the current of the gravel; for the jetee being low, and pointing directly acrofs its courfe, the gravel was driven over it, and round it, and the jetee itfelf in a great meafure deftroyed. Had the materials employed in erecting this jetee, been applied to carry out the pier head in its N. E. direction, this, flanking the current with a confiderable obliquity, would have probably had a perceptible good effect, and had the head been lengthened to its full height, to as great a length in addition as that of the jetee, in all probability this would have in a good part relieved the complaint. I fay in a good part relieved it, for the upper pier being too much in fhallow water, to fully anfwer the purpofes of bufinefs, I am of opinion that nothing can more effectually tend to relieve the complaint, and anfwer the purpofes intended, than a removal of the pier, and carrying it out in the direction and extent propofed in the plan exhibited to me; which was, to carry it out from the Bulwark rock, in a direction towards the fort; that is, S. 52° 30' E. to the length of 175 yards, or 525 feet; and afterwards in the direction S. 69° E. to the length of 105 feet: in the whole 630 feet.

To the 2d.—I apprehend the place of beginning will depend upon the fituation of the principal and weighty materials. If they are to be brought or floated by water to the place, then they can be beft depofited at the pier head; and carrying on the work towards the Bulwark rock, to that degree of forwardnefs that the new pier becomes a fhelter, which experience will point out; then the materials of the prefent pier may be removed, to fill up the gap and join the land. But if the materials are principally to be brought from the land, then the beginning from the Bulwark rock and advancing forward progreffively, will be found to be the eafieft and cheapeft mode of proceeding, as well as the fafeft from derangements. And thus carrying the work forward till an equal fhelter is found to what is afforded by the prefent pier; then this laft may be begun to be removed, in order to carry the other work forward; beginning in taking down with the pier head. I think there is this objection to the beginning at both ends, as then there will be two unfinifhed terminations, always lying open to the derangements of ftorms.

To the 3d.—I muft obferve, that a degree of inconvenience will always be outweighed by a fuperior utility. In my difcourfe upon St. Helliers harbour, I have fully pointed out, that where a floping fhore is interrupted by the erection of a wall, as a wall has not that tendency to fpend and deftroy the waves of the fea that a floping fhore conftantly has; wherever walls are erected in the confines of a harbour, it is rendered, in a degree, lefs

quiet

quiet by this means. That there will be a difference is evident and demonſtrative; but degrees of quietude not being meaſurable by a ſcale, it is not always poſſible to ſay what will be the degree of difference. The want of a road to a harbour acceſſible at all times of tide, is a capital defeƈt; but to the remedy propoſed, of taking ſuch a breadth from the harbour as may not be prejudicial, the whole that can be ſaid, is, that as the road muſt be formed and ſupported by a terminating wall towards the bay, while you are gaining a greater breadth, you are wanting a higher wall; and a higher wall when there is water before it, will more ſtrongly reflect back the waves, inſtead of deſtroying them; and thoſe meeting with, and compounding themſelves with others, will moſt undoubtedly render the harbour more unquiet, than if things were, in this reſpect, to remain as they are. Again, the more uſeful this wharf wall is made, for laying ſmall boats and craft againſt it, by a depth of water before it, the more it will tend towards general inquietude; and unleſs in calm and quiet times, as the face of this wall will lay open to the bay towards the eaſt, I believe thoſe boats will not find it very eligible to lie againſt it. But yet a road is ſo great a convenience, that if forty or even fifty feet of frontage be taken from the ſhore, before the moſt projeƈting wharfs (I mean thoſe ſouthwardly towards the pier), as they at preſent ſtand; and the wall is carried in a regular curve from thence to join the projeƈtion where the road aſcends the ſtreet; I ſhould apprehend, as this can be wharfed up with a wall of a few feet in height, that its influence at the diſtance to which the new pier is propoſed to be ereƈted, will not be ſo material as the convenience; but whatever that influence may be at that diſtance, it certainly will be greater at the diſtance of the preſent upper pier, which is much nearer. There poſſibly might be conveniences and motives in having this road made now, which might make it worth the while to hazard even this difference. But as it does not appear to me, that the want of ſuch a road for two hours a day, for four, or at moſt five days at each ſpring tide, can be of great conſequence; ſeeing that at thoſe times, the low water is in the middle of the day, and the interruptions can therefore be only at mornings and evenings, the reſt of the day being free of acceſs; and furthermore, that in caſe of need, the pier is acceſſible at all times by a road that hitherto has been praƈticable; all theſe things confidered, I cannot recommend to begin the road and wharf, till ſome progreſs is made in the building of the intended pier; or to do any thing that may in its effeƈt be hurtful to the preſent ſituation, till a better one is acquired; and the appearance of the benefit, more ſubſtantial than the riſque.

I am, honourable Sirs,

Your moſt humble ſervant,

J. SMEATON.

*Grays Inn, London,*
*30 Sept.* 1785.

## QUEENSFERRY SHIPPING PLACES.

The REPORT of John Smeaton, Engineer, upon the Shipping Places at Queensferry, in the Firth of Forth, from a View taken the 15th of August 1772.

IT feems fcarcely neceffary to fay any thing upon the importance of this ferry : it being the fhorteft and moft commodious place of croffing the Firth of Forth, and thereby muft recommend itfelf to the attention of every one who has concern in and wifhes well to the kingdom in general ; and on occafions is, and may be made more ferviceable for the paffage of troops, could a bridge be built at this place, though at the expenfe of £100,000, I apprehend it would become a national object to conftruct it ; but as we cannot hope to fee this effected, the next beft thing is a commodious ferry ; and if it would be worth fo large a fum to obtain a bridge, it would furely be worth a confiderable fum to obtain the moft commodious ferry poffible. As matters now ftand, there appear to be good veffels well adapted to the purpofe in fufficient number, and very able and expert hands to manage them ; and even in its prefent ftate, from what I have experienced of it myfelf for years paft, I believe it may be faid to be one of the moft commodious and beft attended ferries of the width in Great Britain. The principal defect is in the fhipping or landing places of embarging and difembarging paffengers, carriages and cattle ; what conveniences of this kind there are already are in a great meafure furnifhed by nature, and the difadvantages thereof are fuch, as where nature requires a little affiftance from art.

The principal defects arife from the roughnefs and awkwardnefs of fome of the prefent places, for want of proper fhelter in particular winds and times of tides, but principally for want of any proper fhipping place at all, at or near the low water of a fpring tide ; by which travellers are often detained when the wind is fair, and afterwards further detained by the winds coming foul.

There are feveral places of fhipping on each fide, but moft of all upon the fouth fide, and the fkippers make ufe of thefe as occafion ferve ; and in fact it is more commodious to have various places on each fide than to be confined to one on each fide though ever fo complete, becaufe from the courfe of winds and fet of tides, they can frequently make their way good between two places without tacking, when they cannot do it between others without great lofs of time, and *vice verfa* in other fets of winds and tides.

The

The principal fhippings on the fouth fide are as follow:

The Binks being the weftermoft and oldeft fhipping place; it fronts the weft, and therefore moft proper in eafterly winds; it extends to high water.

The Grey Shipping lies near to low water, and is of ufe only when the water is low.

The Forenefs Shipping fronts eaft, therefore propereft in wefterly winds, and ferves from fomewhat below half flood to high water.

The Weft Hall fronts eaft, therefore propereft in wefterly winds.

The Long Craig fronts weft, therefore propereft in eafterly winds; this is chiefly ufed when the tide is low.

The White Houfe fronts and is nearly fimilar to the former, being further to the eaft.

### North Ferry.
Craig End, the principal fhipping place, anfwers to both eaft and weft.

The Weft Nefs fronts the weft, therefore propereft in eafterly winds; is ufed chiefly when the tide is low.

The Wich or Eaft Nefs fronts the eaft, and therefore beft in wefterly winds; this fhipping ferves at low water, but is at prefent very difficult of accefs.

On communing with the fkippers as well as the principal perfons at South End and North Ferry, I am of opinion that the fhippings which are of the moft immediate confequence to improve are the Grey Shipping contiguous to Queensferry harbour and the Weft Halls on the fouth fide, and the Craig End and the Wich on the north fide.

The Grey Shipping feems to have been formerly attempted or made a place, but has been either deftroyed by ftorms or left imperfect; there is a good mafs of ftones upon the place, but will need a confiderable addition to make it regular and proper; it will be right to make a face both eaft and weft, and then it will be proper for winds either way, and to build it floping from low water, to anfwer fuch time of tide as the Fore Nefs begins to take place, which ferves to high water.

The Fore Nefs fhipping wants fome repair, and it would feem better if it was made ftraight and the jetees taken away near the lower end; but that is not fo immediately neceffary as the Grey Shipping.

The

The Weſt Halls is chiefly a natural ledge of rocks, ſtanding from low water to high water; the rocks here want part facing, part building, and part levelling; the height of this above the ſand ſhould not be leſs than five nor more than ſix feet.

Some little help is wanted to cut down the face of the long craig from whence it has been begun to low water, in caſe funds ſhould not hold out after the work at Weſt Hall is done.

The craig end at North Ferry has had the moſt improvement made thereon of any of the reſt, and indeed ſeems the moſt important; but the work wants to be extended to the turn of the point, by which means near two feet addition of water will be gained. This ſhipping however will not extend to the low water of a ſpring tide, it is therefore of importance to improve the Eaſt Neſs ſhipping, which has plenty of depth at low water, and is very commodious in weſterly winds, and in all other winds that do not overblow. The principal thing wanted here is to ſmooth down the rough rocks in order to form a road which in its preſent ſtate ſeems quite impracticable for carriages and horſes; this may be done by blowing away the irregularities of the rocks, or by bolting down timbers to take the wheels of carriages in the manner of a rail road.

The great irregularity of each of the places that want improvement, makes it not eaſy to get ſuch meaſures as would be neceſſary to form any eſtimate that can be depended upon without the whole was lined out upon the ſpot, which would take up much more time than my occaſions could poſſibly allow. But upon a general view of the thing, I am of opinion that if the ſum of £500 were raiſed and diſtributed amongſt the three places mentioned, viz. the Grey Shipping and Foreneſs, the Weſt Halls and Long Craig, and the Craig End and Eaſt Neſs at North Ferry, the improvements that would be made under prudent management would be ſuch as would be of real uſe to the public, and demonſtrate that ſuch further ſum as might be neceſſary to make the whole complete, would merit the aid of thoſe who are enabled to give it.

*Edinburgh*
26th *Auguſt* 1772.                                                J. SMEATON.

P. S. The length to be made at the Grey Shipping is 96 yards.
    That at the weſt halls    -    -    -    - 142 do.
    The addition at Craig End, about    -    53 do.

And a buoy anchor and chain would be very uſeful to be laid at a convenient diſtance off the craig End, to enable the veſſels better to haul off and get under way in contrary winds.                                                        J. S.

## EDINBURGH BRIDGE.

The REPORT of Messrs. SMEATON, ADAM, and BAXTER, concerning the Bridge of Edinburgh, addressed to the Right Honourable the Lord Provost, Magistrates, and Bridge Committee of Edinburgh.

HAVING been called upon by you to give our opinions concerning the bridge of Edinburgh, in confequence of the unhappy accident that happened thereto upon the third of this prefent month, we have in confequence each of us minutely examined the ftate thereof with all the attention we are able, we have alfo carefully confidered every matter of fact that has been laid before us in way of evidence concerning it, and we have fully heard and confidered every thing that has been offered and propofed to us by Meffrs. Mylnes towards the effectual reparation and reftoration thereof ; and upon the whole matter we are of opinion as follows :

That not only the particular accident which has been the occafion of this enquiry, but every other matter of a leffer kind of which we have found this ftructure complaining, has been caufed by an overload of earth upon the upper parts thereof, and therefore in general that though the effects of this over preffure has become evident by fetts in various parts of the building, yet if the whole be eafed of this overload, and properly bonded together, we apprehend that relative to a lefs weight, the remaining parts of the ftructure may be rendered firm and fecure.

We beg leave to obferve by the way, that all heavy buildings are more or lefs obnoxious to fetts, and particularly thofe where great weights are obliged to reft upon fmall areas of ground ; yet we fee buildings ftand for ages under thofe circumftances much more than thofe cafes where they can be relieved of the preffure which has originally occafioned thofe derangements, and we muft further obferve that though this bridge were to be taken down and rebuilt with all the fkill of Europe, yet it cannot be enfured, but that fomething of this kind would appear.

We alfo beg leave to take notice, that the particular accident upon which we have been called, is of a very uncommon nature, viz. the falling in of the three vaults in the fouth abutment : we can infer in the general that this has been occafioned by an over preffure,

3

but

but by what particular circumſtance it has happened that the arches of thoſe three vaults have given way in ſo ſudden and ſo alarming a manner, we cannot at preſent inveſtigate, but which perhaps will be more fully diſcovered when the rubbiſh is removed ; this however we infer from it, that as it does not appear to us that the three great arches with any of their connections northward, have ſuffered any alteration by the ſudden and even momentary deſtruction of the ſouth abutment, thoſe arches muſt in conſequence be very well poiſed upon their own legs.

Such being the general cauſes, and ſuch the general remedies, we come now to the particulars, with the method of reparation pointed out to us by Meſſrs. Mylnes, with our obſervations thereupon.

To remove the great overload of earth, Meſſrs. Mylnes ſuggeſt may be effected by ſhaving off two feet of ſolid earth from the whole ſurface of the bridge.

This will certainly have a very good effect in the reſpect mentioned, and we ſee no objection to it, ſave that it will cauſe the declivity to be in the whole two feet greater ; but we apprehend it will ſtill be ſufficiently eaſy ; and that it will in ſome meaſure obſtruct their view ſideways, but we may ſet againſt this, that the ſame two feet relative additional height in the parapets, will make a ſkreen to the violence of the wind, which may be expected in ſome ſeaſons of the year to be too ſeverely felt.

1ſt. Reſpecting that part of the north abutment which is filled ſolid with earth, we obſerve that the wall on the weſt ſide is ſomewhat rounded, which we imagine has been occaſioned by the lateral preſſure of the earth within it relative to the great height of the wall, and as this has doubtleſs happened by laying the earth againſt it, while the wall was green, it may poſſibly go no further ; but to make it ſecure, Meſſrs. Mylnes propoſe to raiſe up outſide buttreſſes, as well againſt the eaſt wall as the weſt, which we approve.

2d. Reſpecting the vaults in the north abutment (which are ſimilar to thoſe on the ſouth which fell in, and were the immediate cauſe of the accident above mentioned) we obſerve that the outwalls not being bonded in with the arches, are ſeparated therefrom about an inch, at a medium, at the crowns, and the croſs walls from which the arches ſprung being loaded with a conſiderable thickneſs of earth above them, are thereby preſſed with a greater weight than the ſide walls, and have therefore ſettled, by the difference of from a quarter to half an inch more than the ſide walls, ſo as to break the bond between them. Theſe effects are doubtleſs owing to too great a quantity of earth above theſe vaults, preſſing the

F f 2

walls

walls that fupport the arches downwards, and the out wall fideways; but as independent of this preffure, we fee no reafon why the walls fhould fall, fuppofing they had originally been built in this manner, Meffrs. Mylnes are of opinion it will be proper to lay the greateft part of the fpace above thefe vaults hollow by arch work; and alfo to confine the two fide walls together by chain bars from fide to fide, which we approve; and we beg leave further to offer, that we look upon it, that the beft mode of arching in this place, will be fimply to raife crofs walls upon the former; and upon them to fpring new arches over the prefent, to rife as high as the road will admit.

3d. The fmall open arch on the north fide feems perfectly found and entire, Meffrs. Mylnes therefore think it will be fufficient to lay chain bars from fide to fide to prevent the lateral preffure of the earth from fwaying them outwards, till the whole is hardened and compacted together; now as we underftand, there are already relieving arches over that open arch, we alfo fuppofe this to be a remedy fufficient, provided the two feet reduction of the height of the road be approved of by the Magiftrates, otherwife we think fomething fhould be here done by way of avoiding an addition of weight.

4th. The three great arches appear to ftand firm upon their legs, as they were built, there are however fome fplits in particular parts, chiefly in the rubble work of the foundation of the piers, which though proceeding from the unequal bearing that inevitably attends that fort of work, yet would probably not have appeared, had they not been charged with an extraordinary preffure; in order therefore as well to relieve the piers with their foundations as to prevent the lateral preffure of the earth againft the fpandrill walls from the chance of producing any ill effect, we cannot but recommend any method of arching in the fpandrills between the great arches that will effectually fave weight in lieu of chain bars acrofs the bridge at thefe places.

5th. We come now to the fmall arch on the fouth fide, concerning which the whole of the difficulty lies we underftand that previoufly to the falling of the three vaults contiguous thereto, there was a fett or rend of the abutment wall from the top to the bottom, which extended itfelf acrofs the fmall arch as far as the impoft of the pier on the north fide, which rend not only in the abutment wall but in the arch opened confiderably wider than before upon the fall of the contiguous vaults; in this arch the rend divided into two branches, the former almoft directly acrofs, and which now affects the piers in a fmall degree, the new one diagonally towards the north weft fpringer. This original rend we apprehend to be caufed fimply thus, that the matter of the natural foundation of the abutment wall, being lefs firm weftward than eaftward, when the whole became loaded with

the

the fuperincumbent weight of earth before mentioned, the weft fide fettled, leaving the eaft fide ftanding firm, and thereby occafioned the rend as firft mentioned. The falling of the vaults we look upon as occafioned by fome failure of a different kind; though ftill owing to the general caufe of over preffure, it is not therefore to be wondered that this rend before made, fhould become worfe by the great concuffion of ground that muft have attended the fall of the vaults, with all the matter above them. In this ftate of things, Meffrs. Mylnes are of opinion, that it is abfolutely unfafe to take down this fmall arch, left the lateral preffure of the great arches fhould by overfetting their piers, bring the whole into ruin; they therefore fuggeft that it muft be remedied by the following artificial means.

To infert a ftrong fpring arch or courfe of aifler ftone, diagonally from the lower part of the bafe of the fouth pier, to the upper part of the bafe of the abutment wall, by which means (as a brace in timber work) the over preffure upon the foundation of the abutment wall will be difcharged upon that of the pier, and thereby the dependance made upon the two together. They proceed then to take down the abutment wall, to a proper height below the crown of the arch, and to cap it with fome courfes of large hewn ftone, firmly bonded together with chain bars, fo as to prevent any further feparation of the abutment wall, and upon this cap as a bafis to raife a wall of a fufficient height; to fpring a frefh arch over the former, fo as to occupy the whole fpace between the abutment wall and the next great arch, by which means the prefent arch will have nothing to do but to preferve the uniformity of the appearance, then to draw the fractured and disjointed ftones to make all fair, and bond the whole together with chain bars in the fecureft manner.

To this propofition we muft be candid enough to fay, that if very carefully executed we muft expect it to fucceed; but we think ourfelves equally obliged in juftice to all concerned to declare, that in our opinion it would be more eligible, and what we would attempt were the cafe our own, to rebuild the fractured arch, and as much of the abutment wall, as can be taken down with fafety.

To prevent therefore the lateral preffure of the great arches from taking effect, we would place an additional row of fhoars in a fimilar pofition to thofe now there, which we would ftep upon the next fet off, on the abutment wall below the prefent, and infert them into the folid of the pier at about the height of the impoft of the great arch, thofe fhoars to be of whole timber, and to be fo many in number, that the fpaces will be only equal to the timbers. This being carefully done and well keyed home, we would take down the fmall arch, leaving as much of the foffite ftanding on the north fide as will fupport itfelf without

a center,

a center, and as much of the folid work above as this will fupport, then taking down the abutment wall within two courfes of the bafe, we would lay on two ftring courfes to be of aifler within and without and ftrongly bonded together with chain bars. Upon this as a frefh bafe the abutment wall is again to be raifed to its proper height, and the arch fprung over as before, and then finifhed with an arch of relief as fuggefted by Meffrs. Mylnes· This method to the beft of our opinion may be put in execution with perfect fafety to the great arches, yet we cannot take upon us to fay we can abfolutely enfure it.

6th. Refpecting the rebuilding of the fallen vaults there is the leaft difficulty of all, for if the walls of thefe vaults are raifed fo high that the crowns thereof do properly clear the road, we perfectly agree with Meffrs. Mylnes in opinion, that they will be attended with all poffible fecurity, as the caufe of failure will then be taken away, that is, the great weight of earth above them, due attention being had to the ftate of the foundation.

7th. The walls of the fouth abutment fo far as they were filled folid with earth, feem to ftand perfectly found and firm, and being of lefs height than the correfpondent ones on the north fide are the lefs in want of help, however for the fake of fecurity, as a further addition of earth will be wanted there, Meffrs. Mylnes in like manner intend to ftrengthen them on both fides with external abutments, which we entirely approve.

We cannot help adding by way of relieving the apprehehenfions that may occur hereafter, that we do not expect thefe or any other methods that can be taken, will put the building in fo equal a poife, but that from the compreffibility of the matter fome joints will appear (efpecially of the upper work) to open, as has happened to all large bridges confifting of feveral arches hitherto built, and probably will happen to all that will be built.

Thus we have endeavoured in the cleareft and moft candid manner we are able, to fet forth the true ftate of this work, as it refpects the contracting parties and the public, and we do not doubt, that notwithftanding the accident that has happened, the reparation will be attended to with fuch diligence and care, that the work itfelf will remain a lafting monument to pofterity.

*Edinburgh*,
22d *Auguft* 1769.

J. SMEATON.
JOHN ADAM.
JOHN BAXTER.

To the Right Honorable the Lord Provost, Magistrates, and Bridge Committee of Edinburgh.

The REPORT of John Smeaton, Engineer, upon the several Matters referred to him by the Minutes of the said Bridge Committee, the 17th January instant.

HAVING carefully confidered the anonymous plan, together with the feveral matters tranfmitted to me by Mr. Stuart along with the faid minute, I find myfelf at a lofs to come at an accurate judgement upon the whole affair for want of the following materials, videlicet.

A true fection of the ground for the whole length of the bridge from the high ftreet to the oppofite intended landing upon the natural ground.

A copy of the original defign and contract thereupon.

An accurate drawing of what has been really executed in confequence thereof. And

A Report of the condition in which the foundation of the crofs fide walls of the fouth abutment arches, appeared when the rubbifh was removed fo far as to enable Mr. Mylne to build upon them as it is faid ; which materials feem neceffary to compare with what is now laid before me. Neverthelefs, that the accomplifhment of this great neceffary work may fuffer no delay from me, from what I have feen and know of the affair, and from what I can collect from the feveral papers tranfmitted to me, I fhall endeavour to deliver myfelf with all the precifion I am able.

Having perufed and re-perufed all the papers above referred to, I find no reafon to deviate in opinion from what is contained in the report of the 22d of Auguft laft, wherein I have the honour to be joined with Meffrs. Adam and Baxter, but find many reafons to confirm me therein, and to more ftrongly enforce the neceffity of the moft material particulars therein contained.

Refpecting the anonymous plan, it feems in many refpects a delineation of what is pointed out in the before mentioned report, which, for the fake of diftinction, I fhall join in calling the Engineers report, the following particulars however excepted.

1ft. He

1ft. He propofes to build a divifion wall and to lay floors in the fouth abutment by way of tying the crofs and fide walls together, and to take down the north abutment, in order to rebuild it upon the fame plan as the fouth.

2dly. He propofes to raife the building four feet higher than it was laid out for at the time of the accident, to fill up this fpace with earth or rubbifh, and reduce this additional load by conftructing hollow cylinders over the piers, which from the draft appear to be intended twenty feet diameter.

In regard to the matter contained in the firft article, as the anonymous gentleman pretty ftrongly infinuates, that the public will not be fatisfied with the judgment of thofe who have fpent a great part of their lives in the ftudy of the poife and ftrength of materials; as I expect the gentleman knows their mind, I hope they will be fatisfied with what he has now propofed, and as Mr. Mylne makes no objection to the execution thereof, I fhall not object it, as it can do no harm. I only defire to fay, that I think it unneceffary, efpecially in works to be carried up anew. Had he propofed to have run a divifion wall through the north abutment, and to have only taken down the walls below the fpring of the prefent arches, in order to have tied them afrefh together by good bond ftones, and then to raife them high enough to turn frefh arches at a proper height, inftead of turning frefh ones over the prefent as propofed by the engineers, he would have made a judicious improvement; to take them down to the ground can do no more, and to take them up entirely, after having been compreffed with double the weight they will again be fubject to, in order to lay them on new bearings, may render them obnoxious to cracks and fets as before.

To what is contained in the fecond article, namely, the raifing the whole building four feet higher, I entirely object: for though this in his opinion may relieve the fears of the public, it will greatly increafe mine.

From what appeared on view, it was my opinion, and feemed that of the gentlemen with me, that the piers of the great arches had complained of the load of earth that then had been upon them, but not in fuch a degree as to be abfolutely unfafe, and therefore could they be lightened there could be no danger of their failing. What Meffrs. Mylnes propofed concerning the fhaving of two feet from the whole furface, might poffibly be underftood in different fenfes by the gentlemen to whom delivered; and though we concurred in the fame words, it is poffible we might have different ideas affixed thereto, (fuch is the ftate of manhood); I did not for my part underftand that Meffrs. Mylnes propofed to fhave off two feet from the body of the earth then upon the bridge, which if I

remember

remember right, was about that quantity below the height, which would have been neceffary to have completed his then plan; but to fhave off two feet from the plan as propofed in its completion; for had I underftood that he meant to reduce the earth upon the bridge, then two feet too low, to be two feet lower, the relative heights of the parapets and declivity would have been four feet, and not two feet greater as obferved in the report. In fhort, my opinion, and I underftood that of thofe with me was, that the piers of the great arches could not fafely carry a greater load than then lay upon them, and with fome reduction would be fafe; with this view the 4th article was drawn up, which I defire in this place may be read.

It is faid in remarks upon the anonymous plan, article 5th, that by Mr. Mylne's plan, the top of the cornice or level of the caufeway, was to be nine feet above the foffit of the arches. The ground was then, or was to be, reduced by fhaving to feven feet above the foffits; the anonymous plan adds to Mr. Mylne's plan four feet, that is 13 feet above the foffits, and is fo delineated in the plan before me; confequently the anonymous plan propofed to lay upon the bridge fix feet additional height and weight more than was thought prudent by the engineers' report, amounting to upwards of 4000 tons, upon the bafes of the four detached piers, and I beg leave to obferve, that 20 feet cylinders, or the value thereof, are practicable over the piers, when the road is at the height of feven feet above the foffits, this 4000 tons is all fheer addition to the engineers' plan; fome fmall deductions for the different height of the parapets above the road excepted.

In order to be more diftinct upon this fubject, we will confider the difference that will arife to one pier alone, fuppofing one of the piers fupporting the centre arches, which will be fubjected to the greateft preffure; it is evident, that one of the piers fupports not only the matter above it, but one half of each adjoining arch, and the matter above them. Six feet addition will therefore lay on a load of 1200 tons upon the bafe of one of the centre piers more than by the engineers' plan, and 600 tons more than they have yet been loaded with, even after deducting the cylinders propofed; and 350 tons more after deduct-ing the greateft cylinders practicable in the given dimentions: but without purfuing the anonymous plan any further, I will now take it up as acceded to by Mr. Mylne, and fup-pofe it to be executed to the greateft advantage. Now according to the reduction propofed by Mr. Mylne there ftill will be an addition of four feet to the engineers' plan, and though that height will yet admit of cylinders larger in diameter by two feet than marked out in the anonymous plan, yet the load on one of thefe piers will ftill be greater by 684 tons than by the engineers' plan, and greater by 86 tons than it ever yet has carried, that is, fuppofing the ground never made up within two feet of the top of the cornice. The bridge committee may, if they pleafe, try this experiment; and I won't fay it will not fucceed; but I can by no means recommend it.

It remains therefore to point out how the bridge is to be made acceffible in the moft eafy manner that prefent circumftances will allow, confiftently with the reductions propofed in the report of the engineers.

From the eighth article of the explanation of the anonymous plan, it is faid, *the building is raifed four feet higher to make the line of declivity one in 16, as was originally intended, and to end at the breaft of the fouth abutment :* but the plan before me makes a declivity of about 1 in 14½, and the remarker, article five, obferves, that to recover the loft height will require a rife of fix feet ten inches; taking this for granted it follows, that at the height affigned by the engineers there will be a lofs of eight feet ten inches; now I find the length of the flope contained within the compafs of the plan before me is about 360 feet, and from the tendency of the lines there feems to be about 106 feet wanting of the fummit in the high ftreet to complete the whole length : it will follow therefore that this addition of eight feet ten inches will reduce the flope from 1 in 16 to 1 in 12; and I beg leave to obferve that in hilly countries, where the flopes in turnpike roads can be reduced fo as not to exceed 1 in 12, they are efteemed fully to anfwer the intentions. The afcent of the arches over Weftminfter Bridge was originally laid out to be 1 in 20, but I apprehend from the eye, that the afcents at each end, are at leaft 1 in 12, and it is a kind of rule in thefe parts that in laying out roads and bridges, if the afcents do not exceed three inches per yard, they are no ways objectionable. The flope by me originally laid out for the afcent of the weft abutment of the bridge at Perth, is 1 in 12, and of this may be produced a multitude of examples. In fact, fince the whole afcent between two given points muft always remain the fame, where the length of the declivity happens to be fuch, that horfes in carriages cannot conveniently trot up, they may as well be reduced to 1 in 12, at which afcent they can conveniently walk and draw, becaufe the length is lefs than if more floping.

If the declivity marked out in the plan before me is to be adopted, then by extending he flope a little beyond the middle of the firft great arch, this flope will cut the level marked out in the engineers' report; and to reconcile thefe two planes without fo great a difference in the height of the parapets, as would be the cafe if the top of the parapets preferves it level to the fouth abutment, I would make no fcruple (were the cafe my own, and I were left to do what I pleafed) to make the parapet conform to the road; but yet, I would not fuffer them to meet in an angle, but reconcile them by a curve, fo that the mafonry

should

should seem gradually to conform itself to the ground it stands upon, and to that to which it is to be connected. The line of the top of the parapet is, according to this idea, marked out with red ink upon the plan *; the height of the road being also distinguished by a dotted line, either of the above methods, as they leave the *stability* of the bridge *without a doubt*, I think far preferable to any increase of its load, which in raising it can scarcely be avoided; so far from it, I would sooner propose to reduce it two feet more, that is to five feet thickness from the soffit of the arch to the top of the road; for, as I understand the penstones are from three feet high, there will still be a two feet cover, which, in my opinion, is amply sufficient, and even then, the slope from the south abutment will meet the horizontal plane before you are over the nearest pier of the center arch, and this plane will still be higher, though not much above the level of the point from whence the north abutment commences; and in my opinion the reducing the whole to one level after the first descent will have a better effect upon the eye than by a second angle bent the contrary way, in leaving the breast wall of the north abutment, and will at the same time greatly relieve the vaults and retaining walls thereof.

I beg leave further to add, respecting the fractured arch on the south side, that I have no reason to differ from what is contained in the fifth article of the engineers' report. I think it may be repaired as Messrs. Mylnes have proposed, and I think it may be taken down and rebuilt with safety, and this I advise. I think the abutment wall may be effectually tied together, if taken down within two courses of the base, and that the preference of taking it down to the surface of the base can amount to little more in reality than taking away the remaining *appearance* of a crack, as the foundation will then remain fractured as it is, nor do I think it any ways eligible to take it quite up, for since the pressure it has had has undoubtedly brought all to a solid bearing, the taking it up and laying it upon new bearings, may produce greater derangements than it is intended to cure; lastly, I don't think the method of shoaring the great arches proposed, by timbers to be laid horizontally, is so efficacious against a sett of the great arch as the shoars recommended in the engineers' report. At the same time it is my opinion that either will do, and that neither is absolutely necessary: my reasons for them were rather to make that sure, which might easily be done, and to relieve the apprehensions of Messrs. Mylnes and the public, than any absolute necessity: on the contrary, I am of opinion, that the great arches would stand upon their own legs, *if unloaded of the earth upon them:* but as there is no necessity to run this risque, I do not advise the experiment to be tried.

*Austhorpe,*
27th January 1770.

J. SMEATON.

* This plan is not among the papers of Mr. Smeaton. Ed.

# EDINBURGH WATER SERVICE.

The REPORT of JOHN SMEATON, Engineer, upon the State and Improvement of the Water Service of the City of Edinburgh.

THE fupply of the city of Edinburgh with a fufficiency of good water being a matter of very great concern, and the Town Council of the faid City having done me the honour to confult me upon the propofed improvements relative to this bufinefs, when I was laft at Edinburgh, in the month of September, I have been folicitous that no inveftigation or enquiry fhould be neglected that might contribute to the perfection and certainty of this great object; and this I was the more induced to do, finding that Baillie Cleghorn, who attended my viewing of the premifes, had taken this matter very properly up, and needed only that kind of information which might be expected from me profeffionally, and which my particular ftudy and application to this branch of hydraulicks, I truft, enable me to furnifh.

Since the view above mentioned, which was upon the 6th of September laft, I have received from him not only copies of the two memorials delivered by him to the Town Council, and of the regifter from April 1778 to March 1779 inclufive, and of the neceffary printed papers, but alfo a plan and fection of the Hareburn and Swanftone pipe from the fpring to the Coniftone refervoir, which I fuggefted as neceffary to be taken, in order to come at proper data for the calculation of what may and ought to be done there; fo that the matter being now fully before me, I fhall do myfelf the honour of laying before the Town Council the refult of my calculations and animadverfions on this occafion.

It is to be obferved, that a deficiency of water in the city of Edinburgh arifes from two caufes, firft, from a failure of fprings collected into the refervoir at Comingfton, in the dry months of the year, and, fecondly, from an incapacity of the main pipe of conduct, which conveys it from thence to the town's refervoir on the Caftle Hill, to bring a fufficiency to anfwer the demands of the increafed number of houfes and inhabitants, who are generally wanting the winter feafon, when there is the greateft plenty at Coniftone, if the pipe would but bring it. Doubtlefs, the firft and moft obvious expedient would be that of enlarging the main pipe; but this being a work of great expenfe and not foon effected, Baillie Cleghorn very judicioufly confidered that an increafe of declivity, or total fall between the refervoirs, by increafing the velocity of the water's current, would in event bring more water as certainly as an enlargement of the pipe; he therefore propofed and

executed

executed what immediately admitted of being done, that is, to let in the water iffuing from the main pipe into the refervoir at Caftle Hill, at the bottom inftead of the top, as it ufed to be, by which empty an addition of fall is gained equal to the depth of the refervoir, which being 7½ feet, this added to 44, the total fall or difference of level between the two refervoirs (in a length of 2¼ miles), making together 51½ feet, inftead of 44, has fo far increafed its velocity, when there is little water in, there is little water in the Caftle Hill refervoir, and when it is moft wanted by the town, that as it appears by the regifter it is now capable of difcharging 200 pints per minute, inftead of 160, a difference equal to one full fourth of its former produce, and amounting to the additional quantity of 112 tons per day, which alone is a very valuable acquifition.

Obferving the good effects of this alteration, he has propofed by way of ftill further increafing the quantity of water to be brought by the fame pipe without enlargement of its dimenfions, to conftruct a new refervoir in the garden of Heriot Hofpital, which being upon a level with the ftreet of the Trone church, is confequently above the greateft part of the city, new town, and fuburbs of Edinburgh, at the fame time that it is 119 feet below the level of the refervoir at Comiftone, inftead of 51½, which is the prefent greateft fall to the Caftle Hill refervoir ; confequently if thofe parts of the town that cannot be commanded by the lower refervoir, are continued to be ferved from the Caftle Hill, a part of the 24 hours may be appropriated to collect water in the Caftle Hill refervoir, fufficient for 24 hours fupply for that part of the town, that diftrict being ftill dependent on that fource, and a much greater portion of the 24 hours can then be applied to collect water into the new refervoir, which, from the greater fall, will be delivered at a much greater rate, and much more water will be brought for the 24 hours fupply of the reft of the town, than if it were all obliged to afcend to the Caftle Hill refervoir ; this he has proved experimentally by obferving that the difcharge of water at the cleanfing cock in the grafs market, lying 170 feet below the level of Comiftone refervoir, was 160 pints per minute more than would have been difcharged at the Caftle Hill refervoir, though the orifice by which it was difcharged was much lefs than the bore of the pipe, and he alfo further confirmed this, by trying the difcharge at the fecond cleanfing, each above a mile from Comiftone, which though only 131 feet below the level of the faid refervoir, yet difcharged 200 pints per minute more that could be difcharged at the Caftle Hill, and it is to be prefumed, that this cock alfo was lefs than the bore of the main pipe.

This propofition for a new refervoir at Heriot's Hofpital, however judicious and effectual for bringing an additional quantity of water from Comiftone when there is a furplufage there, yet when from the other caufe, a failure of the fprings, there is not more water than the prefent capacity of the pipe will convey, no more can be had ; and though this

chiefly

chiefly happens (as appears from the regifter) in the months of Auguft, September, and October, when moft families are out of town, yet as the prefent furplufage in May, June, and July does not appear confiderable, in every event an additional fupply at all times is a thing moft defirable; and this Baillie Cleghorn has not only in part done, by opening fprings and preventing unneceffary waftes before the arrival of the water from the feveral fources at Comiftone; but having obferved that a confiderable quantity of water furnifhed by the Hareburn runs continually wafte that cannot be received into the Swanftone pipe, and thereby conveyed to Comiftone, which the 12th of Auguft laft, in a dry feafon, amounted to no lefs than 160 pints per minute more than was received into the pipe, has propofed to lay a new pipe, of a fufficient bore to convey the whole water of this burn in dry feafons as above, which he propofes to be of four inches, in aid of the other, and to avoid an overload of expenfe, he propofes it to be of wood.

Thefe, I think, are the propofitions laid before me. On viewing the premifes I obferved that the whole of the Hareburn might be taken up upon confiderably higher ground than the Swanftone well, where it is now taken in; and being informed that the water furnifhed by the Swanftone fpring itfelf is but an inconfiderable part of the water now conveyed by the Swanftone pipe, (which, at the utmoft, appears from the regifter to be 76 pints per minute) it occurred to me that it might poffibly come out from a proper calculation, that by lengthening the Swanftone pipe to the rife fo as to take in the water of the Hareburn with that of the next fpring above, would add fo much to the fall (leaving out the Swanftone fpring, as inconfiderable in proportion to the acquifition of the Hareburn) that the prefent pipe might then be able to convey the whole of the Hareburn water; but this could only be known by a furvey of the length and profile, fhewing the difference of level that there now is and may be further obtained in the courfe of this pipe. From the furvey and level now before me, it appears that the fall in the pipe itfelf is much greater, and the additional fall to be gained to the next fpring above much lefs than I expected; I fhall therefore infert the following table, fhewing the refult of the different calculations I have made, checqued with the water actually given in pints per minute, where opportunity has offered; from whence may be inferred what degree of aid it is to be given to thofe where that opportunity has not offered.

For in the prefent length of the Swanftone pipe, viz. 2145 yards, there is no lefs than 168 feet of fall; and to get 40 feet of fall in addition we muft increafe the length 280 yards more, which, though giving upon the whole a greater power, yet is not fufficient increafe to carry the neceffary addition of the Hareburn water.

TABLE

TABLE of Results of Water given and expected by Calculation on different Water-Pipes belonging to the City of *Edinburgh*.

| No. | Pipe calculated upon. | Bore | Length in Feet. | Fall in Feet. | Water expended by calculation. | Water given by Measure | Difference. | Proportional Difference. |
|---|---|---|---|---|---|---|---|---|
| 1. | Swanftone pipe as it now lies | 2½ | 6435 | 168 | 116 | 76 | 40 | $\frac{1}{3}$ |
| 2. | Do. with 280 yards to the fpring above, and 40 feet additional rife | — | 7275 | 208 | 122 | — | — | — |
| 3. | For the firft 1000 yards, in which is a fall of 30 feet | — | 3000 | 30 | 72 | — | — | — |
| 4. | Do. for the principal defcent, exclufive of the firft 1000 yards from the Swanftone well | — | 3435 | 138 | 145 | — | — | — |
| 5. | A new pipe to convey the water from the prefent Swanftone well to Comiftone refervoir | — | 6435 | 168 | 236 | — | — | — |
| 6. | The main pipe from Comiftone to the Caftle Hill, before the alteration | 4½ | 14637 | 44 | 159 | 160 | 1 | $\frac{1}{139}$ |
| 7. | Do. after the alteration | — | — | 51½ | 173 | 200 | 27 | $\frac{1}{6}$ |
| 8. | The main pipe water drawn at the Grafs-market cleanfing cock | — | 13985 | 170½ | 346 | 360 | 14 | $\frac{1}{24}$ |
| 9. | The main pipe water drawn at the fecond cleanfing cock | — | 5359 | 131 | 492 | 400 | 92 | $\frac{1}{5}$ |
| 10. | The main pipe water drawn at a refervoir in Heriot's Hofpital garden | — | 13550 | 110 | 280 | — | — | — |

The deficiency of No. 9. in giving ⅕ lefs than it ought to have done by calculation was, doubtlefs, owing to the bore of the cock not being large enough to let the water freely out, and it may be inferred in refpect to No. 8. that it was fufficiently large; but in regard to No. 1. it appears that the Swanftone pipe is confiderably defective in giving its proper quantity, which may arife from an injudicious ufe of its air-cocks, or their not being air-tight when clofed, or from the pipe itfelf not being air-tight; for if any of the cocks are fuppofed to be open, or to take in air as much as the pipe will want, then it appears from No. 3. that the Swanftone pipe gives as much water as by calculation under thefe circumftances it ought to do, or the deficiency may arife from its not being fully 2½ inches as reported, for if really only 2⅓ it gives nearly its proper quantity. It does not however appear, that, if rendered perfect, it is likely to convey the waters of Swanftone fpring and Hareburn, viz. 76 pints per minute that it now brings, together with 160 pints in addition, in the whole 236 pints, but it appears from No. 4. that a pipe of 3½ inches bore would fully anfwer the end.

Doubtlefs a wooden pipe of four inches in aid of the prefent would be more than enough, nor can there be any fufficient objection to the ufe thereof; but as all the reft of the

water-

water-pipes belonging to this work are of lead, which for this purpofe is not only the fweeteft but moft durable, it appears to me likely that if the prefent 2½ pipe be taken up and recaft of 3½ inches, the addition of labour and weight of lead may be done for nearly the fame expenfe as the laying a wooden pipe.

The weight of a molded pipe of 2½ inches bore is feldom lefs than 30lb. to the yard, or more than 36 ; in the prefent cafe I fuppofe the medium weight, 33lb per yard ; this, for the length of 2145 yards, will amount to 622 cwt.

It appears from Mr. Leflie's fection, that this pipe is almoft upon a continual defcent, there being no rife but from a point which is 58 yards from the fpring head to another at a thoufand ; in which fpace the pipe rifes 26 feet perpendicular, fo that the pipe being conftantly open at both ends this is the only part that fuftains the preffure of a perpendicular column ; for were the pipe laid upon one continual defcent, and both ends open, the water would lay no ftrain upon it, but run through it merely as it would through an open channel, and the preffure of 26 feet that particular part fuftains is fo inconfiderable that it may be laid out of the queftion, as it will be neceffary to make the pipe ftronger merely to preferve its figure from being preffed inwards than would be neceffary to fuftain the column of water tending to prefs it outwards.

The cafe of the main pipe to the Caftle Hill is very different, for it appears from the general fection, that from the Grafs Market cleanfing-cock to the top of the Caftle Hill refervoir, whither it ufed to rife, is an afcent of 126½ feet, and fo much preffure of perpendicular column of water, that part of the pipe muft fuffer in croffing the bottom of the Grafs Market.

For thefe reafons, the Swanftone pipe needs no other ftrength than what will be prudent to give it, to preferve its figure againft the preffure of the earth, the accidental croffing of carriages, together with plows and animals over the ground ; and for thefe purpofes I look upon a 3½ inch pipe as fufficiently guarded (it being carefully bedded and covered) if made of fheet lead turned and burnt, of the ftrength of 11lb. to the fquare foot, which is exactly $\frac{3}{16}$ of an inch in thicknefs, which method I alfo look upon as more fafe from leakings than molded pipe.

A pipe of this conftruction will contain a foot fuperficies in a foot running, fo that it will be the fame weight as it is computed the prefent pipe is of, viz. 33lb. to the yard, fo that the prefent pipe recaft will make the new one, with the addition of fo much lead as will make good the wafte : and from a careful eftimate, I find, that this pipe fhould be

8

recaft,

recaft, the wafte made good, carriage to Edinburgh and out again included, for 3s. 9d. per yard, which, for 2,145 yards, amounts to the fum of 402l. 3s. 9d., whereas by a fimilar eftimate I find, that an elm pipe of four inches bore, including carriage out, the fum of 360l. 11s. 3d. exclufive of the expenfe of opening and filling the ground, which I fuppofe to be nearly equal in both the cafes; the difference, therefore, being no more than 41l. 12s. 6d. I apprehend will not be an object in the determination, and in cafe the recafting of the lead pipe is determined upon, I fhall be ready to fupply fuch further directions for the cafting and laying of it, as fhall enable it to produce the whole of the effect expected from it.

We come now to the propofition that relates to the erection of a refervoir in Heriot's Hofpital Garden; and to give it fome elevation and advantage, I have fuppofed it to be elevated nine feet above the ground, that is, the water to come in at 110 feet below the Comiftone refervoir, at which depth may be expected to be got 280 pints per minute, as per No. 10.

It is laid down by Baillie Cleghorn, as a certainty drawn from experience, that when the water comes into the prefent refervoir, at the rate of 200 pints per minute, that this is fufficient to ferve the city of Edinburgh well, that is to fay, 288,000 pints in 24 hours. It muft follow, that 288,000 pints in 24 hours iffued from a refervoir in Heriot's garden, muft ferve every part of the town that can be ferved thereby very well.

Now 288,000 pints coming in at the rate of 280 pints per minute, will be received in 17 hours 8¼ minutes, there therefore will remain 6 hours 51¾ minutes for the main to run into the Caftle Hill refervoir, which will ftill thereby be capable of ferving nearly upon $\frac{7}{14}$ of the inhabitants as amply as they are now ferved, at the rate of 200 pints per minute. But if $\frac{7}{14}$ of the inhabitants are fupplied from the Caftle Hill refervoir, this will fpare $\frac{7}{14}$ of the water brought daily into the new refervoir, to be diftributed to the fouth fide of the town and extended royalty, beyond what is fuppofed to be comprehended in the fervice of 200 pints per minute.

It therefore now only remains to fix the dimenfions of the new refervoir in Heriot's garden, fo that it may hold a quantity of water equal to the fervice laid down.

I take it for granted, that by far the greateft part of the water ufed in the city of Edinburgh is drawn in the compafs of 14 hours per day, and confequently that this may be looked upon as the average time of the daily expenditure. Now if the water to be drawn from the new refervoir were to come in at the fame rate it is drawn out, no refervoir would be neceſſary, but only a ſmall ciſtern to receive it. But if that be taken out in 14 hours, that requires 17 hours $8\frac{1}{4}$ minutes to come in, then there will be required a reſervoir capable of holding 3 hours $8\frac{1}{4}$ minutes water to begin with in the morning, ſo that after 14 hours draft upon it, the contents of the ciſtern, with what comes in during the time, will be ended with the cloſe of the ſervice each day.

But a reſervoir that muſt contain 3 hours $8\frac{1}{4}$ minutes water, at the rate of 280 pints per minute, muſt contain 52,794 pints, which at $103\frac{4}{10}$ cube inches to the pint, will amount to $3159\frac{1}{10}$, ſay 3160 cube feet, and if we ſuppoſe five feet depth of water to be filled and emptied therefrom, then, if ſquare, the ſide of the ſquare will be 25 feet $1\frac{1}{4}$ inch, or any other equivalent dimenſions, as for inſtance, 24 by 27, allowance being made in depth, ſuppoſe half a foot, to prevent its running over the brim, inſtead of the waſte pipe, and what may be left at the bottom.

I would, therefore, ſuppoſe it convenient, according to this eſtabliſhment, not to diſpenſe any water from this reſervoir before ſeven in the morning or after nine at night, but to let the water run from the main into the reſervoir till 10 at night, then to turn it upon the Caſtle Hill till five o'clock in the morning, when being again drawn into the new reſervoir by ſeven o'clock, it will have got three hours water ready to begin the day's ſervice. Experience, however, of the wants of the inhabitants muſt ultimately regulate the matter; and though I have ſhown that on a ſuppoſition the wants of the inhabitants, that muſt neceſſarily be ſerved from the Caſtle Hill, will be ſupplied by ſeven hours water from the main (but like the other to be diſtributed in 14 hours), that a reſervoir of the dimenſions given will be fully ſufficient; yet there is nothing but the charge to prohibit its being made of any larger ſize, that ſhall be thought convenient, becauſe that will always be a treaſury of water to anſwer any ſudden exigence, as in like manner will that at the Caſtle Hill, which ought always to be got and kept full whenever the ſupply from Comiſtone will admit thereof.

J. SMEATON.

*Auſthorpe,*
12th *February* 1780.

Elevation for a Bridge over the Tweed at Coldstream.

Scale of Feet.

Plan.

One of the 5 Ribs of the Centers.

Scale of Feet.

Foundation for a Pier.

# COLDSTREAM BRIDGE.

(See Plate 10.)

To Mr. Pringle.

Sir,                                                       Edinburgh, 31ft July 1763.

This comes to acquaint you, that the Truſtees of Coldſtream Bridge having had a meeting, I laid before them the different plans, with the reaſons thereof, when the low plan was unanimouſly choſen, as being ſufficiently elegant, and at the ſame time better adapted to the funds, than either of thoſe elevated to a ſemi-circle. According to this plan the ſide arches are 58 feet, and the two adjoining the centre 60 feet 5 inches, and the centre arch 60 feet 8 inches. They are all parts of the ſame circle, and the leaſt arches are one-third, ſo that the others riſe higher, as the additional ſpan admits. The height of the piers above low water, including the impoſt of two feet, is 12 feet preciſely; ſo that the ſhaft of the piers below low water muſt be raiſed 10 feet before the impoſt courſe comes on. I hope I ſhall be able to ſend the deſign for the centre, with the fair plan. I have ſatisfied the gentlemen relative to the foundation and method of proceeding.

Your's, &c.

J. SMEATON.

Mr. Foreman to Mr. Smeaton.

Sir,

I trouble you with this by order of a general meeting held this day for our bridge, who took into conſideration the centre you propoſed for throwing over the arches, and the one you ſaw Mr. Reid have a draft of, and which they had all ſeen before, and which you had no objection to, except that you thought a cheaper one might be made, that would anſwer the purpoſe. But on the wright's (carpenter's) eſtimate of both, yours proves to be the deareſt by about eight pounds. Out of deference to your judgment, they would not, however, fix upon it till they had conſulted you, and therefore ordered me to write to you, ſtating the different eſtimates, and begging to know if the centre propoſed by you is preferable to the other, without regard to the difference of expenſe, for they chooſe to be determined by your opinion.

Coldſtream,
March 27, 1764.

Your's, &c.

JOHN FOREMAN.

* This is the plan here given.

Hh 2

To

To Mr. Foreman.                                             Aufthorpe, 3d April 1764.

I am obliged by the confidence the truftees are pleafed to honour me with, though in
the prefent cafe there was no need for hefitation.   The draft Mr. Reid fhewed me being
upon the fame principle as the centres whereon the arches of Weftminfter Bridge were
turned, there can be no doubt of its anfwering the end propofed, if juftly and properly
executed.   But as I look upon thofe centres, when originally defigned by the ingenious
Mr. King, as intended to let the boats, &c. upon the river pafs under the arches even
while they were building, there was a neceffity of making them hollow, and of diftributing
the fupporting piles on the fides only ; and on this account the middle part being geo-
metrically fupported, induced a neceffity not only of ftronger timbers, but of greater care
in framing the bearings thereof, than if each part of the weight were more directly fupported.
It therefore occurred to me that when this was the cafe, not only flighter timbers would
ferve, but a much more fimple and eafy conftruction.   What I had therefore in view was
to diftribute the fupporters more equally under the burthen, preferving at the fame time
fuch a geometrical connection throughout the whole, that if any one pile, or row of piles,
fhould fettle, the incumbent weight would be fupported by the reft.

With refpect to the fcantlings, I did not fo much contrive how to do with the leaft
quantity of timber, as how to cut it with the leaft wafte ; for as I took it for granted the
centre would be conftructed of Eaft Country fir, I have fet down the fcantlings fuch as they
ufually are in whole balks, or cut in two lengthways, and as I think the pieces will fuffer
lefs by notchings in the middle interfection, and being cut into fmall pieces, it will remain of
value for common building, after it has been done with as a centre.   If therefore the value
of the timber each way, after it has been done with for centres, has been taken into the
account, and there ftill remains againft me a balance of eight pounds, or even eight
fhillings, I have not that partiality for my own fchemes, to prefer them where they are
not applicable.   Yet I muft obferve that had I a lefs opinion of your carpenter than I
have, I fhould prefer my own for this reafon, that as the conftruction is more obvious,
and lefs exactnefs required in the handling, I fhould expect to get a good centre made by
fome in this way, that would make but bad work of the other.

                                                                    Yours, &c.
                                                                    J. SMEATON.

To Mr. Pringle.
    Dear Sir,                                              Aufthorpe, 27th June 1764.

I am forry to find that the Tweed's gravel lies fo very open ; had it been wrecked up
with fand and matter, as is moft common, you would have completed the excavations with
all eafe and pleafure ; but as it is common in water-works for things to turn out contrary

                                                                              to

to expectation, change of circumstances requires change of measures, and nothing but time and patience can overcome the difficulties that often arise ; and you may remember this as an undoubted maxim, that the less the business is put in a hurry, the sooner it will be done, provided the power is steadily applied towards the end proposed.

Piling foundations is generally a very tedious and expensive job; I was therefore desirous to avoid it by going down to the rock ; but since the nature of the gravel makes the drainage likely to turn out so very expensive, of two evils it is best to choose the least ; I have therefore enclosed a design for piling the foundation at its present depth, viz., as I understand it, three feet below the bed of the river. It is common in cases of this kind to lay down a grating and build upon it without any piles at all ; and this is certain, that so long as the matter keeps under the grating undisturbed, nothing bears weight better than gravel ; but in so rapid a river as the Tweed, we are far from sure that the floods may not take away the gravel round the pier, below its present surface, and lay the grating bare ; in which case the matter will move from under it, first near the sides, and then farther in, so as to let down the pier, and damage the arches. The common defence is to build starlings round the piers, terminating above low water ; but as these considerably contract the water way, they are the cause of its acting more violently upon them towards their destruction.

In the enclosed design (see the plate), the main dependence is upon a sheet of plank piling AAAA, encircling the matter whereupon the pier is built, and which will hold it in, though the external gravel should by accident be taken away three feet below the tops thereof, that is, six feet below the present bed of the river; but to prevent this accident, the excavated space between the sheet piling and the coffer dam is to be filled with rough rubble-stone B, as it comes out of the quarry ; which, if any accident happens to it, may be repaired by an addition of the same kind of matter ; but which can never happen after it is consolidated by the gravel washing into the same, unless the gravel without the boundary thereof should happen to be disturbed, and which, if filled even with the bed of the river, by throwing in rubble, will stir no more in the same place.

The sheet piling is to be driven as close to one another as may be, and the heads spiked to the frame CCC, and afterwards cut level with the surface thereof; the bottom of the plank piles being canted off as represented at D, will keep them close at the foot. This kind of plank piling when intended to hold water is grooved or dove-tailed one into another ; but as these are only intended to retain the gravel, they will do plain, which will save a good deal of time and labour ; for to drive into gravel, they ought to be of double the thickness if of fir ; or else to be made of elm or beech to prevent their splitting, which

7                                                                                                    last

laft mentioned woods I apprehend your fituation could not readily afford. The piles reprefented under the frame CCC, are to be firft drove to the level of the prefent bottom, and cut to a level, and the frame to be faftened down with pins thereupon. The tye-beams EE, &c. being dove-tailed to the outward frame, are chiefly intended to prevent the frame from flying outward by the preffure upon the matter within; thofe beams may be let in the outer frame about $3\frac{1}{2}$ inches; the other piles not reprefented under the frame, are to be drove after the frame is laid, and cut level with the upper fide of the frame; then the whole of the frame and tye-beams are to be underpinned with rough ftones; and the reft of the fpaces to be rough paved and drove down with a heavy paviour's hammer, to be worked by two men, and the fpaces filled with gravel; upon this foundation (which will be a kind of artificial rock) you are to build your mafonry, but note, that the outfide ftones muft all be headers, and long enough to reach upon the pile heads of the interior row, and alfo to project about nine inches beyond the fheet piling, and to be let down three inches thereupon, as fhewn in the little fketch F, which will prevent the fpikes from drawing or flying from the frame. In the courfe that brings you above water, you muft lay in block ftones in rows agreeably to the tye-beams, and thereby again cramp the work together from fide to fide. With refpect to the interior piles, if care is taken to lay a found ftone upon each pile head, this will anfwer better than planking; becaufe it gives leave for every part of the work to be brought into contact upon the paving; fo that the matter of every part of the area will come to an equal bearing, and fave expenfe. This method being carefully and judicioufly executed, will I think be little inferior to a foundation upon the rock itfelf, efpecially if the whole has fufficient time to fettle before the arches are turned.

As your large piling engine will, I fear, be too heavy and cumberfome for this job, I would advife you to get two light ones made in the cat's-tail way, to be worked with 10 or 12 men each; the fcantling of the ftuff need not be above $3\frac{1}{2}$ inches, fo that the men that work it, will eafily move it; three, $3\frac{1}{2}$, or at moft 4, cube feet of timber, will make as large a ram as thofe men can work, but it had better be under than over-weight; it is probable that the piles will require no fhoes, or if they do, the bearing piles being fharpened to a flattifh point about two inches broad, and covered with a piece of rolled metal, fuch as is made for cafk hoops, turned up and nailed on each fide like a faddle, and a broader one of milled iron plate put on in the fame way upon the plank piles, is the whole we have ufed in the many thoufand piles we have drove upon the Calder, which is in general a hard gravel mixed with ftones. I would not advife you to drive longer than while you find the piles have got a firm feat upon the rock, left by too much driving you cripple them, and render them ufelefs.

The

The piles reprefented to be under the frame and tye-beams may be got near enough their places, by lines ftretched from fixed points upon the coffer dam, without taking off the water; and may be driven down below the water's furface by fetts, made as fhewn at Z in the plate, the little iron pin at the bottom going into a hole in the pile head will keep it fteady thereupon; the upper end fhould be convex, and the lower concave, to prevent fplitting the pile, and two handles ftanding at right angles to each other will be better than one, as two men will thereby keep it more fteady than one. This done, the water muft be pumped out while they level the pile heads, and lay down the frame; then the reft of the piles may be drove by the fide of the frame, by fets as before; a little practice will render the whole eafy; laftly, the water muft be pumped out while the pile heads are levelled, the underpinning and paving is performed, and the pier brought above water; the ftones being all ready prepared and got as convenient to the place as poffible in the mean time; as to the rubbling on the outfide, it may be thrown in at convenience, as well when the water is in as out; but muft be done before the coffer dam is removed. I would advife to throw into the water for about 30 feet round the coffer dam, a light fprinkling of corn mould earth, this may poffibly choak up the pores of the gravel, and eafe you of part of the water; as much as will amount to three fourths of an inch thick, will be fufficient.

If you will pleafe to let me know whether the rock be of fuch a nature that the piles may be driven five or fix inches therein, I can contrive you an eafier way of laying the next pier.

I am, Sir, your moft humble fervant,

J. SMEATON.

P. S. The piles that are reprefented round, need not be made fo if your ftuff is fquare; if you have not a fufficiency of ftone for the outfide of the firft courfe, as above directed, you may interfperfe headers and ftretchers, but the more headers the better.

To Mr. Reid.                                   Aufthorpe, 27th May 1765.

The making of a defign for the caiffon has not yet come into the courfe of bufinefs: however, that you may be proceeding to get proper ftuff, as many two and a half inch deals as will reach round the bafe of the caiffon by the fum of their breadth, and whofe length is one foot more than the common depth of the water in the place where it is to be ufed, will make the covering of the fides; and fo much Riga balk as reduced into three inch planks when laid crofs and crofs, fo as to form a folid bottom of fix inches thick, about one foot on every fide larger than the bafe of the pier, will form the bottom of the caiffon: other timber for braces, &c. will be wanted.

I am

I am much furprifed at your account of the water rifing three feet perpendicular on the weft fide of the bridge, more than on the eaft in time of floods, and think you muft have made fome miftake; but if the cafe be as you ftate, it will be highly neceffary to take away the ground, and make a clear paffage through the firft arch, as well as to remove all the gravel thrown out of the foundation, and every other impediment to the free paffage of the water: and alfo carefully to examine from time to time whether the rubble at the foot of the piled piers keeps its place without derangement; if not, timely to renew the fame with heavier ftones, or by extending the quantity. But I do not apprehend this need be done till you find fome call for it.

Your's, &c.
J. SMEATON.

From Mr. Pringle.                                                    Lees, November 22, 1766.

The Truftees for Coldftream Bridge hearing that you have infpected it very narrowly, beg you will be fo good as to let them have your opinion of it in writing; particularly if you are perfectly fatisfied about the fufficiency of the work, and if you think it has been carried on with proper expedition and economy.

Do you approve of laying a gangway on each fide for foot paffengers? And is there gravel enough laid on for this winter, to protect the crown of the arches from the weight of any carriage whatever?

Your's, &c.
J. PRINGLE.

To Mr. Pringle.
    Dear Sir,
    Your favour of the 22d came to me at a time when I was fo much engaged (and which has indeed been the cafe ever fince I had the pleafure of feeing you at Coldftream), that I could not poffibly give you a diftinct anfwer to your queftions concerning Coldftream Bridge fooner than the prefent; for of all things I cannot make time.

1ft. As to the fufficiency of the work? In anfwer to this I can fay with great pleafure and truth that I have not feen better, and in this refpect I fhould have thought myfelf very happy if every defign of mine had received equal juftice in the execution; nor can I apprehend that any material part will be fubject to failure in a very long courfe of years, provided that a little attention be paid to the two piers, that are founded on piles; to fee that the furrounding rubble ftones lie in their places; or if removed by the great rapidity of floods, to be replaced by others as foon as may be; it will be neceffary to

attend

attend to this point for fome years, though it is what moft probably may not ever happen; there being all the fecurity in the cafe that can be.

2dly. As to the length of time? According to the account I have received, it was begun the 7th July, 1763, and opened for all kinds of carriages the 28th October, 1766. A fpace of time comprehending three years three months and three weeks. This bridge confifts of five principal arches, which with the four intermediate piers extend $353\frac{1}{2}$ feet between the abutments; befides two abutment arches of 20 feet each; and including abutments, extends 568 feet in the whole, being in breadth, including parapets, 25 feet. I doubt not but a much greater quantity of ftones may be put together in the fame time upon the dry land, or in a fituation not expofed to fo many hazards: but it is confidered that the River Tweed fwells by rains and fnows frequently 14 feet perpendicular, and comes down with great rapidity; fo that the work could not be purfued with any tolerable profpect of fafety more than fix months in the year, and thofe frequently interrupted by fudden land floods as aforefaid; thefe things confidered, I do not know whether there is any example of fo great a work attended with the fame difficulties that has been done in fo fmall a time.

3dly. Refpecting the economy that has been obferved in the conftruction thereof, I think you informed me that the neat expenfe of building this bridge did not or would not when fully completed exceed 6000l. Things of this kind are beft judged of by comparifon; we have feen inftances of bridges as long, as broad, and as high-built for a lefs fum; but either they have been built in a more gentle river, or their piers have not been laid fo deep below the low water furface, or below the bed of the river; or they have been laid upon a more uniform ftratum of matter, or lefs fcrupuloufly fecured, or they have been built with fmaller ftones, or lefs exactly put together; fo that taking in the whole I believe there has been no example of any bridge being built for fo fmall a fum where circumftances have been equal or fimilar: and it is much to the credit of Mr. Reid, the furveyor of this work, that every thing that has been tried has fucceeded; and the whole has been done without lofs of life or limb to any one concerned: nor does there appear any fettlement, falfe bearing, or fiffure in the whole work, fave a very minute one but juft difcoverable in the common fpandrell walling above one of the piers, or one fide, and in the joining of the work of two different years, and this wholly occafioned by laying a weight of gravel in the fpandrells of the arches that reft on this pier when none could be laid in an adjacent fpandrell to balance it; the whole not worthy of mention, but to fatisfy thofe who may have thought it of more importance than it really is.

So much as to the fufficiency of the work and the conduct thereof : you next enquire, Whether there is a fufficiency of gravel to protect the crown of the arches during this winter from the weight of any carriages whatever? I am told there is 18 inches thick above the crowns of the arches, and if fo it is very fufficient; if it fhould happen to cut deep, which is not to be expected, the rutts fhould be occafionally filled : and I apprehend it would do very well if never filled higher; but for greater fecurity it will be proper to make an addition as hereafter mentioned.

Laftly, You afk me if I approve of laying a gangway on each fide for foot paffengers? I think it will be very proper to cover the fpandrell walls with tolerably thick fcapelled flags or paving ftones, which will not only be a caping for the fpandrell walls, and thereby prevent the wet from foaking down into them, but at the fame time make a walking path on each fide, and as there will ftill be 16 or 17 feet clear, this will be very fufficient for the meeting of carriages : the gravel then may be made up within nine inches of the furface of the walking paths, and laid a little rounding in the middle like a turnpike road. N. B. It will be proper to lay a ribband of ftone nine inches thick next the road, to prevent the grazing of the wheels, or elfe to make the flagging nine inches thick, efpecially on the outward fide next the gravel. This I think, Sir, anfwers all your queries.

Your's, &c.

J. SMEATON.

To Lord Home.

My Lord, Aufthorpe, 5th September 1770.

The very day your letter is dated I was viewing the bridge at Kelfo, and in my journey to Woolerhaughhead had paffed within fight of Coldftream Bridge, and where I fhould with great readinefs have called had I had intimation that any thing had been amifs with it. In my way north I paffed over it in the beginning of July, but as I heard nothing I did not enter into any particular examination ; but obferving the ftones lying as ufual upon the falient points up ftream, I took it for granted all was right. It was a requeft I many times repeated to my late worthy friend Mr. Pringle, and to Mr. Reid, that great care fhould be taken for fome years to maintain the rough ftone work round the two piers founded on piles, and from time to time to repair the fmalleft derangement till the whole fhould become fettled : and I am forry to fee that this, as well as many other public works, has been fo cramped in the finifhing, that many effential things are left incomplete ; and that it is not left worth the while of the perfon who has had the principal hand in the execution, and who muft

4 be

be beſt able to ſee and judge, and apply timely remedies, to intereſt himſelf therein.

Something of the kind now wanted is always neceſſary to be applied in gravel bottoms, and the caſe is noways difficult if done with judgment : and no better method can be taken than what you have already done, that is to throw in large rough ſtones, ſo as to form a ſlope up againſt the baſes of the piers, and when this is done, to throw a conſiderable quantity of gravel upon them, or ſmall rubble, in order to waſh into and fill up the chinks of the larger ſtones. The leſs this outwork is extended, beyond what is neceſſary to ſecure it a ſufficient natural baſe, the better, as all obſtruction to the free courſe of the river through the arches, is hurtful. It is poſſible that as the work will not be all of a piece, ſome of the ſtones may be deranged by the action of the flood, but thoſe being made good again ſeldom fail a ſecond time, unleſs it be occaſioned by a failure in ſome other place. All piles and wood-work are to be avoided, as the failure of the wood proves ſooner or later a failure in the ſtones. The piles and boards there are the remains of the coffer dams, which it was then thought not prudent totally to eradicate, for fear of diſturbing the gravel; but now that the rapidity of the flood has, I apprehend, taken away the ſurrounding gravel, and let down the wood-work, and thereby ſapped and taken away the gravel from under the ſtones, the latter muſt neceſſarily go down; let therefore the whole be made up with rough ſtones, and as they can no more be ſapped by the running away of the gravel from underneath them, you will find they will lie, and this the better, if thoſe which can be laid by hand are ſo diſpoſed as to lean down ſtream, as I formerly directed ; and if it be found, that after a great flood the ſtuff to be laid upon the larger has diſappeared, partly by waſhing into the interſtices, and partly by waſhing away, it muſt be renewed till the interſtices appear filled. It perhaps may be a queſtion, as I expect it is only ſome part of the ſurrounding works that has ſuffered, whether thoſe which have not failed ſhould remain till they do, and then be made up in the ſame manner as before directed, or whether the piles ſhould be now drawn, and the whole made up at once ? But this is a matter which, without inſpection, or the Report of Mr. Reid, ſetting forth the condition thereof, I am not enabled to décide upon. But this I am clear in, that if this buſineſs be carefully and properly performed, after all dependence on the wood-work is gone, and ſuch derangements as may happen in the courſe of two or three years repaired, till the whole comes to a bearing, this part of the work will laſt for ages, and in caſe of any future misfortune can always be made good in the ſame way. I beg leave to obſerve that round ſtones will not do ; rough ſtones from the quarry are beſt, the rougher they are the better.

In

In regard to forming a dam below the bridge, it is in this place totally unneceſſary, as the piles on which the piers reſt are conſiderably below the water's ſurface in the loweſt ſtate of the river, the work mentioned would be ſtill neceſſary, and a dam would coſt as much as would do the work in the way I mentioned four times over.

I am, &c.

J. SMEATON.

To the ſame.

My Lord,                                                         Auſthorpe, 16th Auguſt 1781.

I have been informed by a friend who paſſed over Coldſtream Bridge laſt month, when the water of the Tweed was remarkably low, that he obſerved the rubble laid for the defence of the piers to be much ruffled, and a conſiderable part taken away, and left in the Tweed about forty or fifty yards below the bridge, and that in his opinion the rubble that had been laid, was too ſmall and thin.

On this occaſion, if any thing be neceſſary to be added to my letter of the 5th September 1770, it is, that I would recommend the rubble ſlope againſt the piers to extend thirty feet up ſtream above the ſalient points; and if a bank or ſhoal of rubble were made about thirty or forty yards down ſtream, the place where the preſent rubble is carried to, it would greatly tend to keep the defences about the piers in their place: and as this bank or ſhoal need not be raiſed ſo high as one foot below the ſurface of the water, in the ordinary ſtate of the river, there can be no objection on account of the ſalmon fiſheries; it ſhould be a competent breadth at the top to reſiſt the action of the floods in paſſing over it.

From what has happened, I am apt to think with my friend, that the rubble made uſe of has been too ſmall. As he is not a man of my profeſſion he may be miſtaken, yet as he is of a general good judgment of things, I could not omit giving you this intimation for the public ſervice, as Your Lordſhip had formerly honoured me with your correſpondence on this ſubject. It was the original intention to build all the piers upon the rock; but Mr. Pringle finding the expenſe of pumping run very high, begged that the two piers in queſtion might, if poſſible, be founded on the ſtratum of gravel which laid above the rock; which having been gradually worn away by the action of the floods, has occaſioned the derangements that have ſince happened to the defences; which, however, being carefully attended to and kept up, the piers will reſt with equal ſolidity upon the central matter as at firſt.

I am, &c.

J. SMEATON.

REPORT of ROBERT REID on the State of Coldſtream Bridge.

To the Honourable the Truſtees, appointed by Act of Parliament, for the Roads and Bridges in Berwickſhire.

IT evidently appears from experience, that the ſecuring of the foundation of the piers of Coldſtream Bridge is an annual expenſe, and notwithſtanding all the vigilance and care of the truſtees living near it, the expenſe is ſtill multiplying. The reaſon is obvious; a great many of the guard piles are gone that formerly ſecured the rubble ſtones which were laid about the foundation of the end piers. And immediately after the great flood in the year 1782, they were obliged to lay a conſiderable quantity of large ſtones, which exceeded the limit allotted to them, which was ten feet from the piers which theſe guard piles pointed out. This being the caſe, the water is ſomewhat confined in its paſſage, and when it begins to riſe it runs with ſuch rapidity that it has thrown out the gravel from between thoſe two piers founded upon wood, in great quantities, eaſtward from the tails of the piers. In ſome parts it has gone the depth of four feet, at other parts eight feet, and five feet; and if it ſhould continue to work at that rate, it will bring out the gravel from between the piers, and leave theſe two piers ſtanding upon ſtilts; and may be attended with bad conſequences.

The queſtion will now ſtand, how this is to be remedied? This I have pointed out to be by a dam-head, with a row of piles through the river, and a quantity of ſtones on each ſide of them, about 100 feet or thereabouts below the bridge; and whatever gravel ſtones, or other ſtuff the river brings, it will leave it there, and ſo go on till it has raiſed it as high as the dam-head; and in place of taking away any thing from the foundation, will be conſtantly laying to it; which muſt ſecure your foundation as long as the dam-head ſtands, the piles being in the middle of it, will ſecure it a long time.

Perhaps ſome objections may be made about the ſalmon fiſhing: to obviate theſe, there will be ſome openings left at proper diſtances for the paſſage of the fiſh.

Now this method I am perfectly ſure of being effectual, as I helped to put one of theſe dam-heads into execution, at a bridge built by the Government over the river Tay, in the year 1735, and has been the ſafety of that bridge ever ſince. Before the maſon work was perfectly finiſhed, the river was making whirlpools at the tail of the pier; and after

observing

obferving that, all hands went to work to make the dam-head, which remains there to this day.

Another thing would require attention: as the fouth arch is for the moft part of the year dry: formerly, when converfing with Mr. Smeaton about this bridge, he faid it would be of great advantage that the water were brought through this arch, that it might give relief to the reft; and at the breaking of a froft there is commonly a mountain of ice lying in this arch, for want of water to carry it off.

ROBT. REID.

---

OPINION and DIRECTIONS of JOHN SMEATON, Engineer, for the Prefervation of Coldftream Bridge, upon the Condition thereof, as ftated in the Plan and Report of Mr. Robert Reid, dated October 19th 1784; tranfmitted to him in the Letter of Mr. James Gray of 3d May 1785.

To the Honourable the Truftees, appointed by Act of Parliament, for the Roads and Bridges in Berwickfhire.

Sirs,

I obferved, in my anfwer to the letter of my Lord Home in the year 1770, upon a derangement of a fimilar nature to the prefent, that I had many times repeated to my late worthy friend Mr. Pringle, of Lees, and alfo Mr. Reid the executive mafon, that great care fhould be taken for fome years to maintain the rough ftone work round the two piers founded on piles; and from time to time to repair the fmalleft derangement, till the whole fhould become fettled; that fomething of the kind then wanted, is always neceffary to be applied in gravel bottoms; and that the cafe is no ways difficult, if done with judgment.

Since the year 1770, much greater and more violent floods have happened in all the northern rivers, than had been before pointed out to my experience: notwithftanding which, the fame kind of directions having been punctually and carefully attended to at the bridge at Perth, no derangement has happened to produce any alarm.

In the year 1770, the fpeats of the Tweed had then in fome meafure deftroyed the pile fences, originally driven to form the coffer dams for draining off the water while the piers were founding; and which it was thought improper to draw, for fear of

disturbing

disturbing the surrounding gravel; the interior space being filled with rubble stones, to defend the gravel immediately surrounding the piers within the said fences; but those giving way, let down the stones contained within them. As it thus appeared that the pile work was not defensible by its own strength and stability, but that if reinstated, must be externally defended by the deposition of stones, I thought proper to advise, that the whole should be made up with rough stones; observing at the same time, that it was a matter of question which I could not decide without inspection, as the whole of the surrounding works might not have suffered together, whether those which had not failed should remain till they should fail, and then be made up in the same manner as those already directed; or that the remaining piles should be drawn, and the whole made up at once.

It appears however from the present report, that by pursuing the methods then directed, the immediate foundations of the bridge have been so far preserved entire: but as the stones deposited for the safety of the foundation have hereby enlarged their base and slope, and in consequence somewhat straitened the water-way; and proving too stubborn to be removed by the rapidity of the stream, its action has been exerted in removing the gravel between the piers remaining undefended; and has pooled the holes, as particularly set forth and described in that report.

For my own part, I generally find it right to apply the strengthening plaister immediately to the weak part, and would begin the work by filling up every pooled part, even with those parts that remain of four feet deep: and in consideration of the increased rapidity of the floods within the last 15 years, I would advise all these fillings to be performed with stones of a larger size and weight than before directed; (that is to say) with stones from a ton to half a ton weight, and to chock in the interspaces with lesser stones, of such sizes as to make the whole a tolerably regular surface, extending as far as the pools exceed the depth of four feet; and then by way of further security, I would begin an artificial shoal to act in the nature of a dam, by laying a row of large stones across the river, at the distance of 100 feet below the bridge; and then depositing other stones above this, bringing the work forwards towards the bridge, and a little increasing in height, to the breath of 30 feet; and so as to pen the water $1\frac{1}{2}$ or 2 feet deeper under the bridge at low still water, than it now is; and then to slope it upwards to the further breadth of 20 feet; making 50 feet in the whole breadth: this will make so easy a slope, that the salmon will make their way over it without interruption. I would make the lowest part of it, for the water's ordinary passage in dry times, in the place where it now is; but so near upon a level, that in the smallest speat it will flow over its whole length.

It

It appears on this occafion, when the work of the dam or fhoal is eftablifhed, as it will interfect the deep part of the river, that it will be proper to furround the pier that ftands upon the rock, with a ftratum of large ftones chocked in with fmaller, fo as to form a bed or ftratum of about two feet thick; and to a diftance of fix or feven feet from the bafe of the pier, where the rock is bare on the north fide; and till it meets the ftone of the rubble for defending the north fide of the fecond pier: but at the down ftream end to be continued till it meets the foot of the flope of the dam; by which means the aforefaid bed will be fupported.

The ufe of the depofition of this bed round the pier upon the rock, is to preferve the floor of timber of fix inches thick, that lies between the rock and the bafe of the ftone work of the pier; for though it will not rot, yet the rapidity of the water running by, and acting upon its outward border, will doubtlefs in procefsof time corrode it, and may undermine the outfide, fo as to caufe the outer to feparate from the interior core. I do not imagine that any effential harm can as yet have happened from this caufe; but it feems now time to fet about fuch an expedient, to prevent future mifchief.

When the work is brought to this fituation, it will be proper to make the trial of another winter, in the courfe of which I would expect that part of the prefent bottom that does not now lie above four feet deep, being undefended, will be further pooled, and ought then as foon as perceived to be filled up even with the other, with heavy matter of the fame kind; in confequence of·which the whole will become capable of refifting the action of the water; and when by attention the whole of the foft places are made good, then it may be expected the bottom will remain untouched for ages. I cannot fay I would advife the ultimate performance in this manner juft now, becaufe it will coft more to work away the gravel under water to a fufficient depth, than it will to fill the cavity when made; and if a ftratum of fufficient weight and folidity were depofited over the prefent, it would too much obftruct the water-way.

In regard to the erection of a dam-head, without filling up the pools, though it is very poffible the river may do it, yet it is very poffible that it may not: and though this may be the cafe at Tay Bridge, this is but a fingle inftance; and the operation of rivers is fo very various according to fituation, and I have feen fo many dam-heads where the water has not filled them with gravel, that I cannot recommend it to depend upon this alone: whereas by filling the extra depth of the pools, and forming a fhoal or dam of a lower kind, laying firft a row of the largeft ftones by way of footing to the reft, no piles will be wanted, which are both expenfive and tedious in the execution. An application of this fame kind I recommended at Kelfo Bridge, where notwithftanding a dam that had for years fubfifted below, the piers were become fo pooled and undermined, as to put the

bridge

bridge in imminent danger : befides, in this mode of working, after the ftones are won from the quarry, there will not be the need of a mafon's tool to be laid upon them.

Furthermore, I apprehend it will be of utility and very proper that the fouth arch fhould be deepened to the fame level as the pool when filled as already directed ; and when this frefh furface pools further, to be treated in the fame manner as the reft.

J. SMEATON.

*Grays Inn,*
16th *May* 1785.

---

AT a Meeting of a Committee of the Truftees for the Berwickfhire Turnpike Roads and Coldftream Bridge.

Coldftream, 9th Auguft 1790.

Prefent — Eight Truftees, and Lord Swinton, Præfes.

*Inter alia,* The following Memorial concerning the Bridge over the Tweed, near Cold-ftream, was laid before the Meeting :

About fix years ago the river Tweed being very rapid, formed feveral pools or weels round the pillars or piers of the bridge, particularly the middle piers, which are founded upon wood, whereby it was thought the bridge was in confiderable danger.

That thereupon Mr. Smeaton was confulted, and gave in a very full advice to the truf-tees, which was, in general, to build a dam-head or dike acrofs the river, about feventy feet below the bridge, and of a height about two feet above low water mark.

That this was done accordingly ; but fome time after it was made, upon a fuggeftion of fome of the truftees that if it was continued in that manner it would be an obftruc-tion to the falmon getting up the river, a cut was made in the middle of the dam-dike, about feven feet wide, by way of making a free paffage for the falmon.

Mr. Reid, mafon, who is employed to take charge of the bridge, ftates that the gufh of water paffing through this opening makes a hollow in the channel of the river, both above and below the dam-dike, and particularly it fweeps away much of the gravel which the river brings down from above the bridge, and which would otherwife remain above the dike, fecure its permanency, and prevent the blowing round the piers.

Mr. Reid farther informs the truſtees, that in caſe a paſſage through the dike for the ſalmon coming up the river was neceſſary, it might perhaps be ſafe to fill up the preſent cut in the middle of the dam, and in place of it to make another opening at the ſouth end of the dam-dike, and oppoſite the ſouth arch, which is founded upon rock, and where the river runs upon or near the rock.

The truſtees having conſidered the above repreſentation from Mr. Reid, and having themſelves viſited the river, which happened then to be ſpeat or flood, ſo that they could not ſee the effects of this guſh of water either above or below the dam, they had various opinions concerning this matter; ſome thinking that there was no need of any opening at all for the paſſage of the fiſh, others thinking that the fiſh would be ſtopt except in ſpeats; but all agreed in this, that the dam-dike had been of the greateſt advantage to the bridge, and that the preſervation of the dam was the ſalvation of the bridge; it was fit to conſult Mr. Smeaton what was neceſſary to be done in this matter, viz. whether, in his opinion, any opening was neceſſary for permitting the fiſh to paſs up the river, and whether there was any danger from ſuch an opening, and whether, if there was any neceſſity for it, it ought rather to be on the ſouth ſide than in the middle? and that he would be pleaſed to communicate his advice and particular directions on the whole matter.

And in order that Mr. Smeaton may be the better enabled to give an opinion, they direct ſoundings to be made by Mr. Reid and Mr. Sharp, ſurveyor, in Coldſtream, of the depth of the water both above and below the dam, and particularly in the channel of the opening; and to be ſent to Mr. Smeaton along with the memorial. Extracted from the minutes of the truſtees book.

JOHN TURNBULL, Clerk.

REPORT of JOHN SMEATON, Civil Engineer, upon the State and Condition of the Bridge of Coldſtream, as ſtated in the Memorial to the Truſtees of the Berwickſhire Turnpike Roads, and Coldſtream Bridge; the 9th Auguſt 1790.

When I had the honour of delivering my opinion upon the proper remedy for curing the defects then already felt, and further apprehended, in the foundation of Coldſtream

Bridge,

Bridge, about fix years ago; wherein I advifed a dam-dike to be built acrofs the river about 70 feet below the bridge, and about two feet high at low water; I ftated this height, in view that it fhould be no obftruction to the falmon getting up the river; having feen feveral inftances of even upright dams of four feet high and upward, in falmon rivers, and which were not confidered as an obftruction to the paffage of the fifh; as they are frequently feen to leap to a greater height than that. I could not therefore conceive that at two feet high, upon a *floping* current, the leaft objection could arife, or be thought of. I am very happy to find that the expedient propofed and executed, has fo effectually anfwered the intention.

Had I been confulted concerning the expediency or manner of the cut fuggefted to be made through the body of the dam, I fhould certainly have delivered it as my opinion, that none was in reality neceffary: but if to fatisfy the apprehenfions of thofe who thought themfelves interefted in the fifhery, the truftees fhould think fomething neceffary, I fhould never have advifed the execution to have been in the way it is fuggefted by this memorial to have been done; viz. *a cut of* SEVEN *feet wide in the* MIDDLE *of the dam-dike.*

I think it very lucky that this cut, being in the middle of the river, and confequently where in fpeats the principal current muft be, has not by degrees grown wider and deeper, fo as have deftroyed the dam; and thereby have brought the bridge itfelf into that kind of danger which this work has been intended to guard againft. I therefore recommend that the cut be filled up and made good, as at firft; and in cafe, to fave objections, an opening muft be fomewhere, it will, as Mr. Reid very properly fuggefts, be much the fafeft on the fouth fide of the river, where the river runs the fhalloweft, and upon a rock. Indeed, I fee no objection to remove the dam-dike fo far entirely that the water may flow freely and without obftruction through the whole of the firft arch of the bridge from the fouth; and oppofite to the firft pier begin the dam-dike, to rife gradually fo as to come to its prefent height oppofite to the fecond pier from the fouth. Indeed this flope will be formed in reality by the fubtraction of the prefent matter, but yet it muft be taken care to be caped with fuch new matter or weighty ftones as are likely to lie and refift the current in fpeats.

J. SMEATON.

*Gray's Inn,*
*26th November* 1790.

# NEWCASTLE BRIDGE.

REPORT of John Smeaton, Engineer, concerning the State of that Part of Tyne Bridge, between Newcastle and Gateshead, which is in the County of Durham.

HAVING carefully inspected the south part of Tyne Bridge the 16th of September last at low water, I found it in a general state of disrepair: but as it has been originally ill built, I look upon it as impossible, after standing so many years, to render it perfectly sound, unless the whole was new built, which is not the present proposition; yet by occasional repairs seasonably applied, it may last for many years. I shall therefore take the arches in order, and confine myself to the pointing out of such things as more immediately call for assistance.

The first arch beginning from the south side, is in a great measure blocked up by cellars for convenience of the houses above, and has no current of water through it, when the low water is below the sterlings or jetees, as they are called, which surround all the piers in the manner of London Bridge: this arch seems at present to want no material repair.

The second arch has a passage between the jetees at low water. The aislering of the piers on both sides this arch wants repairs, many of them being loose, and some of them dropped out: the aislering on the north side appears worse than it really is, having been built originally bulging, at least so it seems to me.

The whole or greatest part of the arches of this bridge have been lined with ribs, as was customary formerly, with a view to strengthen them, but it so happens that a great many of those ribs have separated themselves from the arches that they originally were in contact with, and have tumbled down, one of the ribs now remaining in this arch; viz. that on the up stream or west side, is so far separated from the arch, and is in imminent danger of falling, that to prevent mischief to any that may be under it when it happens to fall, it will be proper to take it down: I do not apprehend it to be any ways necessary to rebuild it, because I cannot suppose it has ever been of any real use.

In

In the middle of this arch the ſtone work is entirely perforated by an area of about four yards by ſix; and as the bridge has been ſo conſtructed at firſt, this area had once been covered by a draw-bridge by way of defence, being ſo placed that, if open, the paſſage over the bridge, as it now is, between the houſes would have been ſtopped thereby. This area is now floored with timber covered with earth, and paved at the top like the reſt of the bridge, ſo that when carriages go over this part of it, the vibration of the timber makes it appear to ſhake. The main timbers are pretty ſtrong, but the whole has been very roughly executed, and has all the appearance of a job done in a great hurry. It ſeems alſo to have had ſome repairs occaſioned by the rotting of the ends of the great beams which have been ſupported by piers put under them; ſome of the ſmall wood that is ſupported by the greater appears to be decayed; but while ſo ſupported nothing of great conſequence can happen. In fact, as I don't find the ſtate of this flooring ſenſibly different from what it was when I viewed it in the year 1765, for that reaſon it may be ſuppoſed poſſible to continue for a number of years to come; but as it is a piece of work ſo put together, that one cannot anſwer for it, a failure may happen when it is leaſt expected; and as the lives of men depend upon it, and it is in a viſible ſtate of decay, it appears to me that it ought to be repaired; and as it is very probable it may never be wanted again to ſerve the original intention, while it is doing I would recommend this area to be arched with ſtone, and as the centre may be erected underneath, and every thing prepared for turning the arch before any thing is diſturbed upon the top, I apprehend all may be with eaſe completed in three days ſtoppage.

The next arch has loſt all its ribs, yet ſhows no ſigns of infirmity, except that as the pen ſtones are in a double layer, compoſing an interior and an exterior arch, the former is a little ſeparated from the latter, on the down ſtream ſide of the ſouth haunch. Some repairs are wanted in the ſetting of the jetees of this arch, as alſo more or leſs in all the reſt.

The fourth arch from the ſouth ſide, or ſecond from the draw-bridge arch, is called Keelman's arch: it has originally had five ribs underneath it, of which there is only one remaining; but it ſhews no loſs by the want of them, the up ſtream ſhoulder of the pier on the ſouth ſide of this arch, wants repairs; and together with the reſt a number of ſmall articles which it would be uſeleſs as well as tedious to mention.

As the whole of the repair is a kind of jobbing work, there is no ground upon which to form an eſtimate of the expenſe; for when part of an old edifice is pulled down in order to be repaired, it often diſcovers ſomething unforeſeen, of which a repair is equally neceſſary

ceſſary; for this reaſon (except the arching of the draw-bridge area) it cannot well be done by contract, becauſe a contractor will not do more than originally appeared, and thereby the ſore may be left unbottomed; and if done by day-work, the expenſe will depend greatly upon the honeſty and addreſs of the workmen. But I ſhould imagine the whole, ſtone-arching in the draw-bridge included, may be done as well as the ſtate of the bridge will admit of, for 150l., or at moſt, 200l.

J. SMEATON.

*Auſthorpe,*
*October* 18th 1769

N.B. This part of the bridge belongs to the ſee of Durham. It fell down in a great flood in 17 ; and was rebuilt in its preſent form by Mr. Mylne.

---

REPORT of JOHN SMEATON, Engineer, upon the State of that Part of Tyne Bridge, belonging to the Town of Newcaſtle.

HAVING carefully viewed the north part of Tyne Bridge, the 30th of April 1771, being attended by Mr. Craiſter and Mr. Gunn, I am of opinion as follows.

That the whole of this bridge is founded upon piles ſawn off above low water mark, in the manner of London Bridge, and in the ſame manner, each congeries of piles for ſupporting the pillars, is ſurrounded with a work of defence called ſterlings or jetees, which, reaching from two to three feet above water, each probably one or two feet above the heads of the bearing piles, thereby defend them from decay, and from being ſapped at the bottom by the immediate action of the water upon them. Thoſe jetees, however, being chiefly conſtructed of timber, and being immediately expoſed to the action of the water, are always in a periſhing ſtate, and therefore themſelves gradually come under a courſe of repair, though they ſeem pretty effectually to have defended the bearing piles for ſupporting the ſuperſtructure, as there does not appear to me to be any ſhrink of the foundations which at preſent can be noticed. The jetees from the middle pier, that divides the counties, to the north ſhore, appear to be at preſent in tolerable order; ſome few ſtones are wanting from the upper ſurface, but which will be eaſily replaced in courſe of a repair.

The whole of the ſuperſtructure has more ſenſibly felt the effects of time, having been originally very ill built, and in general of too ſmall ſtones, and not of the beſt kind; this

4

added

added to the great wear and tear occafioned by the frequent running of the keels againft the piers and hips of the arches, has produced derangements that can never be effectually mended, becaufe the new work, which from its nature muft be fuperficial, can never be firmly united to the old, and therefore, thofe repairs cannot be of long duration, as they are eafily difplaced by frefh ftrokes, being chiefly retained by cramps of iron: from thefe caufes almoft the whole of the aifler facing of the piers is loofened from the core of rubble work within, and which appears alfo to have been the cafe for a courfe of years back; the arches refting upon ftones which are very much decayed, and which cannot be replaced without danger of bringing down the arches they fupport; this renders the whole fuperftructure in fo ill circumftanced a fituation that nothing lafting can de done at it.

It is not eafy to fix the duration of a piece of ftone work, as we daily fee inftances of old buildings hanging together in a furprifing manner: and as the infirmities that I have pointed out are not of very recent date, nor any thing happened that is particularly alarming, I fhould expect that with conftant attention to repair its defects as they gradually arife, this bridge may be fupported many years longer; however, as it is not poffible to be affured how long it may laft, its apparent condition is fuch as would readily induce me to recommend the thoughts of a new one, if I could fee a poffibility of advifing any thing to which there are not very manifeft objections.

The moft defirable fituation feems to be a bridge below the prefent one, as being moft central to the town; but the depth of water there is fo great as to make fuch an undertaking exceedingly expenfive.

The moft eafy, leaft expenfive, and what could be done in far the fhorteft time, would be to build a new fuperftructure upon the old foundation; but as upon this plan it could not well be made above four feet wider than the prefent breadth where wideft, and as there would be a neceffity of continuing the prefent jetees, I apprehend this meafure would ftand univerfally condemned.

To put down new piers in the prefent fituation, upon new foundations, fo as to afterwards take away the jetees, would greatly prolong the time, and add to the expenfe and difficulty; efpecially as the places of the new piers would be obliged to conform to the old ones, and thereby increafe the number of piers that would be neceffary or eligible in an entire new ftructure.

To

To build a bridge above the prefent one, would be done at lefs expenfe than in any cafe, if an entire new one, but would be lefs central than the prefent.

I have endeavoured to throw together thefe general thoughts, that it may be clearly feen how the matter hinges. With refpect to the current repairs that are now neceffary, and of which it is not poffible to make an eftimate, being fuch as have ufually happened; it is, however, neceffary to fay fomething refpecting the rebuilding the land-breaft and firft arch from the north fide.

The north arch is certainly in the moft ragged condition of any; but being a fmall arch is in lefs danger of falling. The immediate rebuilding thereof depends upon another matter. If a building is to be erected upon the key-wall, connected with the land-breaft, the land-breaft wall is fo bad above low water, as not to be fafe to make any building intended to be durable upon it. The key-wall cannot properly be rebuilt, without doing that part alfo whereon the laft arch of the bridge refts, being connected together. That part upon which the arch refts, cannot be done without taking down the arch, and when down it will not be proper to rebuild the arch without new doing the face of the pier, whereon its other haunch is to reft. The neglect of which, in rebuilding two or three other arches contiguous, which appear to have been done not very many years ago, will occafion their not being of longer duration than the old ones. In fine, if there is no immediate neceffity of building upon the key-wall, and if there was any profpect of a new bridge, then the rebuilding of the faid arch may be deferred; but if the bridge is to be fupported as long as it can be, then I advife at all adventures to build up the wall and arch in the moft effectual manner.

The great arch, which is next the divifion of the counties, and the fixth from the north fhore, as alfo the arch next it north, commonly called the white arch, or fifth, appears to have been original. Thofe arches are built with ribs, which are now fo much fettled as to bear very little weight: two out of five in the great arch, and three out of five in the white arch are tumbled down, or have been driven down by the keels running againft them; at prefent the north end of the lower rib of the white arch has by a ftroke from a keel been driven fideways feveral inches out of its place; it is poffible it may fo hang for years; but as it is plain from this that it fupports nothing, and as another ftroke in the fame way by another keel will infallibly bring it down, and by the fall thereof lives may be loft, it feems therefore advifeable to take it down.

J. SMEATON.

*Newcaftle,*
6th *May* 1771.

P. S. When

P. S. When the north arch is built, it feems practicable to make a foot paffage on to the quay, which, as it will greatly eafe the prefent narrow avenue to the bridge, will deferve confideration *.

---

## MINUTES of a View of Newcaftle Bridge, 30th April 1771, by JOHN SMEATON, Engineer.

### Great Arch or 6th.

The great arch, being the fixth from the north fide, beginning at the middle of the pier on the fouth fpringer of the faid arch where the counties divide.

The down ftream angle of the faid pier between the town and the Bifhop of Durham is in bad repair, and the face of the aifler wants new pointing, and filling up to high water mark, which is in general the cafe with the whole bridge.

When any of the falient angles are new done, I would advife their being rounded by a fweep of about 18 inches radius, by which means they will be lefs liable to be broken and diflocated by the keels, and that all the mortar ufed within the tides mark be made with pozzolana rather than terras.

The great arch has been originally built with five ribs, of which the two outfide ones are gone, the penftones making the cover being put on joint and joint.

In the down ftream courfe of penftones, a bad ftone near the fouth fpringer ought to be in part fcreeded off, and fupported by a frefh ingrafted ftone. Near the north fpringer of the great arch both on the up ftream and down ftream fides, the joints in the penftones are become open ; thofe moft perifhed to be wedged and pointed.

### Pier North of the Great Arch, or 5th from the North Shore.

Several cramps are loofe, and ftones out of the face, which fhould be repaired and pointed as before mentioned ; in the jetee north fide of the pier, fome ftones are out and want repair.

---

* Nothing was done, the bridge tumbled down in the greater part, and this portion thereof belonging to Newcaftle, was rebuilt by Mr. Wooler, military engineer.

### White Arch, or 5th.

This arch was alfo originally built with five ribs, of which one up ftream and two down-ftream are gone, the penftones near the fouth fpringer on the down ftream fide are fharped off, and fhould be frefh abutted. The north fpringer of the down ftream rib appears from the blow of a keel to have been driven feveral inches fideways out of its place, and as from hence it appears to fupport nothing, and may be driven down and do damage by fuch another accident, it feems beft to take it down, nor can any new rib be erected, fo as to anfwer any good end. The joints of the penftones towards the north fpringer are perifhed and want wedging.

### Pier North of the White Arch, or 4th.

The outfide ftones in a general ftate of decay, but no particular failure; like the reft it wants new pointing.

### The Tower Arch, or 4th.

This arch feems to have been new done fince the original, but appears to be the firft of thofe that have been rebuilt; it is without ribs, and not very fubftantial.

The penftones at the fouth fpringer on the up ftream fide are broken, but may be eafily repaired. The penftones at the north fpringer both up and down ftream are a good deal galled by the keels.

### The Tower Pier, or 3rd.

The aifler on both fides that fupport the arches are a good deal wafted; the down ftream point of the jetee has been damaged and wants repair.

### The Arch North of the Tower Pier, or 3rd Arch.

This arch has been rebuilt without ribs in man's memory, is in tolerable good order, but is fprung from bad bearings.

### The Pillar North of the Tower Pier, or 2nd Pillar.

The up ftream point wants repair, and feveral ftones are out of the faces, which need replacing.

### The firft Pillar from the North Shore.

This pillar feems in general in as good order as any, but its aiflering depends altogether on cramps, and being out of the common road of the keels is not fo apt to fuffer thereby.

The

### The firſt Arch next the North Shore.

This arch is at preſent in a very ſhattered condition, though part of it ſeems to have been rebuilt within the compaſs of a few years, the oldeſt part of it does not appear to have been original, but when rebuilt has been haſtily done with bad workmanſhip and materials; the new part which is on the up ſtream ſide, has been very ill abutted, for it reſts upon a baſe of about 18 inches broad at the north abutment, and is gained on upon the old work till it comes to the breadth of five or ſix feet.

The abutment of the north arch is upon a part of a quay wall, extending both above and below the bridge, the whole of which above low water is in a ſhattered condition, and ſeems very improper to build upon without firſt rebuilding the wall itſelf, at the ſame time taking down the north arch, and rebuilding both at the ſame time.

When this is done, I adviſe to clear away the north face of the firſt pier, to at leaſt four feet in breadth, quite down to the foundation pile heads or platform; and if any thing there appears deficient to make it good with new, then to raiſe up this face of the pier with large and good aiſler, in order to reſt the ſouth end of the arch thereon, by this means this will become a part of a new ſuperſtructure, and the reſt being gradually ſerved in the ſame way, will in time produce in effect a new erection.

It ſeems practicable to increaſe the breadth of the bridge ſo as to make it about four feet wider than at preſent, that is, about 20 feet wide within the parapets; 16 feet may be allowed for the carriage-way, which will admit two carriages to paſs freely, and leave four feet for a raiſed walking path.

It is alſo practicable to make a foot-paſſage from the bridge through the old building to the quay, and from thence immediately to the ſand hill, which will greatly relieve the preſent narrow avenue to that end of the bridge.

N.B. The jetees being chiefly timber, are in a general ſtate of decay, but ſeem at preſent in tolerable order; what is moſt amiſs has been particularly pointed out.

J. SMEATON.

*Newcaſtle,*
9th May, 1771.

A RE

## A REPORT relative to Tyne Bridge.

THIS bridge having been erected fome centuries ago, upon the principles that were then in common ufe for fuch edifices, as may be feen from many of the old bridges over great tide rivers in England, and alfo in feveral parts of Europe, namely upon ftilts, furrounded with a kind of bafement called ftarlings; has always had the fame inconveniences and imperfections attending it, to which all others of this kind are fubjected.

The ftilts are a great number of large piles driven down into the bed of the river, fupporting a timber platform or framing under each pier, a little above the loweft water of the river, upon which the mafonry is eftablifhed, and thofe ftilts being furrounded with a row of clofe piling at the diftance of feveral feet, and the intervals filled up with gravel and other ftone materials, thrown down at random; and the whole connected with proper timbers, and paved down at top with large ftones, a few feet above low water mark, form the footing or bafement for the fecurity of the pier, which is called a ftarling.

Upon thefe principles this bridge has been hitherto fupported, though the river is fubject to rapid floods and drift ice in the feafon, without fuffering any thing very confiderable, except in one or two of its arches, within the memory of man.

The total breadth of the river between the quays or abutments being 539 feet, the piers or folids of the bridge took up or occupied a third part of it, and when the tide fell below the top of the ftarlings, they took up fomewhat more than one half of the free waterway, immediately above bridge; and this great contraction of the water-way has at length contributed its effects towards bringing about its ruin, by increafing the rapidity of the ftream paffing under it, fo much as to gull away the ground under the arches, fweep away fome of the ftarlings, undermine the ftilts, and throw down or render incapable of repairs, five of the arches out of ten whereof this bridge was formerly compofed. The arches of the bridge itfelf are likewife too low, feven out of the ten being totally filled up with water above their crowns or key ftones, by the late flood of the 17th of November laft; which flood produced this, and many other unhappy events of the kind upon this and the neighbouring rivers.

This flood was three feet fix inches higher at this bridge than the memory of man or tradition has ever taken notice of; and the ground on the eaft or lower fide of the bridge is gulled away fo as to produce a depth of twenty-feven feet fix inches of water at low

water,

water, when at a medium there is only feven and eight feet depth in the river immediately above the bridge, or on the weft or upper fide of it, and this circumftance is now mentioned to fhew the danger the four remaining arches ftill continue expofed to, by ftanding as it were upon the *edge of a precipice.*

The inconvenience juft now mentioned, arifing from contraction of water-way by this manner of conftruction, and which at length has been fatal to the edifice itfelf, is far from being the only one which this mode of proceeding occafions; the trade now carried on upon this river, fo neceffary to the kingdom in general, requires many hundred of keels with coals, and other fmall veffels, to pafs and repafs under this bridge almoft every tide, both in the day time and in the night. The ftarlings which project from fix to eight feet from the piers, under each of the arches, when covered a little by the tide, expofe them unavoidably to frequent damages and loffes, and often to utter ruin and deftruction. And of fuch events, gentlemen converfant in the trade are fufficiently acquainted with number-lefs inftances, fo as to make it totally unneceffary to mention any further, than that the ftarlings themfelves frequently fuffer in the rencounter, and inceffant blows contribute to their wanting frequent and expenfive repairs.

Other inconveniencies attending this bridge muft now be mentioned, which are fufficiently known to every perfon who has had occafion to pafs over it; one of which is the fteep afcent from the fouth end of the bridge, rifing one foot in feven or eight for a confiderable fpace, with the confined narrow ftreet all the way, which renders the whole extremely dangerous to every paffenger, whether in a carriage, on horfeback, or on foot, and fre-quently occafions great interruptions and confufion, when carriages or a number of paf-fengers happen to meet upon it; and being the principal road of communication between Scotland and the fouthern parts of the kingdom, expofes every individual to thofe hazards, who has occafion to make ufe thereof. The other inconvenience is the narrownefs of the bridge itfelf, which within the parapets is 15 feet wide only, and in feveral places reduced to 9 feet by buildings and other obftructions, and equally productive of the laft mentioned hindrances, dangers and delays.

After this detail of circumftances relative to this bridge, the next confideration, of courfe, muft be the manner of reftoring it to public utility; and, in order to form a proper judg-ment hereupon, many trials, in order to afcertain the nature and quality of the ground clofe to the piers and ftarlings of the bridge, have been made by borings into the bed of the river, in moft of which it was found clogged or filled up with loofe ftones or gravel, to the depth of 12, 15, or 17 feet below the bed, and afterwards a loofe fand

with

with fmall gravel for 10, 12, or 15 feet more, and at a medium beween 30 and 40 feet deep from the furface of low water before a folid bottom was met with, which on the fouth fide of the river appeared to be a ftiff clay, and on the north fide a hard compact gravel.

Under thefe circumftances, the difficulty of reftoring the part of this bridge which is fallen to the former ftate, *without the ufe of ftarlings*, will be fufficiently apparent to every perfon converfant in thefe matters; and though it may not be an impoffibility in point of art to do it, fuppofing that time and expenfe were unlimited, yet it is one of thofe things that do not feem capable even of an eftimate: and if ftarlings fhould be admitted, and the former method of conftruction allowed of, the difficulty of eftablifhing them in a folid manner, as well as the foundations of two of the piers, which are totally overturned in fuch a heap of ftones and rubbifh, not only fallen down at prefent, but which appear from the borings to fill up the bed of the river to the depth above-mentioned, would be fo great, and the execution fo tedious, that it is difficult to form a judgment what the expenfe would amount to; and, after all, the bridge would be eftablifhed with every inconvenience that now attends it, and which have already been enumerated, and at laft remain equally expofed to the danger of being overturned by the next great flood, as well from the defects in the manner of eftablifhing the foundations as the contracting of the water-way, and the want of a proper elevation of the arches.

In order to afcertain the nature of the ground under the bed of the river, two fets of borings have been made acrofs it, the one about 50, and the other 100 yards above the prefent bridge, the latter from the place called the Javel Groop, to the oppofite fhore, from both of which a mafs or body of loofe fand mixed with fmall gravel from 12 to 30 feet in thicknefs or depth from the bed of the river, appears to cover a ftiff clay, in-clining from the faid fhore to near the middle of the river, and a mafs of the fame fand from 30 to 40 feet in depth, covers a hard compact gravel from the middle to the north fhore, conformably to the borings made at the bridge itfelf, which have already been mentioned.

As the laft mentioned place, or near it, feems to be the moft convenient fpot for erecting a new bridge whereby the inconvenience attending the old one might be removed, and, at the fame time, preferve the beft practicable communication from the town to the oppofite fhore, it neceffarily remains to procure by art the means of eftablifhing a folid foundation upon this great mafs of fand and gravel; but as a detail of this part would run to too great a length, it may at prefent fuffice juft to mention the outlines of this

propofition,

propofition, which are to fpan the river with feven arches only, whofe piers or folids may not take up more than one-fifth part of the water-way, to lay the foundation of the piers and abutments feveral feet below the bed of the river upon a fufficient number of ftout piles driven down under them, and to circumfcribe each with a border of plank piling driven clofe to the foundations, and thus at once to remove the danger and inconvenience to the navigation arifing from the ftarlings, and at the fame time to admit of four-fifths of the breadth of the river to be free water-way: to give this new bridge more elevation than the old, to conftruct the piers and arches with large blocks of the hardeft and moft durable ftones the country affords, to make the breadth of the bridge within the parapets 30 feet wide at leaft, with a paved way for foot paffengers on each fide, and to fuffer no erections or buildings whatever to occupy any part of that fpace, to keep its landing on the fouth fhore as much elevated as poffible, and to bring the road to join it 40 feet wide at leaft, from the top of the fteep part of Battle Bank near the Half-moon Lane in a winding courfe round the hill, and by this means to reduce the defcent or afcent to about 1 foot in 19 or 20, or thereabouts, which will not be very perceptible; to open the communication from the north end of the bridge to the *Clofe*, and after allowing a fufficient area or void fpace at each end of the bridge, its communication both ways will be rendered as convenient to paffengers and carriages as it is capable of being made.

In order to communicate the road through Pipewell Gate, a dry arch can be made in the abutment; and to enable the inhabitants of that place, and the lower parts of Gatefhead, to go over the bridge in the readieft way, a flight of ftairs can be erected from Pipewell Gate to the top of the bridge on one or each fide of the fouth abutment, if thought neceffary.

It is, no doubt, an objection to the new placing of the bridge, that it will be detrimental to thofe who have houfes in the thoroughfare down the Battle Bank, but when buildings happen to be erected in a place in fuch a manner as to be totally inconfiftent with public accommodation, thofe partial inconveniences muft either give way to the greater convenience of the public, or the public muft fubmit to the having them entailed upon it to perpetuity: but to fhew this in a clearer light, let us fuppofe a new bridge built in the old fituation, whofe fuperftructure is raifed at the fouth end as much as the pofition of the prefent piers (which at all events it feems neceffary to conform to), will admit of, then if a gradual flope be continued from the brow of the Battle Bank (where it begins to be fteep, that is about 70 yards north of Half-moon Lane) to the bridge, it will form a flope of about 1 in 12½, which is not to be called an eafy one; and in this

cafe

cafe the furface of the road will be in fome places near the bottom of the Battle Bank, 26 feet above the prefent pavement; fo that the greateft part of all the houfes down the Battle Bank muft be rebuilt; nor can the road through Pipewell Gate be communicated with the bridge for horfes or carriages without forming a flope almoft 100 yards into Pipewell Gate, which will in like manner induce the neceffity of rebuilding there, and the fame of Hill Gate.

The flope may indeed be eafed by beginning at Half-moon Lane, and cutting down the breaft of the bank, but this will not bring the flope under $14\frac{1}{4}$ to 1; the height of the road above the prefent pavement will be nearly the fame towards the bottom of the bank, and the fet of houfes where the ground will be cut, being undermined or left confiderably above the pavement, will alfo need rebuilding to render them convenient, and after all, the narrownefs of the ftreet will be a permanent inconvenience, which cannot effectually be remedied without totally rebuilding the whole of it.

It feems, therefore, that there is no other choice on this matter, but either to rebuild the bridge upon its former principles, or to choofe a new fituation. It is not to be doubted but the bridge may be reftored to what it was by proceeding upon the old plan, at lefs expenfe and in lefs time than the building of a new one; but the inconvenience of paffage over and under before mentioned remaining with the public to perpetuity, can fcarcely be counterbalanced by the faving of any moderate fum of money, nor can the public be well accommodated with a temporary bridge without being at a much greater charge.

It is not poffible to judge completely of the expenfe of building a new bridge, till a juft plan thereof is formed, and from thence an eftimate made by a regular induction of particulars. But as this alone is in reality a work of time, to enable the gentlemen concerned to form their judgments of preliminary matters, we judge from the execution of fimilar works, that a bridge may be built at or near the Javel Groop to the oppofite fhore, including the extra expenfe on account of the whole bed of the river being a mafs of fand, for the fum of 50,000l., from fhore to fhore, that is, exclufive of land-works and purchafes. A bridge of this kind will probably take feven years to execute; and we are of opinion, that to build a new bridge in the old fituation, and on the fame principles as before, cannot be executed for lefs than half the money, and will take half the time.

From

From this propofition, the neceffity of purchafing a confiderable quantity of buildings and other private property for the ufe of the public, will be very apparent, in order to make the accefs to the bridge convenient on both fides of the river, and if all this were not wanted for the above mentioned purpofes, a great deal would be abfolutely neceffary to be had for eftablifhing yards and fheds for the workmen for performing the many different operations the edifice itfelf would require.

As the time neceffary for completing this work will be fo confiderable; a better communication than the ferry already eftablifhed, may probably be judged unavoidable during fo long a fpace of time. In this cafe, a temporary bridge of timber materials, erected upon the ftumps of the old piers, fo as to join the four remaining arches with the abutment remaining on the fouth or gatefhead fhore, will be attended with the leaft expenfe and inconvenience of any other method, its whole length being only 320 feet, which is about $\frac{4}{7}$ parts breadth of the river. For effecting this, a proper defign with a conftruction or detail of every particular, is now given in, and the expenfe computed at about 2400l.

Before this work can be performed, the fingle arch remaining towards the middle of the river muft be taken down, and a little time will fhow whether this may be attempted with fafety to the workmen, when it can be well afcertained, that the opening or rent therein has not increafed any more for any confiderable fpace of time. It is further recommended to take up the remains or ruins of the three arches and piers next gatefhead fhore, leaving the platforms or remains of the piers for eftablifhing the legs of the temporary bridge upon; the performing this as fpeedily as poffible, will procure more free water-way to the river, and in courfe fome advantage as to the fafety of the four remaining arches of the bridge, in cafe of another flood, by allowing the ftream to pafs more freely on the oppofite fhore. When the new bridge with its communications is completed, the temporary bridge, with the remains of the old bridge, may be taken away, and the river freed from its incumbrances to the navigation.

J. SMEATON.<br>
J. WOOLER

*Aufthorpe, January.*

To Sir Walter Blackett Baronet, the Aldermen and Common Council
 of Newcaftle-upon-Tyne.

AN Eſtimate of the expenſe of erecting a temporary bridge with fir-timber upon the piers or legs of the old bridge at Newcaſtle upon Tyne, 9th December 1771.

|  | £ | s. | d. |
|---|---|---|---|
| 14000 ſolid feet of fir-timber including the workmanſhip and ſcaffolding at 2s. per ſolid foot  -  -  -  -  -  - | 1400 | o | o |
| 70 Cwt. of iron bolts and plates at 36s. per Cwt.  -  -  - | 126 | o | o |
| 80 Do. of Do. and piles ſhoes at 28s. per Cwt.  -  - | 112 | o | o |
| Walling up the South Pier of the remaining fourth arch from the Newcaſtle Shore, in order to ſupport the end of the temporary bridge, repairing the damaged ſterlings, rooming up the work for ſecurity during the execution, and covering the top of the bridge with gravel, computed  -  -  -  - | 770 | o | o |
|  | £2408 | o | o |

Feet Inches
539 6 the length of the old bridge, from Quay to Quay.
330 0 the length of the temporary bridge.

N. B. Notwithſtanding this Report, the bridge was rebuilt in its old ſituation.

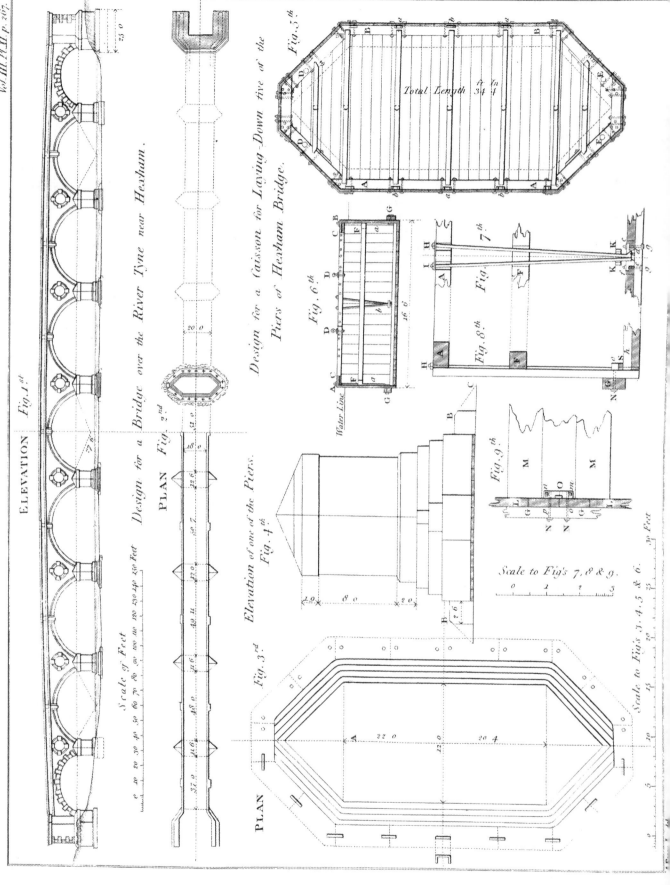

ELEVATION *Fig. 1st*

Design for a Bridge over the River Tyne near Hexham.

Scale of Feet

PLAN *Fig. 2nd*

Design for a Caisson for Laying Down five of the Piers of Hexham Bridge.

*Fig. 5th*

Total Length 34.4

*Fig. 6th*

Water Line

*Fig. 7th*

*Fig. 8th*

*Fig. 9th*

Scale to Fig's 7, 8 & 9.

Elevation of one of the Piers.

*Fig. 4th*

*Fig. 3rd*

PLAN

Scale to Fig's 3, 4, 5 & 6.

# HEXHAM BRIDGE.

(See Plates 11, 12, 13.)

To Mr. Donkin,

Sir                                                          Aufthorpe, 12 January 1777.

I BEG leave to acquaint you, that I have fully confidered the conftruction proper for a
bridge over the river Tyne at Hexham, and have carefully computed every article of
expenfe that will probably attend the erection of it in a fubftantial manner, and fufficient
for me to rifk my credit as an artift upon it. The fituation is that which I particularly
examined at the requeft of Mr. Errington, viz. about fifty yards below the boat, and
nearly in a line with the little bridge that croffes the mill leet; in which cafe it will land at
the north end upon the clofe eaft of the boat houfe, which is fituate between that and the
lane.

The bridge I propofe to confift of nine arches, and to extend between the abutments
518 feet, and including the abutment well, 568 feet; to be twenty feet wide over all,
and about eighteen feet between the parapets. I fuppofe it to be grayelled over in the
manner of a good turnpike road, and the length of the road over the middle arch to
be thirty-one feet above the furface of the river in its ordinary ftate: the height of the
great inundation in the year 1771, being at this place above that furface fcarcely fourteen
feet.

|  | £ | s. | d. |
|---|---|---|---|
| The neat Eftimate of materials and workmanfhip    -    -    -    - | 6036 | o | o |
| Upon the above, I allow 10 per cent. for engines, utenfils, and contingent expenfes | 604 | o | o |
| Expenfe of the bridge complete, with the road formed over it to the extent of the abutment walls    -    -    -    -    -    - | £6640 | o | o |
| To forming the afcents to the bridge at each end, that at the fouth to be wharfed up with walls extending fifty yards with a breaft-work or rough parapet thereon, and fo that the flope fhall not be fteeper than one in twelve; mafonry, filling or forming the road to the extent of the flope    -    -    -    - | 320 | o | o |
| Total expenfe of the bridge complete, materials and workmanfhip    -    - | £6960 | o | o |

It

It is to be noted, that the road is fuppofed to be made up to the height of feven feet above the level of the ordinary furface of the river, and not higher.

It is further to be obferved, that there is here no allowance for a fmall bridge over the mill leet; nor for forming the level road between that and the great bridge, nor for way leave or quarry leave, nor for the fuperintendance or furveyor's falary, which two laft I cannot eftimate at lefs than eight per cent. upon the above amounts, or at about 56ol., making the whole fum 752ol.: fo that, if Mr. Errington undertakes the bridge at a lefs fum than 7500l, with the intent of building it under my direction, it is probable he will be money out of pocket, and therefore lower than this I cannot advife his engaging therein.

I muft obferve, however, that the above is on fuppofition that every thing is to be provided, and therefore whatever the materials that are on hand may be found to be worth to this undertaking on a fair valuation, where they lie, will be fo much to be deducted from the fum to be advanced by the country.

I curforily viewed the fituation oppofite to the Hermitage, where I formerly propofed to build a bridge in 1756, and though I have no doubt of building a bridge there, yet to do it with a probability of fafety againft fuch a flood as that of 1771, which at that time was not thought of, is impoffible there or in any other fituation near Hexham, for lefs money than the amount of the above eftimate.

I am, &c.

J. SMEATON.

To Mr. Pickernell.                                          London, Auguft 8th 1777.
DIRECTIONS for forming the elliptical arches and centers for Hexham Bridge.

The true method of forming an ellipfis is by means of a trammel, with the conftruction of which I take it for granted you are well acquainted.

The dotted femi-ellipfis ABC (Plate 12, Fig 1.) is the medium line upon which the penftones are fet out, by equal parts or divifions, and is 38 4 bafe, or width by 13 7 high, that is,

$$
\begin{array}{ll}
\text{Ad---dC will be} & 19\ 2 \\
\text{dB . . .} & 13\ 7 \\
\hline
\text{Difference} & 5\ 7
\end{array}
$$

and

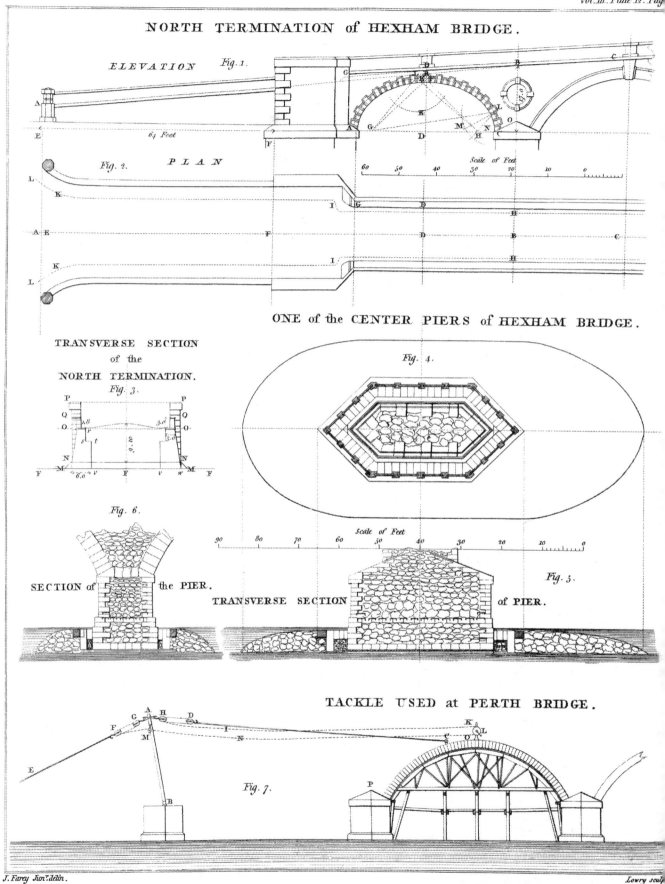

# NORTH TERMINATION of HEXHAM BRIDGE.

ELEVATION *Fig. 1.*

*Fig. 2.* PLAN

64 Feet

Scale of Feet

## ONE of the CENTER PIERS of HEXHAM BRIDGE.

TRANSVERSE SECTION of the NORTH TERMINATION. *Fig. 3.*

*Fig. 4.*

Scale of Feet

*Fig. 6.*

*Fig. 5.*

SECTION of the PIER.

TRANSVERSE SECTION of PIER.

## TACKLE USED at PERTH BRIDGE.

*Fig. 7.*

J. Farey Jun.r del.in.

Lowry sculp.

Published as the Act directs, 1812, by Longman, Hurst, Rees, Orme and Brown, Paternoster Row, London.

and this difference of five feet seven inches will be the length of the crofs grooves, from the center to the ends, fay five feet eight inches, to keep the pins from flipping out; and five feet seven inches will alfo be the diftance of the centers of the two pins that traverfe in the grooves; laying down therefore your crofs upon a floor, big enough to defcribe fomething more than half an arch, fet your tracing point at the diftance thirteen feet feven inches from the neareft traverfing pin, and with this you will defcribe the curve E B F L C at large; and if your ftones will run of twenty inches thicknefs, as drawn, fet off ten inches BE—BF, on each fide the center, and then divide the remainder F L C into 14 parts; but if your ftones will only run about 18 inches, then fet off nine inches on each fide B, dividing the curve F L C into 16 parts; if your ftones will hold but 15 inches, then fet off $7\frac{1}{2}$ or 8 inches for BE and BF, dividing the curve F L C into eighteen parts; and if they will not hold more than 14 inches, fet off feven on each fide, and divide into 20 parts. The points of divifion being thus fixed, fhorten the diftance between the tracing pin and the neareft traverfing pin, to 12 feet 11 inches, and then you will defcribe the interior curve to be of 37 feet fpan, and 12 feet 11 inches high; and if from this you fet out the tracing pin (pencil, or point) two feet, and three feet refpectively, you will then defcribe the curves that will take the extremities of the fhort and long ftones.

To find the directions of the joints, you take the line A d 19 feet 2 inches, and fetting one point in B with the other defcribe an arch, fo as to cut the line A C in G and in H, then will the two points G and H be the two foci of the ellipfis, which have this property; that if to any point of divifion of the curve as fuppofs L, you draw the lines G L and H L, then the line M L, bifecting the angle formed between the lines G and H L, will be the true direction or fummering of the joint at L, and fo of the reft. To avoid confufion, you have another example at the point I, the true fummering of which joint is K I. In this manner the joints of one half of the arch with the key-ftone, being neatly marked out upon the floor, and boards adapted thereto, (the grain of the wood correfponding with the length of the ftone), you will have a fet of molds by which all the ftones in the two arches may be hewn; but you are to obferve, the thicknefs of the penning within need not exceed the length of the fhorter ftones, the longer ones only appearing in front. You are alfo to obferve, that the arch like a femicircle, begins from an horizontal line N O, reprefenting the corner of the falient points of the pier rifing up before the arch ftone, or what may be better united therewith, by proper hewing, the ground joint being in the direction N C.

Refpecting

Refpecting the centers to the elliptical arches, the external curve will be ftruck out by the trammel, about two inches lefs than the fofit of the arch, to allow for the thicknefs of the boarding, and except the making of the curve quite fair, little curiofity otherwife will be required. I propofe nine of thefe ribs, and the curved part may either be fawn out of crooked wood, or lapped together from ftraight, according to the convenience of your materials. The curves may be about feven by $3\frac{1}{2}$, and bafe 10 or 12 inches by $3\frac{1}{2}$ or 4 thick; the ftandards may be made out of any offal ftuff you have, from $3\frac{1}{2}$ to 5 inches thick: in fhort the whole compofition needs not to exceed that of vault centering, when well and firmly done. The bafe may ftand on five bearings, one at each end, one in the middle, and one on each fide the middle, but fomewhat nearer the middle than the fides: perhaps it may be neceffary to put in an upright more in the ribs on each fide the middle one; for I would not have the fpaces near the middle to exceed $2\frac{1}{2}$ or 3 feet diftances at the top, middle and middle.

As the traverfing ruler will be required 20 feet long, and to have a confiderable degree of ftiffnefs edgeways, to prevent its being over bulky or clumfy, it may be proper to obferve that its greateft ftrength will be required at the middle pin, or the pin neareft the tracing point; you may therefore (making the whole $\frac{3}{4}$ thick) let the breadth at the middle pin be eight or nine inches; at the other traverfing pint about four inches, and at the end of the tracing point about three inches.

J. SMEATON.

To Mr. Pickernell.

Inclofed you have the defign for the caiffon, for the five piers of the bridge that remain to be laid, as alfo the defign and elevation for one of the center piers.

By " grooving in" of the fides, and crofs courfe of bottom planks, as expreffed in the explanation, I mean in the middle of the eye of each plank, like a common chamber floor grooved in with Dutch laths.

The reafon why I would have the infide blocks joint more clofe in the firft and fecond courfe than above is, that till the caiffon is fairly grounded, we fhall never be fure of keeping the joints whole; and as the fecond courfe will be fufficient to ground the pier, then all can be made good; and fo long as the upper parts reft upon folid guaged blocks, and then upon one another, it is not abfolutely material whether there were any fittings in the joints of thefe two bottom courfes or not.

3

J. SMEATON.

## EXPLANATION of the Defign for a Caiffon for laying down five Piers of Hexham Bridge.

Plate 11. Fig. 5. is the plan of the caiffon, being 16 feet within, by 22 feet length of the flat fide. The bottom is fuppofed to be made of three inch plank laid crofs and crofs; and the fides of three inch plank only, ftanding upright, and fupported by three ribs of wood, two upon the infide and one upon the out. The uppermoft AA, BB, to be ftrongly fecured with bolts and plates at the angles, and likewife fecured by bolts upon the fides, as fhewn in the plan. The pieces CC, &c. are ftretchers, but which being moveable, will not be in the way of lowering the ftones. The fides not refting upon the bottom as has been ufual, fo as to admit the ftone-work to be flufh with the bottom when the fides are removed; and thereby to let the girdle of large ftones flip down to their own bearing, without being interrupted by any projection, a particular method of faftening the bottom and fides is adapted to this purpofe, and the plates a, a, b, being 16 in number, each fhewing a double nut and bolt, fhew the places and difpofition of thofe faftenings.

DD, EE, fhew the places of iron ftuds, ftrongly fcrewed down, over which a chain being caft the caiffon is thereby moored to a large pile, to be driven at the diftance of 30 feet upftream of the bridge, before the point of each pier, to be kept fteady by another fmall pile, about as much below.

Fig. 6. is a tranfverfe fection of the caiffon, wherein is fhewn that the ground tier of planks in the bottom run lengthways; AB fhew the upper rib, the dotted lines CC fhew the ftretchers, DD the iron ftuds for mooring the caiffon, FF the middle rib infide, and the dotted lines fhew the moveable ftretchers of fix inches fquare, which, when the ftone-work is got that height, are to be removed, and the fides fupported by fhort ftuds againft the mafonry, if need be.

GG fhew the fection of the lower rib, which is put outfide, to prevent its interference with the ground courfe of ftone; this is four by eight; and as this and the middle ribs have no other ufe but to keep the upright planking fteady they need only be fpiked together; and on this account, if thefe ribs were of elm, or any other kind of wood than fir, they would be lefs apt to fplit by a multiplicity of fpikes; a a fhew two fets of the faftening bars feen fideways, and b fhews a fet feen in front; of which follows a more particular defcription.

Fig. 7.

Fig. 7. fhews the upright of a pair of faftening bars to a larger fcale, as they would appear from without if the outfide planking was fuppofed to be removed or tranfparent ; wherein A fhews a part of the upper rib, F a part of the middle rib, and the dotted lines G fhew the place of the bottom rib; c d e f fhews a piece of caft iron let down flufh into the edge of the bottom, and which is held down by the two bolts gg. This piece of caft iron contains a dove-tail notch, in which h H is the dove-tail bar ; and thereby fufpending the bottom upon the nut H; to which i I is the bar compofing the key, which is to be driven in ; but by turning the nut I will draw it out, and releafe the other, and by the fame means all the faftenings in like manner being flacked, the fides of the caiffon will be releafed from the bottom. KK is a ftaple whofe ufe will be fhewn hereafter. But as it will be neceffary that the fides of the caiffon fhould be under fome confinement towards the bottom to preferve a water-tight joint between them,

Fig. 9. is a plan of a part of the bottom, wherein LL reprefent the fide planking, GG the lower rib on the outfide, and MM the crofs planking of the bottom. Now m n o p is the plan of the caft-iron dove-tail notch piece ; which notch being dove-tailed with refpect to its plan, as feen here, as well as with refpect to its upright, feen in fig. 7, and the bars being adapted thereto ; by this means the bars will be confined from flying out fideways, as well as from drawing; and the ftaple ONN, fhewn by dotted lines paffing through the bottom rib, being tightened by the nuts NN, will draw and confine home the fides to the bottom ; there is yet another thing wanting, and that is, that there being nothing to confine the fides from going below the bottom, there will be fomething neceffary to counteract the nut H, fig. 7; now fig. 8, is an enlarged fide view or fection of the fide of the caiffon and bars, wherein the fame letters denote the fame things as the preceding, and the bar h H is fuppofed to be the fame bar as marked with thofe letters fig. 7; now, if between the under fide of the ftaple and the caft metal notch piece there be driven an iron key a little wedge-wife, whofe fection is reprefented by the little fhaded part at S, the fide can be fo regulated as not to go below the bottom, and fo as to keep every thing fufficiently fteady ; nor is any degree of ftrength in this matter neceffary, as the weight of ftone in the caiffon, when afloat, does not in reality hang by the faftenings, but is fupported by the counter preffure of the water under it upwards.

As the bottom courfe of ftone is to be clofe home to the fides of the caiffon, grooves muft be cut in the ftones where they come againft the faftenings, with full liberty to draw them out.

The

The notch pieces neceffarily go along with the bottom, but one fet of faftenings will do for all; and it will be proper to have the dovetail part of the bars well fitted to the notches with a file; care being taken to have the notches caft fo near alike that fitting one they will fit all.

Though it is not mentioned in the defcription, it will be proper that all the joints of the planking and the fides, and the crofs or upper planking of the bottom, be grooved in about three-fourths of an inch in depth; and a lath one and a half inch broad, and about three-fourths or one-half inch thick inferted, and let in with tar, that the pinning of the joints by drought may not induce the neceffity of caulking; and to fecure the joint between the bottom and fides it will be proper to nail on a ftripe of thick flannel and tar three inches breadth upon the out edge of the upper tier of floor planking, and this joint may be further fecured, if feen neceffary, by a flight caulking infide.

Fig. 3. fhews the plan, and fig. 4. the elevation of one of the middle piers, with the plan and fection of the girdle of the ftone round the bafe. The bafement of the pier is fuppofed to confift of two courfes of ftone, both of which are to be made entirely of blocks; the interior blocks need not, however, be jointed as to their upright, or fcrupu-loufly fitted to each other, but to be brought to a thicknefs anfwerable to the courfe, and to be fet upon their bigger bed as clofe to one another as may be, and the interftices carefully filled, and walled with beft mortar.

The two upper courfes reaching to the low-water level muft alfo be interfperfed with blocks gauged to a thicknefs, but need not be fo clofe to each other as defcribed for the two ground courfes; if in thefe upper courfes the gauged blocks occupy two-thirds of the folid within the outfide courfe it will be fufficient. The firft courfe of the bafe or chamfered plinth being above water is to have three chain courfes go acrofs it, that is three rows of blocks, one acrofs the middle and one near each fhoulder, to be ftrongly cramped together, and to the ftones in the outfide courfe, or joined by one continued bar acrofs; and all the outfide ftones in the circumference of the chamfered courfe to be joined by one continued chain bar. In other refpects, the mafonry above the low-water line to be compofed in the fame way as the piers already executed above that line.

N.B. The fhoulder angle of the girdle courfe, and of all the other courfes under water is centered at A, in the plan, and the dotted line in the upright, denotes a flope of rock-work formed by a depofition of rubble, ftone, and cement, called beton.

*Aufthorpe,*
*17th Dec.* 1777.

J. SMEATON.

From Mr. Pickernell.                    Hexham Bridge Office, July 24, 1778.

I can now give you a particular account of the damage we have received from the high water, which came on us laſt Tueſday night and Wedneſday morning.

The pier which I laſt founded at the ſouth ſide is undermined at the weſt point, and ſettled down conſiderably. I meaſured the water and found the gravel to be taken away about 18 inches below the bottom of the caiſſon. The foundation of the fourth pier from the ſouth abutment is taken out at the ſouth end, and nearly half along the pier, ſo as to let the weſt point, or cut-water, ſettle about 18 inches. The gravel is taken out in ſome parts, as nearly as I can find, two feet ſix inches lower than it was when the beton was laid on the bed of the river, and ſeveral ſtones are taken out of the cut-water part of the pier. The water came with ſuch velocity as to undermine the girdle-ſtones, and force up betwixt them and the pier ſo as to force the girdle-ſtones two feet from the pier. In ſome parts the iron cramps which were laid into the baſe courſes have confined that part together, or elſe, I believe, all the cut-water part of the pier would have been entirely ſwept away by the violence of the ſtream. Though the water was not very high, I never ſaw it ſo ſtrong.

I cannot ſee that the fourth pier from the north has received any damage, except that the bed of the river round the ſinking ſtones at the weſt end of the pier is in part taken away; though, I believe, had it been as high as the other it muſt have ſuffered the ſame fate. The third foundation from the north abutment has ſuffered the ſame as the ſecond from the ſouth abutment. The gravel is taken out from under the weſt cut-water about 18 inches, and the point is ſettled down. Indeed I had not the good fortune to have the girdle courſe laid round the weſt part of the two foundations, as we could not get them at the quarry faſt enough. None of the foundations which were laid laſt ſeaſon have received any damage.

I am, &c.

J. PICKERNELL.

To Mr. Donkin.                          Auſthorpe, 29th July 1788.

I have received a particular account from Mr. Pickernell of our very diſagreeable misfortune, and find it no ways more flattering than I had reaſon to expect from the contents of your letter. What I moſt blame myſelf for is, in not ordering every pier to be ſecured, not only with the girdle courſe complete, but with the ſlopes of rubble ſtones brought up againſt it, before another pier was laid on the bed of the river; becauſe, notwithſtanding what has happened, had the foundations been ſecured, and finiſhed off

in

in the manner I had intended and propofed, before winter, I am fully fatisfied this difafter would not have happened. As it was, I was willing to pufh on the work in the prime of the feafon, and confiding in the natural ftrength and compactnefs of the gravel bottom, I could not have fuppofed fo much derangement would have happened in the courfe of two or three fummer frefhes. I hope I fhall not find matters irretrievable.

---

DIRECTIONS for fecuring and repairing the under-wafhed Piers at Hexham Bridge.

NOTWITHSTANDING the natural hardnefs of the upper cruft of the gravel bottom, it being now found not capable of refifting the action of the water in fudden and rapid floods, it appears neceffary to inclofe the feveral foundations by a fence, or cafe of fheet piling, to prevent, at all extremities, the gravel from wafhing from under the bafes of the piers; a plan for the doing whereof accompanies thefe directions.

The work may be begun upon any of the piers that are found convenient, and as foon as may be, the bays compofing the two fheets that form the falient point before the weft end of each pier ought to be completed, together with one bay of the return upon each fide of the pier; and as foon as this is done the cavity wafhed below the bed of the river fhould be filled up rather higher than the original bed of the river both infide and out. The outfide to be filled with rough rubble ftones from the upper bed of Oakwood Bank quarry, which are exceedingly well adapted for this purpofe, to be of promifcuous fizes, the largeft not exceeding half a cube foot, or about 70lb. weight. The infide to be filled with the fame kind of quarry rubble of promifcuous fizes, the largeft not to exceed that of a large double fift, or about 12lb. weight, and for every two bufhels or meafures of rubble ufed within the cafe one bufhel of clean fand muft be thrown upon them.

This done, the defirable thing would be to proceed thus far with all the four piers that have been laid down in caiffons, by which means all the piers will be guarded from further damage, after which the completion of the cafing can be done more at leifure, nor ought this to be omitted with refpect to the north pier of the center arch, which, though it has fuffered no material derangement from the late floods, yet experience fhewing that the gravel bottom ought not to be trufted, the fame means fhould be applied as a guard to its fafety as to the reft. However, if the piling is found to go on

with

with readinefs and facility, and the moving of the tackle from pier to pier a work of labour, then, at the difcretion of the furveyor, the piers may in turn be furrounded wholly before the tackle is moved.

As foon as the gage piles are driven, the fcrew clamp may be applied upon the furface of the water ; or indeed if the piling of each pier be gone on without removal, this clamp may originally be applied as an outward frame for directing the driving of the gage piles, and for retaining them in their places while the fheet piles are driven ; and for greater facility, certainty, and exactnefs, in driving the fheet piling, fo as to render them water-tight, I would recommend the long fides to be divided into four bays, and the falient point fides into three each.

When the cafe is completed, reaching above low water, it may be tried whether the water can eafily be got out, and if fo, the under pinning, where wanted round the fkirts of the caiffon bottom, may be done by men's hands ; but if not, the whole of the interftice between the cafe and the pier muft be fitted with rubble and fand, as before directed for the weft ends, about fix or eight inches higher than the bottom of the caiffon, and after-ward driven down by a fett of about one foot fquare at the lower end, acted upon by one of the hand-rams : this will caufe the matter to fpread under the fkirts of the caiffon bot-tom, and the cafe hindering it from fpreading outward will render all tight and firm. This done, the interfpace between the cafe and the fides of the pier muft be chocked in with blocks of the thicknefs propofed for the girdle or funken courfe, which being fcap-pelled to a tolerable fquare, and adapted to their places, they will not need cramping ; but yet, to prevent any violent flood from turning any of them out of their places, it muft be obferved to chamfer the upper or leading edge of each crofs joint, as fhewn in a de-tached figure, as alfo to pin them faft with wedges, ftones, and pebble mortar, or beton, which may be eafily done if the cafe can be drained, as I expect it may, fo as to dry the upper furface of the funken courfe. This being done, the whole of the cafing is to be driven down with fetts, or otherwife cut off to the level of the funken courfe, and the fcrew clamp ultimately fixed, fo as to be rather below the tops of the piles, which will effectually confine all clofe home together.

The work being thus fixed, the outfide of the cafe is to be guarded with a flope of rubble, which, that it may be the better grounded by a competent body of matter, it will be advifeable firft to let the work ftand the effect of a flood ; and then not only filling up the excavation that may be expected on the outfide of the cafes, up to the general level of the bed of the river, but forming a flope extending to the diftance of about fix feet upon the bafe, at the height

of

of the bed of the river, and to reach as high as the clamp. After this, the flopes muft be examined at every flood, and fupplied where found deficient till the matter appears to be at reft, which fooner or later will be the cafe, when the river hath formed itfelf fuch a channel between the piers as is natural to the new fett of the ftream, that the interpofition of the piers of the bridge muft neceffarily occafion.

In cafe by repeated floods while the work is going on, any of the foundations fhall appear to underwafh more than they have done, it will be proper to throw round the weft end and weft fhoulder of fuch pier a competent body of rubble, the largeft not weighing more than 25 or 30lb.; for through rubble of this fize piles may afterwards be driven nearly as well as through large gravel.

The length of the piles fhould conform to the depth of the water. I would not wifh the fheeting piles round the weft end and firft bay of the return on each fide to go into the ground more than about ten feet; and if they do not drive kindly, muft be contented with lefs; from thence each bay may be gradually lefs depth into the ground, fo that round the down ftream pointing feven feet will be fufficient.

If the gage piles drive more kindly, they may be longer by 18 inches or two feet than the fheeting; but if not, they need not be above one foot longer: and to make them drive kindly, as well as fix fafter, it will be proper to point them longer than heretofore, that is, to about three feet, but not fhouldering, as with a regular taper, but curved, beginning from nothing at three feet.

The length of the piles covering the weft end and returns of the fifth pier from the north, being that moft underwafhed, the fheet piles fhould be at leaft 15 feet long, and the gage piles in proportion, and the reft proportionable to the depth of the water; and refpecting this pier, as the lowering of the water in the cafe will be more neceffary than in the reft, it will be proper before any rubble is put in outfide the cafe, to put a layer of earth or loam upon the bottom, amounting to about fix inches thicknefs, that is, allow a cube yard to fix yards fuperficial, this will choak up the pores of the gravel in the deep places, and retard the percolation of the water through the gravel, fo as to give a better chance of getting out the water, for new founding the weft end, if deemed neceffary.

What regards the repairs of the mafonry I fhall defer to a future opportunity of delivering, in due time.

It

It will be proper, in driving the gage piles, to try to ufe the great ram, fitted with a moveable difcharger, fo as not to be lifted above fix or eight feet above the pile heads ; by which means I apprehend it will be found to drive the piles much more kindly than if lifted its full height, and perhaps more kindly than the hand ram.

J. SMEATON.

*Hexham,*
5th *Auguft,* 1778.

### Additional Inftructions.

When the caiffon bottom is fully underpinned, according to the method defcribed, let the point be taken down one courfe as far as the fhoulder, fubftituting other ftones of fuperior thicknefs, fo as to raife the point rather higher than it was originally, by $1\frac{1}{2}$ or 2 inches ; reconciling, however, the new with the old work at top, and cramping every ftone to its neighbour, and to the block ftones of the old work, with common mafon cramps, run in with lead at each end, to bring them to a bearing, or otherwife fixing them faft with wedges. If you fhould have any difficulty to get them dry enough to run in the lead, heating the cramp, and putting a little oil into the holes, will render the difficulty lefs.

The taking up one courfe will do for the third and feventh piers ; but for the fifth you may take down and cramp as before directed, as far as you can.

I have alfo confidered, that if you can but get large ftones for the up-ftream pointings of the girdle courfe, or funken courfe, the reft may be done with aifler, whofe breadth and length is fufficient ; and if you can get the water off, as I expect you will, fo as to point and fill with the very beft mortar, which when ftruck to a breadth will make the open joint upward, and more capable of being filled. I would not, however, have any thing pointed from the down ftream fhoulders to the down ftream point ; only chocked faft with ftones or wedges, leaving interftices there for what water may get in at the up ftream part of the pier, to make its efcape freely below. Such of the old aiflers as will anfwer to the length and breadth will effectually ferve the purpofe without any new drefling.

J. SMEATON.

9th *Auguft* 1779.

EXPLANATION of the Defigns for the Diving Machine for Hexham Bridge.

(See Plate 13.)

To Mr. Pickernell.

Sir,                                                    Aufthorpe, 16th September 1778.

IF the cafes would have enabled us to reduce the water fo low as to be even with the very bottoms of the caiffons of each pier, I take for granted you would have thought it no difficulty with broken rubble, beton, ftones, and fhort blocks of wood cut a little wedge-ways, to have crammed and wedged full the cavity underwafhed, under the wooden bottoms; fo as to have been equally refifting, and capable of bearing a weight with the original gravel, and particularly when this new body of matter is fupported, and even jambed tighter into its place, by filling up the vacancy between the pier and the cafe, a little above the wooden bottom with rubble, and then driving it tight down by a fetr with the ram. It therefore now remains, that I defcribe, and make you mafter of a piece of machinery, that will put you nearly into the fame condition, as if the water could have been reduced to the caiffon's bottoms as before mentioned; and this is by means of an air cheft or diving veffel, which being let down will exclude the water down to the very bottom of the river, if you pleafe; and therefore as low as the underfide of the wooden bottom, which in the prefent cafe is as low as will be neceffary or ufeful; and the cheft or veffel being large enough to give liberty for a man to work therein; being furnifhed with a pair of boots, he will at mid-leg deep in water, do his bufinefs with almoft as much facility as if the water were pumped out to the fame level.

The principal part of this machine will confift of a ftrong cheft, fuppofe three feet fix inches in length, about four and a half feet depth or height, and as wide as to give free leave for its going down between the cafes and the piers, which I fuppofe will be about two feet wide infide meafure, as the other meafures are alfo fuppofed to be. Now you know very well, that if you pufh a drinking glafs, or any other fimilar veffel with its mouth downwards into the water; that it will exclude the water, leaving the veffel full of air, as it was before it was thruft into the water; in like manner, if this cheft, being loaded with a fufficient weight, be let down into the water mouth downwards, the air will exclude the water to the bottom fkirt of the cheft, and if let down fo as to reft upon the bottom of the river, a man may ftand dry therein, and do any kind of bufinefs, the fame as he could do in the fame fpace in the open air. But to continue this for any length of time, two things are obvioufly neceffary, and thofe are light, and a circulation of frefh air. The former might on occafion be fupplied by a candle; but here we may have the advantage

of

of day-light, by putting in two or three ftrong round panes of glafs into the bottom of the cheft, which will in its inverted fituation in ufe be the top; a fufficiency of light will enter, this top of the cheft being fuppofed above water.

Refpecting air, you will conceive that any quantity might be forced in by a ftrong pair of bellows; but thefe made of leather, would be cumberfome and unhandy. I therefore fubftitute a kind of forcing air-pump, made of thin hammered copper, that will throw in a gallon at a ftroke; which will not only continually refrefh the workman within, but whatever air efcapes out through the joints or pores of the air-cheft, will be replenifhed, and the overplus go out at the bottom or fkirt of the cheft, and boil up on the outfide.

The quantity of weight that will fink it mouth downwards, will be the fame as placed therein (bottom downward) would fink it the fame depth, and as this cheft I propofe to be fufpended by a tackle, and to go down by its own weight, I compute that it will take 16 pigs of lead to fink it to the bottom of the river, and keep it fteady; I propofe that the lead may be as much out of the way as poffible, to place them upon the ends of the cheft, endways upward, that is, four in a row below, and four above, and the fame at the other end; making in the whole 16 pigs, which are to be faftened on with fcrews, either by cleats fcrewed on, or punching a hole through each end of each pig.

At one end of the cheft there is to be a board fixed acrofs for the man to fit upon, and a cleat nailed on each fide, to fet each of his feet upon, fo that while the machine is letting down or hoifting, he is totally dry, and when let down low enough, he ftands upon the bottom of the river, without any more water than the height between the fkirt of the cheft, and the bottom of the river, which may be more or lefs as is found convenient, I fuppofe never more than a foot deep; becaufe wherever the ground is taken out more than one foot below the under-fide of the caiffon's bottom, I would propofe to fill it up with rubble previoufly to that height or depth; nor can it be of ufe to let down the fkirt of the cheft much below the caiffon bottom, becaufe the fide of the cheft will then diminifh the room you will have, to get the matter for underpinning under the caiffon bottom.

The foregoing will I believe be fufficient for explaining the general principles and outlines of the method I mean to purfue in underpinning, and re-fupplying what is underwafhed from the bafes of the piers, and which I dare fay you will now fee to be entirely practicable.

What

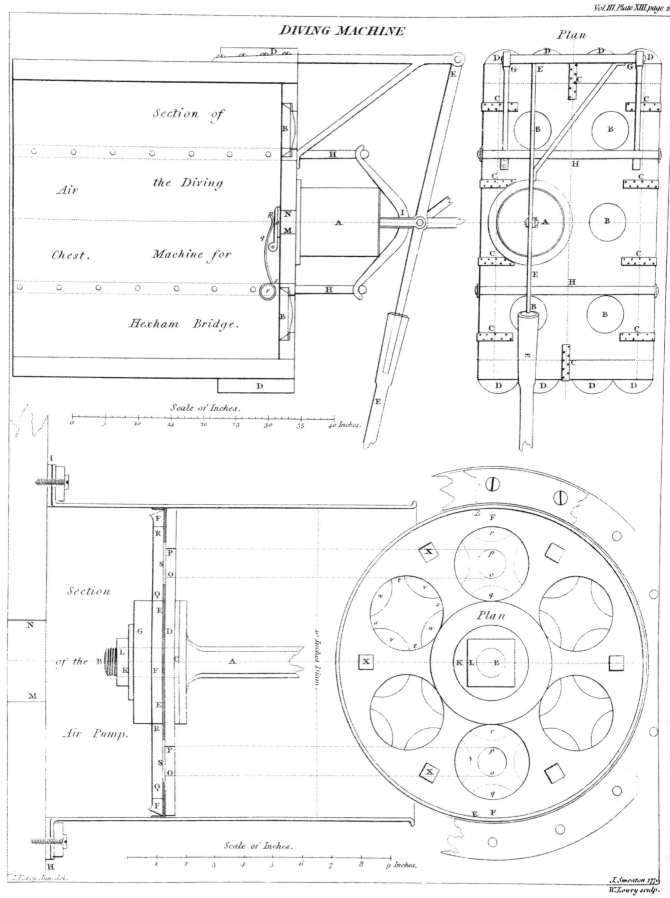

# DIVING MACHINE

Plan

Section of the Diving Machine for

Air Chest.

Hexham Bridge.

Scale of Inches.

0   5   10   15   20   25   30   35   40 Inches.

Section of the Air Pump.

Plan

21 Inches Diam.

Scale of Inches.

1   2   3   4   5   6   7   8   9 Inches.

J.Smeaton 1778

W.Lowry sculp.

Published as the Act directs,1812, by Longman,Hurst,Rees,Orme and Brown,Paternoster Row,London.

What you are therefore immediately to put in hand, is the air-cheft, of or about the infide dimenfions before mentioned; I believe the two flat fides will do very well, if of good red wood deal fhot clean of fap, the two ends and bottom (or in *ufe* its *top*), it would be well if they could be got of fingle planks of elm, beach, or plain tree, as they would hold the nails better; I fancy $1\frac{1}{2}$, or $1\frac{3}{4}$ thick for the fides, and $2\frac{1}{4}$ or $2\frac{1}{2}$ for the ends and bottom, will be fufficient; they fhould be well jointed, and put together with white lead and oil, and the infide joints ftroked with white lead and oil, as the effort will not be of the water to enter, but of the air to efcape from within.

Were I with you when it is put in ufe, I fhould be the firft to go down in it, as there is no more danger (all your tackle being firmly fixed,) than being let down into a coal pit by a rope; and if it fhall happen that all your mafons are too fine fingered, I fancy a couple of colliers to take turn and turn, will find it a very comfortable job; a particular encouragement muft however I expect be given.

I will give you more particular directions in my next: as to the air-pump, all that will be wanted from the copper-fmith will be a cylindrical pipe of copper, 10 inches diameter and 12 inches high, wired at top, and a flanch at bottom of about $1\frac{1}{4}$ inch broad, by which it is fcrewed down upon the top of the air-cheft; the copper to be about the thicknefs of a halfpenny; if you have no neat handed copper-fmith, that can hammer it ftraight and fmooth infide, it may on occafion be made of ftrong tin.

I am, Sir, your moft humble fervant,

J. SMEATON.

The materials will be got into the cheft by letting them down in a fhallow bucket or box, that will go under the fkirt of the air-cheft; which fhould be let down upon blocks to keep it fteady, while the air pump is worked; you will do well to try it, firft in fhallow water, and deeper by degrees, from whence you will find the nature of its working.

Explanation of the Section of the Air Pump, Plate 13.

A—The pifton fhank and flanch terminating in

B—A fcrew by which the whole is compreffed together.

C—Leather to keep the joint air tight.

D—The upper plate (to be made of boiler plate), $\frac{1}{10}$ inch lefs in diameter than the copper barrel.

E—A flat middling piece of fhoe upper leather, turned down upon the border to make a tight joint with the barrel, and which alfo compofes the valves.

F—The under plate (alfo of boiler plate), the edge being a little raifed round the border, and about $\frac{2}{10}$ lefs than the barrel: the leather D being held tight between thefe two plates.

G—A piece of wood by way of butt or ftop upon

HI—The upper furface of the plank of the air-cheft's bottom, in its inverted fituation.

K—An iron ring, and

L—The nut that fixes all faft.

MN—The opening through the plank, by which the air paffes into the cheft, and is fhut by a valve or clack on the underfide.

OP—Shews the opening through the upper plate.

QR—Ditto through the lower plate.

S—The folid part of the leather that fhuts the hole OP when the pifton is forced down, and gives leave, by means of its four arms, for the air to enter when it is drawn up.


Explanation of the Plan of the Air Pump.

FF—Shews the under plate, and under face of the pifton.

EE—The leather.

q r—The holes through the fame.

o p—The holes through the upper plate, as they would appear if the leather were removed.

At

At *s t, s t,* the leather appears in place ; and *v w, v w,* thofe fpaces being cut away, give leave for the air to pafs in going downward from above, while the pifton is afcending ; but not to efcape from below upward, while the pifton is forced down.

N. B. the reft of the letters marked upon the plan refer to the fame things as the fame letters refer to in the fection.

XX—Shew the fquare heads of fmall fcrews tapped into the oppofite plate, in order to hold the plates clofe together near the border, as the nutt and fcrew do in the middle : the heads, however, will be better above.

### Explanation of the Plan of the Air Cheft.

A—The air-pump.

B—The fky lights 6 inches diameter each, to be made of window glafs knobbs, if plate glafs is not to be had.

C—Clamp plates of iron to confine the top and fides ftrongly together.

DD—Pigs of lead, end upward.

EE—The lever for working the pump.

GG—The axis and brace for fteadying the lever.

HH—Two bows for hoifting the cheft.

### Explanation of the Section of the Air Cheft.

The fame letters referring to the fame things as in the plan.

I—A ftrong crooked iron to lay hold of the bows to which the main rope or tackle is to be fixed.

MN—The opening from the pump to the air cheft.

*o p*—The valve ; whereof *o* is leather, *p* wood ; to be fhut by a wire fpring, *q r s,* a little more than fufficient to overcome the weight of the valve.

To

To Mr. Pickernell.

Sir,                                                    London, 1ſt April 1779.

As I look upon it that the judiciouſly ſetting on of the baſe courſes upon the damaged foundations is a very material point of conſideration, it was for this reaſon that I was willing to take the whole of the winter fully to make up my mind about it; and having now ſeen the firſt of April, I ſhall no longer defer ſetting you at liberty.

You will readily comprehend, that if, inſtead of beginning with maſonry from the bottom, two or three large detached pieces of rocks had been depoſited in the river, of a ſufficient ſize to build our piers upon; theſe ſerving as feet to ſtand upon, and each capable of ſupporting a proportion of the weight; thoſe rocks, though unconnected below, yet if they are firmly connected at and above the ſurface of the water, by a cap of ſtone well cramped and united in one; this cap would ſerve as a new baſement that could not ſeparate, whereon to raiſe the reſt of the ſuperſtructure.

Our piers in ſome degree anſwer the above idea; and it will now be our buſineſs to make a cap of the two baſement courſes, bonded together with a degree of firmneſs that would not have been neceſſary, had no derangement ever happened.

I propoſe, therefore, that the two baſement courſes be filled in with block ſtones, gauged to a proper thickneſs in reſpect to their beds; and ſo far ſtruck off upon their ſides, as to give opportunity to cramp them to one another, both inſide and out. The outſide ſtones I would adviſe to be cramped to one another with cramps from $1\frac{3}{4}$ to 2 inches broad, and from $\frac{5}{16}$ to $\frac{3}{8}$ thick, all well leaded in, and of ſufficient length to get at leaſt ſix inches fair hold of each ſtone. The whole to be bedded and worked in with terras mortar, of which I will give you the compoſition.

In the next place, as the ground could not be conſolidated with piles under the firſt foundations, I always expected theſe to ſettle to the value of an inch more than the piers that have piles underneath them; but as in conſequence of our derangements, our bearings will be rendered in ſome degree unequal, a greater ſettlement may be expected to take place at the weſt, than at the eaſt end; for theſe reaſons I would have the baſement courſes ſo ſuited to the preſent work, that the eaſt end of the baſement ſhall finiſh an inch higher than the finiſhed piers; and the weſt end two inches higher; alſo that the weſt point and ſhoulders be carried $1\frac{1}{2}$ inch further weſt than the finiſhed piers, ſo as to make the piers $1\frac{1}{2}$ inch longer than the former. In raiſing the ſhafts, I would gradually loſe this $1\frac{1}{2}$ inch in getting up to the impoſt, by ſetting on about the value of $\frac{1}{8}$ of an inch at each

courſe;

courfe; fo that the impoft will be of the fame length as the former, but ftill finifhed one inch higher at the eaft, and two inches higher at the weft ends. Now if afterwards the weight of the arches and fuperincumbent matter fhould not prefs down the piers in the proportion allowed for, this will make no apparent fault ; becaufe it will rather give addition of eleva-tion to the middle of the bridge ; and if the fettlement fhould be double to what is allowed for, which I think is as much as can poffibly be expected, it will ftill be fcarcely vifible. The main thing, therefore, that we have to guard againft in carrying up our piers, is to do them with that care and folidity, that though the bearings may be fomewhat un-equal, yet that the whole may go together, without fhewing any fetts or rends in the outfide aifler; and which, though it may happen without any material detriment to the real folidity and duration of the work, yet it will difgrace our building in the eyes of thofe very many (in proportion to the whole) who can neither fee nor reafon any further ; on which account I would conftantly keep the weft end fome courfes higher than the reft, building up as it were a head, and making off the courfes. I would alfo cramp every courfe in the outfide aifler, from the weft point to half way down the long fides, and every other courfe down to the eaft fhoulders, till you come to the next courfe below the impoft ; at which height I would throw three block or chain courfes acrofs, in the man-ner I formerly ordered ; there let the work of each pier ftand as long as we can, and afterwards finifh as has been already done. I mean what relates to the cramping above the bafement to regard chiefly the third and feventh piers : as to the fourth, after the cap is on, as before, I think it will be fufficient to cramp every other courfe from the point to half-way down the fides only ; but in regard to the crofs chain courfes at the height above mentioned, no caiffon pier fhould want them. In regard to the fifth, I muft fufpend all directions till I fee how it is likely to turn out.

The place where I am moft apprehenfive that fetts may appear, is about a yard below or down ftream of the weft fhoulders. I would therefore have you avoid putting in long ftones faceways, near thofe parts ; for though this may caufe a few more cramps, and in common apprehenfion not be fo ftrong : yet, as every joint will give way a little, it will prevent the breaking of the ftones in the middle, nor will the weight above act by fuch long levers to break them.

Inftead of our former terras mortar, take as follows :

Common lime 1, barrow lime 1, terras $\frac{1}{2}$, fand $1\frac{1}{2}$; — this dofe with pebbles what you can, for rough work.

To

To fet ftones under water :

Barrow lime 1, terras 1, very fmall pebbles 1 ; — this being very well beaten, and let grow rather ftiff, put it down in lumps about the fize of a pullett's egg. If you have an opportunity of flat bearings, you may mortar the under fide of the ftone to be let down.

Below you have my defcription and directions concerning the new tackles.

I am, &c.

J. SMEATON.

---

DESCRIPTION of the Shears and Tackle that were made Ufe of in raif- ing the Piers, &c. of the Bridge of Perth, and recommended for Ufe in Hexham Bridge.

(See plate 12. fig. 7.)

AB fhews one of the fhear logs, which are framed into a foal piece being of nearly the fame height and fcantling as the fhears, that is, about fix inches, but inftead of being fixed by back ftays, they are made to overfet by guide ropes commanded by tackle blocks in the following manner :

The guide rope CD is fuppofed to be fixed to a lewis to fome part of the adjacent pier or arch ; the guide rope goes to an adjacent pier if raifed to its height, or the next but one, if the next is not raifed much above water. DH are a pair of blocks with two fheaves each, whofe fall HIK goes to the jack roll or windlefs L, mounted upon a frame in the ufual manner, and fixed upon fome convenient part of the adjacent pier or arch: now as the roll drawing by the fall KIH, tends to draw the fhears towards it, with a power of one exclufive of the purchafe of the tackle, and as befides the fall there will be four parts of the rope acting as a purchafe, the confequence will be, that the whole purchafe will be as five to one, and the roll muft gather up five yards of tackle fall to make the blocks DH advance one yard towards each other.

Again,

Again, the tackle blocks F will tend to draw the fhears the contrary way to the other, and the tackle fall FMNO being attached to the under fide of the roll L, this roller will command the fhears either way by turning it one way or the other; now to keep the two tackle falls thus brought to the roller at any equal tightnefs, or nearly fo, it is neceffary that the two purchafes may be equal ; it therefore appears that if the fall ONMF, acting over the fingle pulley M, which ferves only as a director, were fimply attached to the block F, no power this way applied to the roll at L, would produce any action upon the fhears, but only pull at the fixed part of the guide rope EF without producing any action any where. The blocks FG, therefore, muft contain a purchafe of five, and this will be done by making the block F treble, and the block G double, the ftanding part being fixed to the block G, in like manner as the other ftanding part will be fixed to the block H. If, therefore, one fall is attached to one end of the roll, and the other to the other, while one is winding off, the other winding on, four feet in length, and fix inches in diameter will be fully fufficient ; and the handle being of the fame length as before prefcribed, as one man will generally manage the matter, the handles will beft be fet oppofite that they may fimply balance each other.

To avoid all confufion of the figures, I have wholly omitted the main tackle blocks, which are fufpended from the top of the fhears ; and to avoid the platform of the work from being crowded, I fuppofe the fall for the main tackle to be returned from a fnatch block fixed to a lewis, next the foot of one of the fhear logs, and from thence paffed to a main windlafs or jack roll, and fuppofed to be placed upon the adjacent pier at P, or upon the arch near the other windlafs. Two men in ordinary cafes being fuppofed to manage both windlaffes, that is to affift each other in hoifting the ftone perpendicularly, and then one of them to go to the guide tackle windlafs to overfet the fhears, which done, the firft can lower it by the main tackle.

The overlaying of the fhears fhould be fufficient to clear the ftones from rubbing and beating againft the fides of the pier, while hoifting perpendicularly, and the feet of the fhears may either ftand upon the aifler or juft behind it, fo as to drop fuch ftones into their places as come within their compafs, and that on either fide. The fall of the main tackle may be prevented from dipping in the water, in paffing from pier to pier, by a fmall block being fufpended from the guide rope at I.

The guide tackle blocks may be fuch as thofe made for fhipping, the fheaves about fix or feven inches.

J. SMEATON.

*London,*
27th *March* 1779.

To

To Mr. Pickernell.                                        Aufthorpe, 25th Feb. 1780.

I have made up my mind refpecting the north termination which I now enclofe, and expect you will find it fufficiently intelligible. See plate 12. figs. 1. 2. 3.   It is made conformably to an elevation on the fame fcale, fhewing the elliptical and firft fegment arch from the north, fent to you in a former letter, defcribing the method of conftructing the elliptical arch, where the cordon is defcribed as laid down upon the key-ftones, and to which I find I have not had regard in making out my defign of the tranfverfe fections, which was intended chiefly to fhew the projections, by which means the whole parapet became raifed fix inches unneceffarily, and in confequence the thicknefs of the road gravel upon the top of the arches.

I am, &c.

J. SMEATON.

EXPLANATION of the Defign for the North Termination of Hexham Bridge.

Plate 12. figs. 1. 2. 3.

THE line AB, fhews the furface of the road to be inclined in the proportion of one in 12; and which is to meet

BC, fig. 1. The furface of the road over the firft fegment arch, which, if produced to the middle of the elliptical arch, will go to D.

The height of the road at B, the middle of the firft pier above the fpringer, is as per fection 17 feet, which being fettled, it will pafs about two feet eight inches above the fofite of the elliptical arch; that is, fuppofing it two feet in thicknefs, will allow eight inches for the gravel, and as much more as the ftones in the middle of the arch are fhort of full two feet in the pier; and NB the exuberances, may be fcappeled off, fo as not to hurt the general bearings, and give their all poffible thicknefs for the gravel at F, the termination of the abutment walls, it will pafs at or about nine feet ten inches above the top of the dooming, fuppofed the ground level at the line E, the extreme of the road wall pillars, extending from the extreme of the abutment walls 64 feet, it will

2                                                                                          be

be four feet ten inches, that is nearly upon five feet at the center of the pillars from whence the road slopes away till it meets the natural ground, which, if perfectly level, will be at the distance of about 58 feet beyond the terminating pillars, and in this space the road is raised rampart fashion, being upheld by natural slopes on each side, like a common turnpike road.

The walking path is supposed to follow the line BD, and then keeping parallel to the top of the parapet D to G, there to descend two steps, upon the extended gravel surface between the abutment walls, where there will be room enough.

The same letters in the section, fig. 1. refer to the same parts as in the plan, and wherein the dotted lines HIKL, are supposed to be the hollowest part on each side for conducting the water, where the one in 12 slopes begins towards L, it will be proper to cross-sett this hollow with thin rubble, to prevent the sand and downfall of showers from getting a channel.

In the cross section, fig. 3. upon the line FFF, wherein

MM—Is the height of the springers.

NN—The top of the dooming or ground line.

OO—The main height of the surface of the road, being supposed a little swelled in the middle, and hollowed at the water courses, as before described.

PP—The line of the top of the abutment walls, and

QQ—The same.

Y s t o w x, is a section of the body of the termination walls, to be six feet high including caping : the inside of the caping ranges with the inside of the abutment walls, and the outside of the body of the termination walls, to be built with all the batter they can, so that the sloping face x w, does not go beyond the abutment wall of the top of the dooming.

The method of founding and building walls, and for founding the termination pillars, no part need be dug deeper than through the loose soil, perhaps one foot deep; but the bottom muſt be ſet with flat rubble lengthways croſs upon the foundation trench; and when done to be driven down with paviours rammers: the body of the wall to be built dry, but at every half yard in height to lay on bond ſtones well bedded in mortar, at the diſtance of two yards from each other, middle and middle; obſerving, that in the courſe of bond ſtones next above, to put them at the middle ſpaces of the courſe below, ſo that there will be in event, a bond ſtone in every ſquare yard of face; and as the bond ſtones may not be long enough to reach through the wall in one entire ſtone others muſt be laid to make good the bond quite through in mortar.

The body of the terminating walls to be built with ſomething of care, in the ſtyle of a very good field wall, the cordon to be hammer-dreſſed, and reduced to a thickneſs; the ſtone to be found and laid in with mortar; the parapet to be of hammer-dreſſed ſtuff of promiſcuous courſes, from three to ſix inches both inſide and out; built with mortar, and neatly walled like a farm-houſe. The caping to be of good found ſtone, broached and well jointed, the terminating pillars hewn where octagonal, and not hid by earth, the ſquare baſe walled and hammer-dreſſed on the ſides that will appear, the foundation rubble in mortar.

J. SMEATON.

To Mr. Pickernell.

Sir,                                                                 London, 26th May 1780.

In regard to the mode of finiſhing the ſouth termination from the ſouth abutment, if the ground were as high it would be perfectly the ſame as the north; you are therefore to carry on the work ſo that all above that level ſhall be the ſame; the ſection of the walls therefore under that level will be no more than a continuance of the ſame ſection lines, obſerving to increaſe the thickneſs, not only by a continuance of the ſlope on the face, but by ſetts off on the back ſide; and as this continuance of the ſlope in part may make the foot to project conſiderably before the foundation of the main walls of the abutment, you may get rid of it by turning it a quarter round upon the abutment wall.

The declivity at top will be the ſame as the north end; and in this manner you may be going on till you come to the hollow, where I propoſe to have two arches of 18 feet ſpan, with a pillar of at leaſt ſix feet between. Theſe arches are to occupy the deepeſt part of the hollow, ſo as to vend the water with the greateſt advantage, and then to have an 18 feet arch from the ſame center over the mill leet, but ſeparated from the other two by ſuch walling as in length to ſuit the ground, and after you are over the mill leet, then to

make

make a quarter turn, and terminate the walling with two termination pillars like thofe of the north end. I propofe to make the 18 feet arches to rife four feet, fo that you may be preparing the centering; and as the penftones may be from 18 inches to one foot, as they may happen to run promifcuoufly, you may be getting them prepared; they need no work upon them further than to bring them to a bearing. The work of the pillar and abutment walls of the arches, alfo need no other work than to make them courfe and joint'; but the whole floor of the land arches I propofe to be rough fett with flat rubble 18 inches in height or depth, to be juft enough fhaped with the hammer in fetting to make them pack like a tolerable pavement. This to be extended about 15 feet below, and about five feet above the arches, apron fafhion, and to be kept by ribbands or ftring pieces piled down. I mean to pile under the pier and abutment walls; but as this will be particularly defcribed by the defign, I now mention them to give you fome idea of the materials that will be wanted for that part of the fervice. I don't however mean any piling or wood work but what will be got out of old ftores.

In regard to the foundation of the terminating walls, I mean that part which you are now to proceed upon from the preceding directions, I believe the lefs you fink below the natural furface the lefs you will get into foft matter, fo that you may rife and fall by fteps according to your difcretion; a foot into the ground, where the cover will be the leaft, I apprehend will be quite fufficient; this you are to crofs fett, and drive down with a rammer, as directed for the north abutment; but if the matter on trial becomes foft, and in your judgement unfit to bear the weight, then you muft drive promifcuous piles at about the rate of one in a fquare yard, more or lefs, as you judge neceffary; but more rank near the front than the back fide of the wall, and rough fett, and driven down between the heads, fo that the heads and fetting may be flufh, upon which you are to build, filling the fpaces of the paving with quick lime and gravel, in cafe it beats to a puddle: the moft ordinary ftuff will do for this fervice, that will bear driving; but get the piles down into a harder ftratum, if you can.

I am, Sir,

J. SMEATON.

To Mr. Pickernell.

Sir,           Aufthorpe, 28th October 1780.

I am forry that any thing fhould have difcompofed any part of the outworks of the bridge, though, as it turns out by your letter, it is nothing more than might be expected,

         whenever

whenever a top flood came, as before that nothing was ever ftirred, and therefore the rubble not difpofed as laid by the water ; but as its effects feem to have been principally fpent upon the fourth pier, I think it will be proper now to do with it what I have all along propofed in my own mind, in cafe an effect fhould take place to the degree you have defcribed of that pier : that is, to drive fome of the round oak piles about 18 inches diftant, middle and middle, driving them down by fetts three feet under the common low water furface : the point pile being advanced up ftream about 15, or, if you pleafe, 20 feet above the falient point of the cafing : the line of the piles to be made a little rounding, I think the upper fide is the better figure. The internal fpace between the piles and the cafing to be filled with rubble, as large as you can well get it, and fome large pieces to be chocked in between the fhoulders of the cafing and the fhoulder piles, and then filled up with fmaller at the top. I would alfo lay a footing flope of rubble upon the outfide of the piles, to be reconciled with the footing flope on the fide of the cafes. As it appears from all the arches interfpaces wearing deeper, that it is the natural effect of the waters being more confined than at firft by the interpofition of the piers ; it therefore indicates, that we fhould not fill up more than what is abfolutely neceffary for our fecurity ; for the more we clock up the more tendency the water will have to take away the blocking and deepen the interfpaces. You have not given me the foundings in the refpective interfpaces, nor told me where the rubble, &c. is chiefly lodged below the tail of the piers ; if I were informed of thefe matters, as alfo whether the deepening of the interfpaces was greater at the eaft or the weft fhoulders, or below the eaft fhoulders, I fhould be able to tell how to direct you. You mention however that your foundings by the fide of the fourth pier were eight feet ; now if all lies regular, I fhall not think eight feet too great a depth in the middle between the piers ; and if the rubble flopes do not reach within three feet of the top of the cafings, extending above two feet in bafe to one of perpendicular height of the flope above the cafings, it will be as fufficient as if higher and broader.

Nothing will fo much tend to give an eafy paffage to the water through the bridge, as its fpreading equally through all the arches. The beating jetee on the north fide, greatly tends to prevent the water paffing freely through the north arches, at the fame time that it throws it upon the center one ; that therefore fhould not only be removed, but the fhore made fmooth and regular, the ground and all impediments removed from the north elliptical arch, and the rifing ground below it taken away ; for if the water is ftopt below an arch it is an equal impediment to the waters getting through it as if under it or above ; and whatever is there ftopped muft be thrown upon fome other, and the water naturally tends to find its paffage where it finds the leaft obftruction ;

that

that is, where the channel is the deepeſt. Our buſineſs is therefore as much as poſſible to invite the water through the ſide arches by ſmoothing and clearing all impediments to its paſſage through the ſame.

I am, &c.

J. SMEATON.

To Mr. Errington.

Sir,

In anſwer to your requeſt, eſiring my opinion in writing concerning the late calamitous accident at the Bridge of Hexham, and what is moſt adviſeable to be done upon it; I muſt, in the firſt place, obſerve to you, that from reflecting upon every circumſtance that has yet been communicated to me, with all the preciſion I am able, I am of opinion that the true cauſe of any failure was occaſioned, not only by the great violence where-with the bridge was attacked, but by the great weakneſs of the ſtratum of matter that lies immediately under the bed of the river, and which has been ſaid univerſally to pre-vail in that neighbourhood by thoſe who made trial thereof, between the building of the firſt bridge and that of the ſecond; which weakneſs of the under ſtratum I was not only aware of, but turned my thoughts towards every expedient that could tend to avert the ill effects that might ariſe therefrom; and having obſerved, that in all the attempts of thoſe who had gone before me in this enterprize, they had dug conſiderably into the bed of the river, and thereby rendered that weaker which was already too weak, I did not doubt but that, by a contrary practice, my endeavours would have been crowned with the wiſhed and expected ſucceſs; for as I had read of buildings and bridges that had ſtood upon more weak natural foundations than this appeared to be, and even myſelf had a caſe of the kind, that I had effectually remedied, I did not doubt, but that with the precaution of not weakening the upper cruſt of hard gravel, but building immediately upon it, I ſhould in like manner ſucceed in this place.

The inſtances, however, that had come to my knowledge, though the ſtrata under the foundations might be naturally weaker, yet none of them are liable to be attacked with any thing near that degree of violence that this river now appears to be capable of; had it been poſſible for me to have been acquainted before hand that a flood of this river could come down with ſo much ſuddenneſs, as that, for want of time, for the lower reaches of the river to be filled from the upper, there could be created a fall or difference of level between the up ſtream end and the down ſtream ſalient point of the ſame pillar, of no leſs than five feet perpendicular, which would in effect create a

velocity

velocity of the water of above a thousand feet in a minute; I say, could I have been informed of this single fact, as appeared to be at, and for some time before any degree of derangement was apparent in this bridge, I never could have thought of advising you, or any private gentleman, to have undertaken, at his own risk, a building of so much danger and hazard; and, exclusively of that danger and derangement which might naturally be expected to arise from the mere rapidity of the water, I am further of opinion from what now appears, that the mere difference of the weight of the body of water immediately above the bridge, which could not be counterbalanced by a body of water of an equal breadth immediately below, has, in reality, been fficient to force down the under soft stratum out of its former position, so as to be more inclined to the west, and occasioned the upper stratum, upon which the bridge immediately stood, to follow it; and in both these respects, that is, of rapidity and unequal pressure upon the bottom, the violence would be greater than even in the great inundation of 1771; for though, acccording to accounts, the height of that was greater, yet its rise was far less sudden; and therefore its rapidity and tendency to derangement less.

To the question, What is now to be done? I answer, that though I do not conceive it impossible to re-erect the bridge in the former site, and where much expense is standing towards a completion; yet I do conceive it impossible to be done with any reasonable hope of its proving a permanent or successful undertaking. I am further fully convinced the bridge was perfectly safe against all common occurrences.

I remain, Sir,

Gray's Inn,
6th April 1782.

Your most humble servant,

J. SMEATON.

P. S. The bridge I mentioned to have restored was the Bridge of Dumbarton, about 20 miles from Glasgow, undertaken by government. That part of it that failed was built upon a crust of gravel not above two feet thick; and, without any flood, external violence, or previous notice, one of the pillars went down, with the two adjacent arches, and crushed the centers, then standing, under them. On examining, I found the ground so soft under this crust of gravel, that a bar of 40 feet went down to the head by its own weight.

To

To Mr. Pickernell.

Dear Sir, Aufthorpe, 6th June 1782.

All our honours are now in the duft! It cannot now be faid, that in the courfe of thirty years practice, and engaged in fome of the moft difficult enterprizes, not one of Smeaton's works has failed: Hexham Bridge is a melancholy witnefs to the contrary; yet, after all, I feel much lefs for honour and credit than I do for the actual lofs fuftained by Mr. Errington: it would give me much fatisfaction if that matter were fettled between him and the county. I have heard that his appearance there at the Eafter feffions much inclined the magiftrates to fee the matter in a favourable light; but the difficulty was, how far it was in their power.

I faw Mr. Donkin in town, who acquainted me, that he was looking at it when it firft appeared to give way; his fon being but juft returned from the fouth fide, to fee that the fmall arches there were fafe, which was the only part that they had any doubt about. He was wondering at the poffibility that any ftructure could withftand fuch extreme violence, yet not at all expecting that any thing would hurt it; and remarked that before any thing appeared to give way, the water was up to the top of the doom-ing of the piers up ftream when it fcarcely touched the bottom of the impoft down ftream, which makes a fall of five feet, and it was not above a couple of minutes between firft per-ceiving the mortar dropping out of the joints of the fofit and the fall of the arch, and fix more were down in half an hour, fo that it was fo equally guarded that in a manner it all went together. Could I have known beforehand that there was a poffibility of a flood to come down fo *fuddenly* as to have made a fall through all thofe extenfive openings of five feet, I fhould certainly never have attempted the building a bridge in that fituation, as that fall would neceffarily create a velocity to the water in its paffage of 1100 feet per minute: a velocity that it would require the ftrongeft fluice-floor and aprons to withftand. I am therefore clear that it has forced away the very bed of the river and all before it. The occa-fion of this extreme fuddennefs in the waters coming down, which in that refpect was far greater, according to all accounts, than in the great inundation of 1771, though the total height not fo great, was, doubtlefs, owing to this circumftance, that the afternoon before a very great downfall of fnow happened, fo deep as to cover the ground at an average two feet. This was immediately fucceeded by a vaft downfall of rain, none of which would run off, till the whole body of fnow was faturated like a fpunge, and then, like the burfting of a fnow-ball in the fire, it would come down all at once, and that fo fuddenly, that the lower reaches of the river not having *time* to be filled from the upper they would be comparatively empty; and as the velocity of water depends upon its fall, or *difference* of level at any given place, and not upon its total height, the difference will

I depend

depend upon the fuddennefs of its coming down ; and this being further hurried down by a violent gale of wind at north-weft, the very direction that, upon the whole, would tend to bring it down the quickeft from both the Tynes, it would feem as if all the powers of nature were collected to humble my pride and yours. The news came to me like a thunderbolt ; as it was a ftroke I leaft expected, and even yet can fcarcely form a practical belief of its reality. A flood that could mount up to the top of the doomings of the piers was, however, not a fmall, or even middling flood, in point of height ; and as every object that entangled it would moderate its rapidity, it is not improbable but that the downfall of Hexham Bridge might be the faving of Corbridge ; and by the fpreading gradually over the wide haughs in many places below, was not more than a moderately large flood at Newcaftle. There is, however, one confolation that attends this great misfortune, and that is, that I cannot fee that any body is really to blame, or that any body is blamed : we all did our beft, according to what appeared ; and all the experience I have gained is, not to attempt a bridge upon a gravel bottom in a river fubject to fuch violent rapidity.

I remain, your moft humble fervant,

J. SMEATON.

P. S. I cannot fuppofe any failure arifing from the greennefs of the work, but wholly from the whole bed of the river giving way under it ; as nothing appeared to fettle till it gave way in a manner at once, which is a proof of its being firmly bonded together.

---

A COPY of Mr. Mylne's Opinion and Report, delivered to the Magiftrates of Northumberland, refpecting the Practicability of building a permanent Bridge at Hexham, on the Site of that built by H. Errington, Efq.

Hexham, April 24th 1783.

To the Magiftrates and Juftices of the Peace in and for the County of Northumberland, in General Meeting affembled.

Gentlemen,

BEING requefted by you to view and examine the prefent ftate of the Bridge acrofs the river Tyne, lately built by, and at the expenfe of Henry Errington, Efq., agreeably to the plan and advice of Mr. Smeaton, and to infpect the fite and nature of the bed of

the

the river, whereon the fame was conftructed, I beg leave to report, that it fully appears, this bridge was undermined by the great flood, which happened on the eleventh day of March 1782, and that the foil and fubftance of the bed of the river, of whatever matter and quality the fame confifted, was dug or fcooped out from below the greateft part of the piers, and *that* towards the weft or upper fide of the bridge,—and the advanced or guard works, inclofing a fpace round the piers, for the better fecurity and maintenance of the foil immediately under the piers, whereon the whole weight of the ftructure had been charged, were alfo undermined by the gravel and fand, into which they had been driven, being worn away by the velocity of the ftream; thus circumftanced during the height and greateft rage of a flood, it will appear no wonder that the piers, having a fourth, half, and even to three-fourths of their bafe taken out from below them, the arches fplit in two longways, by fome of the piers breaking acrofs, into two parts, precipitated into ruin themfelves, and the parts conftructed upon them.

All the piers fell towards the ftream, but took different inclinations towards the excavation of the bottom, fometimes made more on one fhoulder han on the other of each pier.

The furface of the water-line is now twenty-two inches lower than in the time during which the bridge was conftructed; I have bored the river at the bridge to the depth of twenty-three feet below the latter water level, in a place where I might not be led aftray by any alteration formed by the faid flood, in the height of its impetuofity; and fkimmed over again in its milder velocity; and I have found under the teftimony and perfeverance of Mr. Wake, that the foil and texture of the bed of the river at this place, is uniformly a compofition or congeries of roundifh and flat ftones, gravel and fand, of equal quality and confiftence in the whole of that depth.

The piers which were founded by means of a batter d'eaux, have ftood tolerably well, and thofe which were laid by caiffons, having no piles directly under the piers, were the eafieft prey to the vaft powers of this flood.

The depth to which all the works in general, whether immediate or prefervative, were carried was far too fhallow, and too little into the bed of the river, which (though hard to the touch of boring and compact to the eye, and feeling of inftruments) is wonderfully loofe, and unconnected in its parts, in fo much, that the bed of the river Tyne feems to fhift and alter its form, extent and fituation, with every flood more or lefs, and tearing up at one time to a great depth, that fair moulded and well laid hollow, which the ftream had laid for itfelf on fome former occafion.

In fuch a fituation, under thefe circumftances, with the additional one of many piers being to be fixed as obftacles to its violence, the foundations could not be laid too low. To what depth they ought to have been laid, and the means to be devifed for that purpofe, lay and remain with the parties engaged in the performance.

Mr. Smeaton, than whom there is no perfon or artift better inftructed, more knowing, and of a more penetrating and correct judgment, muft have been deceived in the collection of facts and materials, on which he eftablifhed his plan of operations.

A great mind is often deceived by its own virtues. Habituated to give, on all occafions, the genuine and honeft productions of its faculties, it often relies too much, and implicitly, on that which appears to be the fame of other perfons.

The exiftence of a fand below, and a fuppofed hardnefs and concretion of five feet or any fuch meafure of the upper parts, feem to have precipitately and fatally determined the plan of operations, at firft fetting off, and appear to me to be equally the caufe of the prefent precipitate opinion, for abandoning the propofition as impracticable.

Art furnifhes the means, I humbly prefume, of going to a fufficient depth, with all the foundations, and that too, on the fpot, notwithftanding the damage which this fite has received.

To compare the eligibility of this fituation with that of any other near to Hexham, and the direction of the Alemouth road, is perhaps ufelefs to difcufs at prefent.—But in my opinion, if nothing had hitherto been done, (as too much has unfortunately been done) I would have recommended a place of deep water, and fixed elevated banks, like to that oppofite to the weft end of the dwarf wall of the Spital Green within Mr. Waftell's inclofures.

To enable me to form the opinions, which I have now the honour to report to you, I have feen and examined all Mr. Smeaton's papers, I have heard all the particulars, and hiftory of his proceedings, and motives for the method of operations which he adopted, I have examined every part of the works, and many perfons concerned, and employed in the detail of its execution; and it was my lot to pafs here in Auguft 1778, when I viewed the effects then produced, in thofe very foundations, by floods previous to that time, making for my own inftruction, minutes of what then I faw.—I mention thefe things to enable you to judge of that reliance, you and all other parties interefted in this bufinefs, may be pleafed to give to thefe opinions, and to this report, which is made by

Your moft obedient, and very humble fervant,

ROBERT MYLNE.

## Mr. SMEATON'S Memorial concerning Hexham Bridge.

WHEN Mr. Smeaton was applied to by Mr. Errington, for the building of Hexham Bridge, it was not till after the total deſtruction of a bridge at the weſt end of Tyne Green, near that place, built under the patronage of Sir Walter Blackett; which, about twelve or fourteen months after it was finiſhed, was totally deſtroyed by an extraordinary flood, that has ever ſince been diſtinguiſhed by the name of the Great Inundation, which happened in November 1771; but as this bridge was ſtanding at dark in the evening, and totally demoliſhed the next morning, no other information could be drawn from this very fatal and alarming accident, but that this river was capable at times, from a certain combination of cauſes, of being ſwelled to a degree of violence far exceeding any thing that had before this been experienced, handed down by tradition, or imagined.

As an evidence of this, amongſt many others that might be given, the water roſe ſeven or eight feet, or thereabout, upon the main ground floor of Mr. Fenwick's new-built apartments at Bywell; which being erected from the deſigns of that eminent architect Mr. Payne, it is not likely that he would direct the main floor to be laid within flood-mark, as it had, at that time, been known, or then thought likely to happen; and yet, Bywell being many miles below the junction of the two Tynes (that is of the North and South Tyne, about a mile above Hexham) and after it had had much room to ſpread over the wide haughs that laid between Hexham and Corbridge, and alſo in the ſpace between Corbridge and Bywell, we muſt conclude, that the riſe of the water was leſs at Bywell than in the neighbourhood of Hexham.

Under this degree of information, and experience of the utility of a bridge at Hexham, (Sir Walter Blackett having choſen rather to forfeit the penalty of a bond of three thouſand pounds, that he had laid himſelf under for the upholding thereof, than attempt to rebuild or re-eſtabliſh the bridge), the erection was taken up by the county; and, for this purpoſe, conſulted that eminent engineer Mr. Wooler, then engaged for the town of Newcaſtle, in the re-eſtabliſhment of Tyne Bridge there, ſo far as the magiſtrates of that town were concerned therewith, and (which through the Fan down the river, and in the tides-way, ſuffered alſo an almoſt total demolition) Mr. Wooler, on faith of borings made by a ſurveyor, a perſon employed by the magiſtrates of the county for that purpoſe, who reported, that a bed of clay laid at no more than four feet under the bed of the river, at a place about fifty yards above, or weſtward of the bridge built by Sir Walter Blackett, formed a proper

Q q 2

design

design for building a bridge upon the foundation deſcribed, upon the principles of piling and planking under the piers, and which was begun accordingly in the year 1774 ; and, after building the north land-breaſt, Mr. Pickernell was recommended in the beginning of the year 1775 to the county, by Mr. Wooler, and employed as ſurveyor under him for the erection of this bridge; who having ſunk the foundation pit for the firſt pier from the north abutment, as directed, to the depth of four feet below the bed of the river, to find the bed of clay; inſtead thereof, came to a ſtratum of a very different nature, which, after examining, he reported to Mr. Wooler, then at Hull, viz. " A quickſand full of " bubbly ſprings, and of ſo looſe a texture, that by hand only, a bar of iron entered " into it forty-ſix feet without meeting any reſiſtance ; and that a trial pile of whole timber " entered twenty-ſix feet, at two inches and a half per ſtroke of the ram without ſloping ; " and that the gentlemen concerned were eye-witneſſes to the facts."

Upon the above report Mr. Wooler declared his opinion, that the attempting to ſet a bridge upon ſuch an enormous depth of quickſand, over a river ſo ſubject to great floods as the Tyne, may be deemed ſo hazardous, as to be next to imprudence itſelf — and again, that this wretched quickſand, rendered the attempting a bridge on ſuch principles (that is piling and planking under the piers) little better than folly — a quickſand, which, from its reſiſtance to the iron bar, cannot be deemed much better than a heap of chaff. For, ſays he, " let it again be ſuppoſed, that a flood like that which overturned the late bridge " ſhould happen, it cannot be doubted, that when the looſe gravel under the bed of " the river (only four feet thick) ſhall be ſwept away between any of the piers, the " quickſand under it will preſently follow like water itſelf; and an excavation may be " made in a few hours, as deep or deeper than any of the piles that guard the piers ; " when a downfall muſt be the immediate conſequence. On theſe principles therefore, " the bridge ought not to be attempted in this ſpot; and, if no better can be found in " any other ſituation, there is but one method of dealing with ſuch ground, which has " ſucceeded where expenſe was not regarded: that is, by carrying a ſolid wall quite " through the river, from ſide to ſide, about ſix feet high; and in this caſe it muſt be " forty-two feet broad." The preceding extracts are made from a copy of Mr. Wooler's letter of the 19th of July, 1775, to Mr. Pickernell; the peruſal of which will more amply ſet forth the grounds and reaſons of Mr. Wooler's opinion; that no bridge under ſuch circumſtance, is likely to be accompliſhed at any limited expenſe; he concludes with ſaying, " I had the honour to mention this method (that is of the ſolid wall) " to Mr. Aynſley, when there was a doubt about the nature of the ground, ſometime " before their ſurveyor found out the ſtratum of the clay; but he then looked upon the " expenſe to exceed their abilities; but however, after all, if ever a ſtable bridge be

" made

" made there, I do not know any other means to effect it — You will lay this before the
" gentlemen for their confideration."

After this, Mr. Pickernell proceeded to fink a well or fhaft in the folid foil of Tyne
Green, near the place where the fouth abutment of the intended bridge was to be; when
paffing through the ftratum of gravel, found the quickfand at nearly the fame depth,
(that is to fay, four feet below the bed of the river) as in the foundation pit for the piei
on the other fide, into which he thruft his iron bar as before, and covered-up the fhaft,
till Mr. Wooler fhould come and examine the premifes.

He alfo proceeded to try the river, by boring in other places; and particularly in the
pool below the eaft boat; that is, a little above the place where Mr. Smeaton afterwards
pitched upon, to build a bridge for Mr. Errington; an account of which boaring being
tranfmitted by Mr. Pickernell to the clerk of the peace, reference being thereto had, will
more fully appear; but which went to prove, that whenever Mr. Pickernell had pene-
trated the bed of gravel, univerfally a ftratum of quickfand was found.

Under thefe circumftances Mr. Wooler attended at Hexham, to furvey the premifes;
and in the prefence of fome of the magiftrates affembled on the occafion, repeated the
trial of the bar, both in the foundation pit near the north, and in the fhaft near the fouth-
end of the intended bridge, and which fucceeded as before mentioned. The exiftence
of a ftratum of quickfand under a bed of gravel in this place, then, does not depend
upon the fimple teftimony of Mr. Pickernell, but is alike witneffed by very refpectable,
as well as competent judges.

The refult of which view and furvey was, that no other place appearing more eligible
and likely, than where the beginning had been made, and being unwilling to go on upon
the principle intended and began, of piling and planking under the piers; and the
magiftrates not giving ear to the folid wall propofed by Mr. Wooler (acrofs and under the
whole bed of the river, from fide to fide, as an artificial foundation whereon a bridge was to
be erected), on account of the expenfe thereof, which was not likely to be uncertain, but
fo great as to be very imprudent for even the county to enter upon, the whole undertaking
was at that time given up or fufpended : and Mr. Wooler having been urged, as too eafily
defifting from his original plan, on going away, he very fagacioufly and prophetically faid,
whoever meddled with a bridge there, would burn their fingers. After this the gentlemen
of the county, unwilling to lofe fight of a bridge at Hexham, an advertifement foon
after appeared in all the Newcaftle newfpapers, as from the bench of magiftrates of the

county

county, importing an invitation to all adventurers to undertake the erection of a bridge, taking the rifque of making a foundation upon themfelves, and taking their own method of doing it, but to build the fuperftructure according to a certain defign to be produced to them, and fecurity for the permanency of the whole for the term of feven years. This advertifement was continued till the latter end of the year 1776, in which interval feveral adventurers had in fucceffion offered, but all of them on a clofer view, before the completion of a contract, ftarted off; feveral expenfive preparations having been made at the expenfe of the county, and the materials lying upon their hands : this work being therefore, as will appear from what is preceding, generally confidered as a derelict fcheme, or at leaft a forlorn hope; fome time in the latter end of the year 1776, Mr. Donkin, agent of Mr. Errington, came to Mr. Smeaton in Mr. Errington's name, to know if he would undertake the direction of building a bridge over the Tyne, fomewhere between the Lowford, and the Eaftboat at Hexham, for Mr. Errington, provided he (Mr. Smeaton) could find a place for the founding thereof which he thought fo fufficient, as that he would rifque his credit upon it as an artift : in which cafe, if it could be done upon a moderate eftimate, he (Mr. Errington) would make a propofal for building it to the county ; urging, that as the county had been fo long baffled in the attempt, as it would be an advantage to his eftate, if it could be done there, it was probable that if it could be done at a moderate expenfe, the county might accept of a bridge there, rather than none, and if he (Mr. Errington) was two or three hundred pounds out of pocket on the above account, he would think it worth his while.

Mr. Smeaton, being fomewhat furprized at the uncommonnefs and newnefs of the propofal, defired time to confider of it, as previoufly to that time, he had ftudioufly avoided having any thing to do with it ; though he had been frequently in that neighbourhood for a courfe of years, comprehending, and even preceding, the time of firft undertaking thereof by Sir Walter Blackett ; but confidering it on this occafion as a great advantage to the public, that if that could be done which they then feemed unlikely to get done ; he began to confider the caufes of failure in thofe that preceded, and in thinking ferioufly of the fubject, there occurred to him a mode of conftruction, that could not only be executed for a very moderate expenfe, confidering the extent of the fubject; but the only mode in which a bridge could be executed, on fuch a kind of foundation as was then generally fuppofed, at any moderate expenfe and with a reafonable profpect of fafety.

This confidered, Mr. Smeaton acquainted Mr. Donkin, that he was willing to examine the fituation ; and if he found it competent, fo that he could hazard his credit as an artift

4                                                                                          upon

upon it, he would be willing to give Mr. Errington an eſtimate. The trial was made, aſſiſted by Mr. Pickernell, who on this occaſion recounted the principal part of what is above ſtated, ſo far as he was concerned ; the eſtimate was made, preſented and accepted, the bridge undertaken, built, and ſuffered the fatal overthrow that has occaſioned the preſent litigation.

The preceding narrative will ſufficiently ſhow, that none of the parties preſſed them-ſelves into this unfortunate, this ill-fated buſineſs, or proceeded in it from any intereſted motives.

Reſpecting Mr. Errington, he neither profeſſed nor could ever have any view of profit from the undertaking, the contingent benefit that it might in that ſituation be to his eſtate, being the ſole motive of the pains and trouble that muſt neceſſarily attend it.

Reſpecting Mr. Smeaton, he neither aſked, expected, ſought, nor received more than his accuſtomed daily hire, and he truſts, that it will not be ſuppoſed, that he could wiſh to undertake this buſineſs for want of employ ; and in reſpect of Mr. Pickernell, if he wanted employ, he would have been much more likely to have met with it, by the bridge being proceeded with, if he had reported a good gravel to an unfathomable depth, knowing or believing there was a quickſand at nine or ten feet under it, than he could expect from reporting a quickſand, at nine or ten feet under the ſurface of the gravel, knowing or believing it was gravel unfathomable ; becauſe being then not at all acquainted with Mr. Smeaton's ideas of the proper method of treating ſuch a ſubject, he muſt ſuppoſe it more likely for a bridge to be undertaken and proceeded with, if the foundation was a gravel unfathomable, than if it was a gravel with a quick-ſand under it.

What remain, therefore, as queſtions material to the county and to all the parties, ſeem to be the following, and what they who undertake to judge of the whole matter, ſhould be acquainted with.

1ſt, Whether Mr. Smeaton from the whole matter before him, at the time of forming his project, did it with that deliberate judgment and reaſonable probability of ſucceſs, that have characterized him in other things?

2dly, Whether

2dly, Whether Mr. Errington was sparing of any thing necessary to give success to that mode of building which Mr. Smeaton had adopted?

3dly, Whether Mr. Pickernell did to the best of his power and abilities execute, to a reasonable and possible extent, what he was directed by Mr. Smeaton?

4thly, Whether under all the experience and knowledge of the subject, as it now stands, the present bridge should be attempted to be re-instated, or a new one built at Hexham?

With respect to Mr. Smeaton's scheme for the bridge, the following matters are worthy of observation:

1st, That from the failure of Sir Walter's bridge in the night, no estimate could be formed, of the fall or velocity that the water had in passing that bridge, at the extreme of the flood, before its failure: for though the marks of the flood were left very visible, which shewed it to have risen many feet higher than any former flood, in point of height; yet this gives no light into the stress laid upon the bridge by the velocity of the water, for no bridge, even tolerably built, ought to suffer from the water's rising any height upon it whatever, if stagnant or rising very slowly, by the counteracting of a rising tide opposing the natural current.

2dly, That it is ascertained beyond a doubt, that at the place where Mr. Wooler began, there in reality existed a quicksand of an unfathomable depth, covered with a bed of gravel, of a very moderate thickness and consistence, and intermixed with large tumbling white stones.

3dly, That not only from the faith of Mr. Pickernell's subsequent borings, but from the proximity of the two situations (not one-third of the breadth of the river asunder), it appeared in the highest degree probable, that the same, or some such a stratum of loose matter, lay under the foundation of the bridge built by Sir Walter Blackett, and was the cause of its destruction; otherwise, its sudden and total destruction, in so short a space of time, was to Mr. Smeaton totally unaccountable; who in the way of curiosity (as other business often carried him to Hexham) had sometimes viewed the operations of that bridge while it was building: that bridge having been in his judgment at that time competently well founded to guard against accidents, considering in what manner

it

it was done, when compared with the manner of founding on gravel ufed by our forefathers; and which, for Sir Walter Blackett, was defigned and undertaken at the rifque
of that ingenious and well experienced builder Mr. John Gott, of Woodhall, in the
Weft-Riding of the County of York, who for feveral years previous to that, had been
undertaker of the building and repairs of the county bridges in the faid Weft-Riding,
and alfo furveyor of the rebuilding and repairs of the navigation works upon the rivers
Aire and Calder; a man rendered truly refpeƈtable to all who knew him, from a long
feries of fuccefsful experience in this kind of arduous undertakings; and who, more-
over, previoufly to this undertaking at Hexham, had with great fuccefs and credit to
himfelf, then undertaken and completed the new bridge at Ferrybridge, in Yorkfhire,
which was done upon the felf fame principles that he afterwards put in praƈtice
upon the Tyne, and which bridge ftill ftands unhurt upon the river Aire (there
alfo united with the large river Calder) to the praife of the fkill of the worthy builder
thereof.

This perfon, Mr. Gott, being perfonally known to Sir Walter Blackett, as well as
his works, Sir Walter pitched upon him as a proper perfon to enfure fuccefs to his
favourite projeƈt; and ftill the more effeƈtually to do it, he joined with Mr. Gott,
Mr. Brown, a very worthy and experienced mafon of the neighbourhood; a perfon
that had acquitted himfelf by many works done for Sir Walter and others, and fome
in the bridge way, and was alfo at that time furveyor for the bridges of the county
of Northumberland: and ftill the further to fecure their care and induftry in this undertaking, he had them bound to him as undertakers for the fum for which they con-
traƈted with him, to uphold their works for the term in which he ftood engaged to
the county; but as a demonftration, that fhews how well he was fatisfied, that the
care and fkill of the undertakers were fully and properly exerted, he, after the
accident and a full examination, gave them up the bond they had entered into
with him, contenting himfelf to pay the penalty in which he ftood engaged to the
county.

And now, as it will throw a confiderable light upon what I have to fay further upon
the fubjeƈt, it will not be loft time to explain the mode of founding, adopted and put
in praƈtice by Mr. Gott, as it appeared to Mr. Smeaton by ocular infpeƈtion, and who
at the time was acquainted with the undertakers, but more particularly and previoufly
with Mr. Gott.

                             Having

Having conftructed large and broad coffer dams of earth to fence off the water, by the help of chain pumps, they fank the foundation pit about three feet into the gravel, then they drove piles over the whole area of the intended foundation of each pillar, from ten to twelve feet long, and from ten to twelve inches diameter in the heads, and tapering according to the natural taper of the timber, proper for driving into gravel of confiderable refiftance. The heads being cut to a level, the whole was covered by a platform, made of whole (that is twelve inches) Riga balks, rabbetted or halved into each other, fo that each could not fubfide without its neighbour going with it, and upon this platform the pillars were refpectively built.

Mr. Smeaton has reafon to believe (though he never happened to be there when any piles were driving), that the undertakers finding their piles go into the ground more eafily than they expected, and the upper part the hardeft, did not in all the pillars make the excavation of the foundation pit quite fo deep as above mentioned, but yet all were founded below the bed of the river: and, in a converfation with Sir Walter Blackett, after the founding the bridge was done, Sir Walter obferving to Mr. Smeaton, that a rumour had gone forth, that the founding of the bridge had not been made fufficiently ftrong, Mr. Smeaton faid, that had they encreafed the circumference with plank or fheet piling, as he had done in all the gravel foundations of the kind, that he had had the ordering of, it was all he fhould have done more than was done; but as the laying a folid platform, and even the piling itfelf, were things that our forefathers had not generally practifed in fuch cafes, and yet we found many of their bridges ftanding after many years trial; it muft be fomething very extraordinary that could hurt a foundation fo laid, far beyond any thing wherewith we were then acquainted.

This ferves to fhew what the opinion of Mr. Smeaton was at that time, before any derangement had happened, fo that it was a matter to him of very great furprize, that notwithftanding the extraordinary height of the water, a bridge fo founded fhould be fo entirely demolifhed in fo fmall a fpace of time; but when the operations of Mr. Wooler were known, his furprize ceafed: looking upon it as a certainty, that the violence of the water having taken off the cruft of gravel, wounded alfo by the excavation for the piers, fo as to let loofe the quickfand, he no longer wondered at the fudden demolition of the bridge.

The third matter to be obferved is, that Mr. Smeaton had at that time, (that is, at Mr. Donkin's application) finifhed with fuccefs two capital bridges in Scotland, over

two

two of the reputed moſt rapid rivers of their magnitude in that part of Great Britain; that is, over the Tweed at Coldſtream, which was finiſhed about the year 1767, and the Tay at Perth, which was finiſhed in or about the year 1770, and which in the interim before Mr. Donkin's application, had ſuſtained many ſevere attacks from floods, but without any injury, except (in ſome ſlight degree) to the rough rubble ſtone depoſited round the piers by way of defence, and which being occaſionally replaced, the whole remained and does ſtill remain unhurt.

Theſe bridges, the firſt being in part, and the latter wholly upon gravel of unfathomable depth, were founded on bearing piles, encaſed with ſheet or plank piles, below the bed of the river, the ſpace being filled up, and the foundation farther defended by the depoſition of rough quarry rubble ſtones: and Mr. Smeaton having experienced the great dependence and power of reſiſtance of ſtones ſo depoſited, not only in the caſes of building the bridges above mentioned, but in a great variety of caſes, preceding thoſe undertakings as well as after, wherein he found them the moſt effectual means, not only of controuling the violence of rapid rivers, but of the ſea itſelf, he was naturally led to place very great confidence in that ſpecies of defence.

4thly, That partly from the report of Mr. Pickernell's borings, partly from the ſimilarity of ſituation of the place propoſed by Mr. Errington, to that where Sir Walter Blackett and Mr. Wooler had worked, being both of them near the bottom of an extenſive pool, wherein the water is kept up by a bed of gravel juſt below them, and forming as it were a natural dam, whereby the motion of the water in the pool above, in the low ſtate of the river, was ſcarcely perceptible; I ſay, from ſimilarity of ſituations, Mr. Pickernell's report of the ground, juſt above the place pitched upon by Mr. Smeaton, and the trials that he (Mr. Smeaton) made himſelf, by driving a ſharpened iron bar from nine to ten feet into the bed of the river in ſeveral places, which was very conſiderably leſs reſiſted, and particularly in the main current, after it was driven down ſome feet, than it was in entering the upper cruſt of the gravel bed, which was apparent to him, by his aſſiſting perſonally in the operation; from all theſe conſiderations he thought himſelf well juſtified in concluding, that at ſome depth, exceeding nine or ten feet, at this place, there either actually exiſted a ſtratum of quickſand, ſimilar to that at the weſt end of Tyne Green, or at leaſt matter ſo little compact or capable of bearing weight, that to drive piles into it would only weaken the ſtratum. The queſtion therefore, that he had to decide for his own guidance was, Whether there was a bed of gravel of ſufficient thickneſs and compactneſs to bear the weight of a bridge, in caſe it was unwounded and unbroken? And the experiment of the bar abovementioned

R r 2

(which

(which was tried in feveral places acrofs the river), determined his judgment, that what he had felt and experienced was fufficient.

It may here naturally be enquired, why Mr. Smeaton did not bore the bed of the river, inftead of driving the bar in the manner defcribed? and he anfwers, becaufe former experience had taught him to have very little faith in boring in gravel, for the purpofe of founding bridges; for the colliery borers, though exceedingly expert in boring for the purpofes to which they are to apply them, yet are no competent judges of the compactnefs of the ftratum for the purpofe of building a bridge; and in the trials formerly made by Mr. Smeaton himfelf, from the continual falling in of the fmaller parts of the gravel itfelf, while the fhank of the inftrument is turning round, thereby occafioning a continual grinding; and if the inftrument is attempted to be withdrawn, the holes immediately filling, made it never appear to him in the light of a fatisfactory operation, convincing to his mind of any certain conclufion: he has therefore, for many years paft, contented himfelf with trials by the bar, which being driven by a hammer, he judges of the compactnefs of the gravel, by the number and ftrength of the blows that it takes to go down; and on the faith of trials of this kind, where the bar went down with a competent refiftance and a near equality, he built the bridge of Perth upon piles encafed with fheeting.

From a mature confideration of the above particulars and circumftances, Mr. Smeaton found himfelf led to the following conclufions, viz.

That to build a folid wall acrofs the river as a foundation for the whole bridge, in the manner propofed by Mr. Wooler, would not only be attended with an enormous expenfe, but, in the place where he propofed it, likely to be in itfelf impracticable: for it did not, nor does it occur to Mr. Smeaton, how this is to be done without draining off the water from the bottom of the very large excavation that would be neceffary to be laid open at once; which muft not only go down to the quickfand, but in reality confiderably into it, to lay the propofed foundation of the wall; that in cafe the quickfands fhould break up and run, as it was moft likely to do, the drainage of this liquid matter would be endlefs; and if any part of it was attended with fo much fuccefs as to get founded, yet the part fo founded would be fapped, when the fand is fo broken up in any fucceeding part.

That though in the place pitched upon by Mr. Smeaton, the bed of gravel appeared both thicker and firmer than where Mr. Wooler had begun; yet, as it appears evidently

dently to him, was likely to partake of the fame quality, the execution of the fcheme of the folid wall, or of penning as propofed by others (to make which effectual muft amount to the fame thing) could not be done upon any limited eftimate; and at any rate, would exceed all bounds of expenfe, that it appeared to him likely or indeed prudent to be gone into by the county.

That to attempt the building of the bridge upon the principles of that of Perth; that is, to fink an excavation pit confiderably into the bed of the river, and in this to pile and encafe, would be, in effect, firft to deftroy the very beft and firmeft part of the ftratum, and then by driving piles into what was likely to be incapable of bearing the weight, would be in reality to repeat the errors, that, as it feemed to him, had been committed in Mr. Gott's erection; and as, laft of all, the fecurity of the bridge in any of thefe methods, muft ultimately depend upon the defences to be made by the judicious and proper depofition of rough quarry rubble; it appeared to him a folly, firft to deftroy the firm upper cruft of gravel that he reported verbally on his trial thereof to be comparatively hard, like the pavement of Hexham ftreets, and then, at a great expenfe, fubftitute fomething not fo much to be depended on, and this ftill want defending by quarry rubble, which in every cafe could be applied: and he muft here beg leave to remark, that a quarry, fituated moft commodioufly to this fituation of the bridge in the eftate of Mr. Errington, offered the greateft plenty of this kind of material, and of the moft excellent quality for the purpofe that he has any where had the experience of.

From the whole of the premifes he concluded, that the fafeft way would be to preferve the upper cruft of the bed of gravel inviolably unbroken even by a pile; and particularly in the main channel of the river, where the diminution of the hardnefs of the upper cruft principally to him appeared; fo that, concluding to build the two land-breafts upon piles, with cafting, and alfo the two pillars next thereto, in the fame method, with coffer dams to drain out the water (he having found that within that compafs the bed of gravel appeared equally hard and compact) the method that naturally offered itfelf was to found the reft of the piers by caiffon; a method the moft eafy and ready, and attended with the leaft coft of any. So that having before abundantly experienced that good quarry rubble would refift the action of a current to a greater degree than any kind of gravel, it appeared that the pillars fo funk, being defended from accidental flood till they could be furrounded by a flope of rubble (which the depth of the water naturally admitted in this place) hence would arife every degree of fecurity that the nature of the fubject would admit of.

He

He concluded therefore to build a bridge of nine arches inftead of feven, that it might have more legs to ftand upon, in confequence of the natural weaknefs of the ftratum; and by way of fecurity to the piers, before they could be properly and fufficiently furrounded by the propofed flope of rubble, as well as after, in cafe of any derangement to the rubble defence, a girdle of ftones in blocks of a ton weight and upwards, was propofed to be let down, and furround the bafe of each pier, to be fitted to each other, and to the pier they furrounded, and to be cramped together.

Upon this idea of conftruction Mr. Smeaton formed his original eftimate; and which, from the fimple mode of it, could be executed for a very moderate fum of money, in proportion to the largenefs of the river and extent of the work; and which, in confequence, was bargained for by Mr. Errington, and the work proceeded with accordingly.

The north land-breaft and the adjoining pier were fuccefsfully built upon piles encafed as propofed; and the gravel being there very fufficiently compact (fo as to afford only a moderate quantity of water) Mr. Smeaton determined to try to go on as far as he could upon that principle, and therefore ordered the fecond pier from the north abutment to be tried with a coffer dam, to encavate and found like the firft; but when the pit was funk but two feet under the level of the water outfide, and not much more than half as much under the natural bed of the river, the water boiled up between the interftices of the gravel ftones, bringing fand along with it, that it required forty men continually at the pumps to keep it down; and it was not without the utmoft difficulty that the pier was founded on that principle at that depth.

The fouth abutment and contiguous pillar were alfo fuccefsfully founded, according to the original intentions; but at the fecond pier from the fouth, the water being much deeper (being in the main channel, and the gravel bottom clean wafhed, like the fecond from the north) Mr. Smeaton judged it in vain to attempt any more pillars by coffer dams, becaufe it would be an ufelefs expenfe to conftruct a coffer dam without the leaft probability of maftering the water.

Early in the fummer of 1778, the remaining five pillars were begun to be executed by caiffons, and Mr. Smeaton attended the execution of the firft that was laid, which was the fourth pier from the north fide of the center arch; and which was done with fo much expedition, eafe, and convenience, that the feafon and weather turning out remarkable fine, the whole body of agents and workmen preffed forward to get as many

of

of them done as poffible while that favourable feafon lafted ; but a number of blocks for the girdle courfe not immediately arifing out of the quarries of a fufficient fize for the purpofe, about the latter end of July four out of the five caiffon piers were grounded upon the bottom of the river, and brought above water, when only one of them had any of the girdle ftones brought and depofited, and this only in part around it ; beginning from the weftern or up ftream falient point, and extending from thence about half way round the pier on each fide : in this ftate of things, after a remarkably dry feafon of fome months, there came a violent rapid flood, not indeed a very high one, but the river being previoufly empty of water, and the rain which occafioned it falling very fuddenly, it came down (being alfo urged by a violent wind at weft) with uncommon rapidity ; the confequence of which was, that the four caiffon piers, totally unguarded except as above-mentioned, were all underwafhed at the weft end, to the depth of about 15 inches at a medium, at the borders, and fome to a greater breadth and fome lefs ; but the pier that had the half girdle courfe round the weft point having been found to fuffer along with the reft, though not in fo great a degree, this induced Mr. Smeaton to think of a mode of defence not merely terminating upon the upper cruft of the gravel as the girdle ftones (which this event fhewed was not fufficient to refift the increafed velocity of the current, when paffing by a new objeft), but inftead of the girdle ftones, to go fome depth into the gravel ; which it appeared practicable to do, by driving a cafing of plank piles to furround every caiffon pier at the diftance of three feet, which would alfo enable him fafely to underpin the parts underwafhed.

To this propofition (though attended with an addition of expenfe unthought of before, of between five and fix hundred pounds) Mr. Donkin very readily confented on the part of Mr. Errington. This was therefore immediately put in hand before further proceedings were gone upon.

Hitherto I have been particular, as it feemed abfolutely neceffary to give an adequate idea of the natural difficulties attending this work, and what a very fmall portion of them were known before it was originally enterprifed by Mr. Gott ; how gradually they unfolded themfelves in confequence of the fteps that from time to time had been taken ; and how very far all were from being aware of the whole, when the work was begun by Mr. Errington under the direction of Mr. Smeaton. Suffice it therefore to fay, that the cafing of the damaged piers, and the underfetting of three of them, was proceeded upon with the greateft alacrity, and completed that feafon ; nor did any other adverfe accident happen to the completion of the bridge.

It

It may be proper to mention that when the firſt caſing was completed, which was about the fifth pier from the north ſide, Mr. Smeaton ordered a trial to be made to pump out the water, which, if practicable, would have afforded the moſt eaſy way of underſetting the under-waſhed parts; but this being attempted, Mr. Pickernell reported, that with four double pumps and two ſingle ones they had not been able to ſink the ſurface of the water within the caſe above an inch below the ſurface on the outſide. In this method, however, the coffer-dam caſes uſed at Perth Bridge were driven ſo as to keep the water out of the foundation pits, when the ſurface of the river (when the tide was in) was from ſix to ſeven feet deep upon the bed of the river, and conſequently againſt the ſides of the caſing, it would therefore have ſeemed that there had been ſome very palpable defect in driving theſe caſes, had not the operations at the ſecond pier from the north ſide ſufficiently ſhewn how extremely open the bed of gravel is to the paſſage of water, and how impracticable every method was likely to prove that depended upon the drainage of the water, for the piers to be placed in the main channel of the river.

Several very rapid and much larger floods than that which did the miſchief happened in the courſe of the ſucceeding winter, particularly one upon the 12th of December, when the water was within nine inches of the top of the inpoſt; when Mr. Pickernell marked a fall of two feet three inches, but without any material damage to any thing, which naturally induced all thoſe concerned to proceed in the way they were then going on.

The ſeaſon of 1779 was begun by new founding the weſtern half of the pier, that the weather prevented from being completed the year before, which was the fifth from the north, and was done without caiſſon or draining the water, by means of an air cheſt or diving machine, that had been very ſucceſsfully and conveniently employed in underſetting the other five damaged piers; and the pier before unbegun (being the ſixth from the north) was the next in courſe founded by caiſſon; but with this difference, that the caſe was firſt drove all except the down ſtream ſalient pointing, before the caiſſon was floated into its place through this opening.

It would cauſe too great a prolixity to deſcribe the particular operations of what followed the diſaſter of Auguſt 1778; nor can they be done without reference to figures; I ſhall therefore proceed to ſay that in the beginning of the year 1779, and alſo afterwards while the arches were throwing, the whole of the caſes were guarded all round by a depoſition of rubble; and to render it the more effectual againſt the torrents that this river then appeared ſubject to, the up ſtream points of the rubble bulwarks were extended to the length of 30 feet above the ſalient point of the caſes reſpectively.

That

That the underfetting of the three piers fo treated, was done fo as to be as folid and effectual as the gravel ftratum which the whole ftood upon, as alfo the new founding of half of the fifth pier, and the original founding of the fixth, which was the laft founded; and that the whole ftood upon one bottom equally capable of fupporting the weight of the fuperftructure built upon it, appears from this, that in the whole of the mafonry, from the time of the accident in 1778 to the time of its total overthrow in the year 1782, there never appeared the leaft crack or fet in any part of the work, not even in the parapets, which to thofe who are well experienced in bridge-building, will appear a remarkable inftance of the foundnefs of the work; and furthermore, that the whole was fufficiently guarded againft every accident that could be forefeen or expected appears from this; that in the year 1779 a remarkably rapid flood happened: which Mr. Pickernell obferved, from marks upon the bridge, above and below, fhewed a fall of three feet nine inches, occafioning a velocity of above 900 feet per minute, which not only paffed without any material derangement; but, on the contrary, fuch changes as had been made in the bed of the river had been for the better, as it had acquired a more equal depth from fide to fide; the fhallow parts becoming deeper, and even the deeper parts, by the depofiting of gravel, had become fhallower, and every flood that happened after, occafioned reports of the fame kind till the laft. In this manner, and with thefe ideas, every thing was fatisfactorily finifhed, and the rubbling completed, according to Mr. Smeaton's directions, at or about Chriftmas 1780; and in the beginning of January 1781, Mr. Smeaton viewed the whole, in the prefence of the magiftrates of the county, who attended for that purpofe, and found every thing done to his great fatisfaction as well as theirs; and as fuch reported it to the county, as being completed according to Mr. Errington's agreement.

After all fucceeding floods, every account was of the moft flattering kind; fo that previoufly to the laft flood, every perfon concerned therein or therewith feemed perfectly eafy as to the fecurity of Hexham Bridge.

On Sunday evening the 10th of March 1782, there happened in that country a great downfall of fnow, fo great as to be a foot thick upon the plain ground; which was immediately fucceeded by a violent hurricane: and as the nature of fnow is to drink up the rain like a fponge, till it becomes perfectly faturated with water, it then burfts at once like a fnowball in the fire, and may be fitly compared to an immenfe refervoir, extending over the whole country, and breaking loofe altogether; and as the hills, and whole face of this country, are fteep, comparatively like the roof of a houfe, the water is capable of coming down very fuddenly; it then meets altogether, by the junction of the two Tynes, a little above Hexham, without having any confiderable flat ground to fpread itfelf upon. The next morning, viz. Monday, March 11th, Mr. Donkin perceiving an extraordinary high flood in the river (which runs from weft to eaft) attended with a very high wefterly wind, he was led to go down to the bridge, to obferve the effects of the water upon it; but without thel eaft fuppofition of any damage happening thereto; when he obferved, that the water was up to the top of the dooming (as there called), that is, the tops of the caps of the falient points upon the piers of the up ftream fide of the bridge, when it fcarcely touched the down ftream fide, which makes a difference in level in paffing the bridge of no lefs than between four and five feet, and which, according to the known rules of hydrauiicks, will occafion a velocity of one thoufand feet in a minute; but yet he was fo far from apprehending any danger to the bridge, that he had juft fent his fon over it, to the fouth end, with two or three mafons, to examine the ftate of the land-arches there, who reported all fafe, but they had not returned above five minutes, before he began to obferve fymptoms of failure, when to his great furprize, " he perceived fome particles of lime fall from the fourth arch, about " the fize of chaff, and the lime coming from thence and no where elfe, he pointed his " obfervations to that part only. That the falling of the lime continued to encreafe in " fize and quantity, for the fpace of a minute; that foon after he obferved a crack acrofs " the bend of the arch towards the upper fide of the bridge, which crack gradually " widened, and in about a minute more, the fplinters from the ftones in the plain part of " the fpandrel, between the third and fourth arch, which he could perceive fhake, gave " way, and the two arches and a pier fell together; that the whole bridge was deftroyed " in half an hour," only two arches remaining whole, and one fell in part in the evening. Such was the fudden cataftrophe of this much commended unfortunate bridge.

The fecond queftion is, Whether Mr. Errington was fparing of any thing neceffary to give fuccefs to that mode of building, which Mr. Smeaton had adopted?

To this Mr. Smeaton can only bear his teftimony, that he was not; nor did he ever hear him find fault with any expenfe, thought neceffary by Mr. Smeaton; on the contrary, he always expreffed his wifhes to have the bridge completed in the moft fubftantial and effectual manner. Nor did he ever appear difquieted by the difafter of 1778, or at the expenfe of what Mr. Smeaton had propofed as a neceffary addition, though at fo confiderable an increafe as five or fix hundred pounds, as has been already ftated.

Thirdly, refpecting Mr. Pickernell, whether he did to the beft of his power and knowledge, execute to a reafonable and poffible extent, what was directed by Mr. Smeaton? And to this Mr. Smeaton thinks it but juft to fay as a witnefs, that he always looked on Mr. Pickernell throughout the whole proceeding, as a perfon particularly attentive to

execute

execute orders and directions given by him, and upon whose capacity for that purpose, he could safely rely, after having shewn him the mode of going about any new operation, and upon whose reports of these operations, he could also safely rely; and as the general workmanship of the bridge has been applauded by many, and discommended by none, it seems there is only one point in which Mr. Pickernell's execution of Mr. Smeaton's orders can be called in question, and that is respecting the driving down of the cases of piles, round the caisson piers, to a proper depth; it is therefore necessary to state this matter particularly.

Mr. Smeaton's written instructions were as follow; " The length of the piles should
" conform to the depth of the water; I would not wish the sheeting piles round the west
" end, and the first bay of the return on each side, to go into the ground more than
" about ten feet, and if they do not drive kindly, must be contented with less; from
" thence, each bay may be gradually less depth into the ground, so that round the down
" stream pointing seven feet will be sufficient. If the gage piles drive kindly, they may
" be longer by eighteen inches or two feet than the sheeting, but if not, they need not
" be above one foot longer."

To the above, in the course of the work, Mr. Pickernell reported, that having driven the piles of the fifth pier from the north, which was the first to which the casing was applied, the gage piles went down very well and entered two or three inches at a stroke, but when he came to drive the plank piles, they could not be got into the ground more than from five to seven feet.

He further reported in the course of this business, in regard to the sheet piling of the seventh pier from the north (or second from the south side) which was the last casing driven of the four damaged piers; that the bed of the river at the south side, is entirely full of large flat stones, such as they got out of Oakwood bank quarry, which have been the ruins of the boats' landings, taken away by floods and ice from time to time, and those stones had obstructed their sheet piling round that foundation, and had occasioned many of them to go out of their places at bottom, so that sundry cavities were occasioned thereby, more than in the last.

These were the representations of Mr. Pickernell, concerning his execution of Mr. Smeaton's orders, respecting the piling; so that if they were driven to a less depth, or in any manner less effectual than as above represented, Mr. Pickernell must answer to it, as Mr. Smeaton was totally unacquainted therewith, nor was any insufficiency in this part of

S ſ 2

the

the work ever fuggefted to him by any perfon whatfoever during the courfe of the work, or fince, till he heard of an oppofition to Mr. Errington's bill for relief from his obligation.

But whether in reality Mr. Pickernell did this part of the work equal to the above reprefentation of it or not, that the ftanding or falling of the bridge may not be wholly left to reft at Mr. Pickernell's door, Mr. Smeaton, in juftice to Mr. Pickernell as well as himfelf, thinks it neceffary to declare, that for the reafons already affigned, (as well as the verification thereof during the courfe of the work, by every flood that happened) fo great and abfolute was his dependence upon the application of Oakwood bank quarry rubble, as an ultimate defence to controul the violence of the Tyne's floods (no part of it laid round the coffer dam foundations, having ever been moved); that provided the piles of the cafings were but driven into the ground, fo far as to fix faft therein, and fo clofe together, that though the cafes might not hold water, they might retain the gravel from being wafhed out through the chinks from under the piers; he had not the leaft doubt of preventing any material damage ever being done to the pile work, by the application of the faid rubble to furround them. This fentiment, however, though it dictated that part of his inftructions, " get the bays of fheet-piling at the weft end of the piers down to " ten feet if you can, if you cannot, we muft be contented with lefs;" and alfo made Mr. Smeaton contented with what was above reported to him, as the moft imperfect part of the performance; yet he never communicated this opinion to Mr. Pickernell, or any other perfon; left the workmen from hearing thereof, might be induced to fatisfy themfelves with doing lefs than otherwife they might be capable of, in the way of getting them down as far as they could.

Whether Mr. Smeaton's opinion, concerning the fecurity of Oakwood bank quarry rubble, was well or ill-founded, will be further examined in the fequel; but this is cer-tain, that the driving of the cafes not being completed before the middle of September, and being then very defirous to take advantage of the fecurity they afforded, to get the piers underfet, if poffible, or otherwife fecured before the heavy winter floods came on, he concluded, that if the experience of thefe floods fhould fhew a need of greater ftrength and defence, it might be added in the courfe of the next feafon.

This autumn of 1778, in reality, afforded the experience of a confiderable number of floods, amongft which, the laft, which was of December 12, was a capital one, and the higheft that had been fince the great inundation of 1771; and the water on the weft or up ftream fide of the bridge, was within nine inches of the top of the impoft of the fecond

pier

pier from the north, when at the eaft end it was one foot three inches below it, fo that the fall was then no lefs than two feet three inches, and which would produce a velocity of above feven hundred feet per minute.

After the water was fubfided fo as to afford a full examination, Mr. Pickernell reported the effect; viz. that there were but very few of the rubble ftones removed from where they were thrown in round the foundation; what were moved, were from the weft fhoulders; but that from the third pier to the north, round which no ftones had been depofited, it had torn up the gravel from the falient point and weft fhoulders, to the depth of three feet; and that from thence to the ftones that were laid round the fecond pier, it had deepened the bed of the river full eighteen inches, (which before was too fhallow); but as to all the other part of the river's bed, he could not perceive it altered in the leaft.

The experience therefore of thefe floods, and particularly that of the 12th December, all concurred in proving, that the Oakwood bank quarry rubble was a fufficient defence againft every violence of the Tyne: fo that it did not appear neceffary to introduce any new mode of defence, but only to apply the rubble in the moft effectual manner; and as the weft fhoulders appeared to be the parts that the greateft ftrefs came upon, Mr. Smeaton ordered that the water might not meet with fo fudden an oppofition there, but be more evenly, flopingly, and gradually brought thereupon, that the weft falient points of the rubble fhould be extended weftward of the falient points of the cafes refpectively, to the length of at leaft thirty feet; which was accordingly executed by Mr. Pickernell upon all the caiffon piers.

4thly, We come now to the fourth and laft queftion, viz. Whether under all the experience and knowledge of the fubject as it now ftands, the prefent bridge fhould be attempted to be reinftated, or a new one built at Hexham?

This queftion is indeed of far the moft material import; for it is of little confequence to the public, in the prefent ftate of things, whether Mr. Smeaton misjudged of the fubject? Whether he was deceived himfelf, or was deceived by others? Or, whether Mr. Pickernell did or did not do his beft, towards a full execution of Mr. Smeaton's orders, in regard to the driving the cafing piles? Nor is it of any confequence to know, that in point of art, but without any regard to, or limitation of expenfe, a bridge is poffible to be built: the true queftion is, Is it fitting for the county to undertake it? Suppofing the Treafurer in poffeffion of whatever fum can be recovered from Mr. Errington, in confequence of his obligation, will it not (like Sir Walter Blackett's three thoufand

4                                                                              pounds)

pounds) be a temptation to the county to fpend ftill much larger fums upon an unfruitful project; and it may reafonably be fuppofed, that the whole county ftock is not an unlimited fum ; nor can it be properly expended in the erection of a bridge at one fingle paffage.

It is now known for a certainty, what was not, and could not have been known before the erection of this bridge, that there is a poffibility of natural caufes being fo combined, as to produce a flood fo large, and of fo fudden a nature, as to produce a velocity of the water exceeding one thoufand feet in a minute ; and whether even this may, or may not, be the uttermoft limit of Nature, is not in the power of any man to cal-culate :—

That the velocity of feven hundred and twenty feet per minute, arifing from a difference of two feet three inches, as per flood of December 1778, was fufficient to tear up and remove the natural bed of gravel, which forms the bed of the river in this place, wherever there was a particular fet upon it, but was not cabable of moving or materially derang-ing the defences compofed of Oakwood quarry rubble.

That the velocity of nine hundred and thirty feet, refulting from a difference of three feet nine inches, in a flood of the 1ft of December 1779, ftill made not the leaft altera-tion in the defences, nor to any part of the bed of the river, fave that the rubble ftones depofited at the third pier where the current had torn up the natural gravel in the flood of December 1778, were now wrecked full and covered with gravel, and reduced to the level of the adjacent parts. Another flood fucceeded this in the compafs of ten days, that rofe within eight inches as high as the former, but in this nothing happened of any kind ; in fhort, the bridge being now erected, as far as it was concerned with the water, all the arches cleared, and the defences completed, after a confiderable number of great floods, and nothing happening in confequence, every one feemed fo entirely fatisfied of the ftability of the bridge, that even the Gilligate people, Mr. Pickernell obferved, ceafed their vifits, who before had conftantly, after every flood, come to infpect, in hopes of finding fome-thing correfpondent to their prayers and wifhes for the downfall of the bridge.

Mr. Smeaton was, however, agreeably furprifed on having this account, that the fall of water had been fo great, and no harm enfued ; for had it been poffible for him to be apprifed of fuch a fall before hand, he never fhould have recommended to Mr. Errington to have undertaken to erect a bridge upon that bed of gravel.

It

It therefore at this time appears plain, that though the Oakwood bank rubble will lie ſtill, and reſiſt a velocityof the water of nine hundred and thirty feet in a minute,yet it is capable of being all removed and carried away by the velocity of the water of, or a little exceeding, one thouſand feet per minute; a velocity reſulting from a difference of forty-four, as it was or upwards in the flood of March 1782; and that the gravel bed itſelf is capable of being torn up by a much leſs degree of velocity; the queſtion then is, How in ſuch ſituation a foundation can be laid and effectually ſecured?

Shall we attempt to build a wall acroſs the bottom of the river, according to Mr. Wooler's propoſition? Experience has ſhewn in the building of the laſt bridge, that the gravel is of ſo open a nature in the main channel of the river, that it is impracticable to drain off the water. Mr. Smeaton means not to put limits to the invention and ingenuity of men; but neither his obſervation, experience, nor invention, has hitherto ſuggeſted any effectual method of founding ſuch a wall, without draining off the water; and the ſame will apply to the penning the bottom acroſs the river.

But for a moment, ſuppoſe the thing done: this wall or this apron, muſt have a termination; and wherever it terminates, experience ſhews, a rapid current will form a deep hole, to twenty, or even thirty feet depth, and upwards; and if the gravel under the foundation gets looſe, the downfall of the whole is the conſequence.

2dly, Suppoſe we attempt to build it in an excavation upon piles encaſed, as was done at Perth, the ſame difficulty ariſes; we cannot get out the water; and if done, as rubble will not lie to defend it, the gravel bed being ſcooped out, beyond all practibiſity of driving piles, the piers being ſapped, the ſame unfortunate circumſtance muſt enſue.

3dly, Suppoſe we attempt it by excavation with ballaſt lighters, and drive down piles even with the bottom of the excavation pits, to found the piers upon; which may doubtleſs be done, without taking off, or drainage of the water; ſtill, if neither the bed of gravel itſelf, nor quarry rubble, is capable of reſiſting the violence of the current, when the gravel bed is deſtroyed or deranged, ſo that the piles are laid bare, the pillars will be ſapped, and deſtruction equally enſue: nay, even ſuppoſe the piles could be encaſed without taking off the water, yet this is only giving the river a little more work to do; for if rubble is carried away, as we find it muſt be, it is no defence; and we do not know the depth to which the gravel can be ſcooped out and excavated by the violence of this river;

the

the foundations therefore, however deep, can be ultimately fapped, and the fame ruin enfue.

In fhort, turn ourfelves which way we will, nothing feems certain in this bufinefs, but a very great expenfe, how commenfurate with the county's funds, muft be left to thofe to judge of who know them : but this Mr. Smeaton will take upon himfelf to fay, that he fees no way of making foundations for a bridge to ftand upon, for the whole fum in which Mr. Errington ftands obligated to the county, that is likely to be attended with any certainty of permanency, much lefs alfo to build a bridge upon thofe foundations, for the fame fum.

THE following is a Copy of a Paper delivered at Northumberland Affizes 1783, by Henry Errington, Efq., to Gawen Aynfley, Efq., Chairman for the County of Northumberland, but which appears to have been originally intended to be delivered to the Grand Jury, which had been previoufly difcharged, as the Words Gentlemen and Magiftrates, in the Original, are fubftituted for Grand Jury.

Mr. ERRINGTON'S contract with the Juftices of the Peace refpecting Hexham Bridge having placed him in a very difagreeable fituation, he, as well on his own account as on the part of the County, wifhes the Gentlemen and Magiftrates, now affembled, as a refpectable body of the County, would be pleafed to take the following fhort ftatement of facts into their confideration.

Mr. Errington undertook to build the bridge according to a fpecific plan, under the direction of Mr. Smeaton, and to uphold, fupport, repair, maintain, rebuild, and keep it in good and fufficient repair for the term of feven years, to be computed from the time it fhould be certified under the hands of two juftices to be fo built.

The bridge being completely finifhed according to fuch plan and direction, was in January 1781, certified for.

The

The plan to which Mr. Errington was confined, was not only a plan of the fuperftructure, but alfo of the foundation, and of the manner in which fuch foundation was to be laid.

In March 1782, the bridge was thrown down; from which circumftance it is evident, that to rebuild it according to the fame plan would be ineffectual, and from the nature of the contract, it is not in the power of the contracting parties to alter or vary it; indeed, Mr. Smeaton, whofe direction Mr. Errington was bound to obey, has declared that it is not in his power to devife a better. If therefore the County fhould infift on Mr. Errington's rebuilding the bridge, a great portion of the feven years would expire before he could complete it; (two years and a half having already elapfed fince the certificate)—and if by favourable feafons it fhould be capable of being upheld till the end of the feven years, he would then be difcharged from his obligation, and the burthen would fall upon the County; Mr. Errington in common with the land owners of the county in general, and by having an eftate in the vicinity of Hexham in particular, feels himfelf interefted in having a bridge built, which by an alteration in the mode of ftructure may be attended with a probability of being permanent, and he has been informed that other engineers lately confulted by the Juftices are of opinion that it is practicable; he, therefore, conceives that it would be more advantageous to the County to accept from him fuch fum of money as fhall be eftimated equal to the expenfe of repairing the bridge, and putting the fame into fuch a fituation as his contract requires, and apply fuch fum together* with the materials, (which are of more than four times the value of thofe received by Mr. Errington from the County), towards building a new bridge according to fome other plan, than to require him to repair the old one upon the former erroneous principles. — And if the Gentlemen and Magiftrates fhould be of the fame opinion, Mr. Errington doubts not the Juftices will, as truftees for the public, readily adopt what fhall fo appear to be the fenfe of the County.

* Value of materials received by Mr. Errington from the County, and acknowledged by him to be eftimated at    -    -    -    -    £ 3000

Which multiplied by 4    -    -    -    -    12,000

Which is 2900l. more than the penalty of Mr. Errington's bond to the County, and many thoufands more than expended.

Mr. MYLNE's Second Report.

Edinburgh, 30th September 1783.

To the Magiſtrates and Juſtices of Peace for the County of Northumberland, &c. &c.

Gentlemen,

A PAPER has been tranſmitted to me by Mr. Davidſon, Clerk of the Peace, which contains a propoſition from Mr. Errington to pay a certain ſum (to be fixed hereafter) in lieu of rebuilding Hexham Bridge; and he requeſts me, by directions of Mr. Aynſley, the Chairman, to ſend my full ſentiments thereon, to be laid before your meeting of the 8th October next.

As it will be convenient for me to attend you at the ſaid meeting, I apprehend it is not neceſſary to ſay much on this occaſion; other than ſtating a few words on ſome matters of fact, which require to be aſcertained, before any one can judge with certainty and preciſion. Another reaſon requires me to be the more conciſe at preſent, as in the event of not agreeing with Mr. Errington, and the ſubject being diſcuſſed at law, every previous animadverſion would in that event be ill timed and premature.

The ground work of the propoſal, and the propriety of the reaſoning which it contains, depend upon the following matters:

1ſt, It ſtates that the agreement was to build according to a ſpecific plan.

2dly, To do that under the direction of Mr. Smeaton.

3dly, That it was completely finiſhed according to ſuch plan, and under ſuch direction. And

Laſtly, That he was confined to ſuch ſpecific plan; which was not only a plan of the ſuperſtructure, but alſo of the foundation part; and of the manner in which ſuch foundation was to be laid.

On the firſt head, it will appear, that all the foundations were, by the plan annexed to the articles of agreement, to have been laid full five feet below the water line, in maſonry or timber framed work; and the two abutments, and the two piers next the abut-

4

ments,

ments, are propofed to be piled in fuch manner as fhall appear neceffary on opening the ground.

In all other refpects whatfoever, either as to the quantity or quality of the piling under thefe parts fo mentioned, and under any part of the other eight piers, or of any manner whatfoever of laying all or any of the foundations, the plan does not hold forth any fpecific manner of laying the foundation ; but on the other hand, leaves all thefe particulars to the judgment and adoption of the contracting party.

On the fecond head, the agreement entered into on the propofals of Mr. Errington and his agents, was certainly to put the whole under the direction of Mr. Smeaton ; and the act of parliament which followed thereupon, confirms it to a certainty not to be fhaken. The reference thus to be held to the judgment of Mr. Smeaton, was in no points more evidently neceffary than in the manner of laying the foundations, which were not fpecified at all, as well as in many other things impoffible to be contained in drawings, written agreements, or in acts of parliament. But, in the unfortunate event of things, I conceive, the operation of Mr. Smeaton's directions in all matters not fpecified as above mentioned, he was led aftray, and that his directions were not followed. Proceedings which were eafy in their nature, were followed up with a fatal rapidity, that laid the feeds of ruin ; and the guard works, which were added, on after confideration, and the experience of the fhallownefs of the foundations, were not executed confiftently with the correctnefs and good fenfe of his orders.

His directions I conceive to have been clear and fufficient if they had been fulfilled ; his candour will not allow him to fay thus much ; a commendable regard to others engaged in the executive branch, fuppreffes what ought to be faid : but I who feel for the character and reputation of fo great an artift, and every man fo peculiarly fituate as he is, muft be permitted to fay, that the purport and effect of his directions were not executed, and of courfe, that he was deceived.

On the third head, it is neceffary to ftate that the foundations were not laid according to the plan, fo far as the fpecification thereof went. To begin at the north end, the butment is not fo deep as fhewn by the drawing by two feet ; the firft pier by three feet five inches, the fecond pier by two feet four inches, the third pier by two feet eleven inches, the fourth pier by feven inches, the fifth pier was ten inches more in depth than the plan, the fixth pier was in like manner eleven inches, the feventh was alfo fix inches, the eighth pier is two feet fhort of its depth ; and the fouth butment is two feet in like manner lefs than it ought to be.

From

From this ſtatement, founded on the beſt evidence I could procure, the bridge was not built according to the deſign agreed on where it could have been ; and that the directions given were not followed literally and effectively.

On the laſt head, I have ſufficiently ſhewn that the ſpecification of the plan did not confine the bridge in manner of laying the foundations, nor in the ſhape and extent of the works. Anything might have been done under the agreement, which a more intimate knowledge of the bed of the river, and the experience of the works themſelves, gave, during the time of the execution : in fact, it is ſhewn, that the parts ſpecified were altered and modified to ſuit the manner adopted for the execution.

The concluſion, therefore, naturally draws me to end with ſaying, that no argument can be reaſonably built on the idea of being confined to a ſpecific plan agreed for, or executed in thoſe parts, to wit, the foundations, on which this queſtion depends. And I am thoroughly convinced that if Mr. Smeaton, poſſeſſed as he is of ſo much ſtrength of judgment and variety of reſources, were to view the work and examine its preſent ſtate, with a view to its repair, he could with ſatisfaction to himſelf undertake to reconſtruct the bridge according to the agreement, with the ſame expectancy of permanent durability as he had at firſt ſetting off with this undertaking.

What remains to be ſaid will come better into the diſcuſſion which is propoſed to be held at your intended meeting ; until which time, I remain,

Gentlemen,
Your very humble and much obliged ſervant,
ROBERT MYLNE.

---

OBSERVATIONS on Two Reports of Robert Mylne Eſq. concerning Hexham Bridge, by J. SMEATON, Civil Engineer.

THERE are ſo many points contained in the two reports of Mr. Mylne, of the 24th April and the 30th September 1783, in which I entirely differ with that gentleman in opinion, that to make the proper obſervations upon the whole, would draw me out to a length that in the preſent ſtate of things I would wiſh to avoid. I ſhall therefore content
myſelf

myſelf with obſerving upon thoſe that I look upon to be the foundation of that difference of opinion; and upon which the merits of the queſtion ſeem principally to depend. I ſhall therefore paſs over the compliments that Mr. Mylne's politeneſs prompts him to pay me on this occaſion; and particularly as they ſeem to be at my own expenſe. For the greater eaſe of reference, I ſhall apply to the printed copy of Mr. Mylne's reports.

Mr. Mylne ſays, page 5th, (298 of this volume) " The exiſtence of a ſand below, and a " ſuppoſed hardneſs and concretion of five feet, or any ſuch meaſure of the upper parts, " ſeem to have precipitately and fatally determined the plans of operation at firſt ſetting " out; and appear to me to be equally the cauſe of the preſent precipitate opinion for " abandoning the work as impracticable."

He alſo ſays, page 4, (298) " Mr. Smeaton, than whom there is no perſon or artiſt better " inſtructed, more knowing, and of a more penetrating and correct judgment, muſt have " been deceived in the collection of facts and materials, on which he eſtabliſhed his plan " of operations."

Now, if in the following detail it ſhall appear that Mr. Smeaton made a proper collection of facts whereon to judge; and did make a proper judgment thereon, ſo far as at the time he was to form this judgment had come to light; then, Mr. Mylne's reflections on this judgment muſt appear unfounded, and will in courſe fall to the ground.

Mr. Mylne acknowledges himſelf to have been made acquainted with the reaſons and motives of Mr. Smeaton; for page 6th, (298) he ſays, " I have ſeen and examined all Mr. " Smeaton's papers, I have heard all the particulars, and hiſtory of his proceedings and " motives for the method of operations, which he adopted:" and Mr. Mylne knows that this intelligence was from Mr. Smeaton himſelf, before Mr. Mylne went down *to ex-* *amine the works and perſons concerned, and employed in the detail of its execution:* and that it was in this conference that Mr. Smeaton mentioned to him, that the ground and reaſon which influenced his judgment in not ſinking the piers of the bridge deeper into the bed of the river was an eſtabliſhed opinion that the greateſt part of the river's *bed* con-ſiſted of a gravel, very hard compacted together and difficult to penetrate at its upper ſurface; but which diminiſhes in compactneſs from the ſurface downwards; ſo that at ſome depth exceeding five feet, it deviated into a ſand, and ultimately into a quickſand or mud. The exiſtence of ſuch a ſand below, as alſo a ſuppoſed hardneſs and concretion of five feet, *or any ſuch meaſure,* of the upper parts, Mr. Mylne explodes in the terms already quoted: but as herein reſides the foundation of that entire difference of opinion that ſtill

subſiſts

fubfifts betwixt Mr. Mylne and Mr. Smeaton, it feems proper now to examine the grounds whereon Mr. Smeaton eftablifhed this opinion, and alfo the grounds whereon Mr. Mylne explodes it; becaufe, if well founded, from what *appeared at the time*, it muft be admitted fufficient in reafon for Mr. Smeaton's determining his plan of operations according to it, though, fince the fatal accident, any thing fhould appear to the contrary; and if on the other hand, from any thing that can be deduced from M. Mylne's operations, Mr. Smeaton's original conclufions ftand uninvalidated; then it muft appear, that Mr. Mylne muft have been drawn into a precipitate opinion, or, as he terms it, *deceived*.

That Sir Walter Blackett built a bridge at Hexham, which not long after it was finifhed, was thrown down by the great flood in 1771, in one night, and the materials almoft totally difperfed, is an event too notorious for Mr. Smeaton to be *deceived* in.

That the foundations of this bridge were let into the bed of the river, well founded upon a great number of piles, and a platform upon the heads of them 12 inches thick, of folid balks judicioufly rabetted together, Mr. Smeaton could not be *deceived* in; having been eye witnefs to many of the operations.

That all the piers, arches, and one of the two abutments were totally deftroyed, and the whole length of one fide of one of the platforms raifed out of the water, half way between the horizontal line and the perpendicular; in this alfo Smeaton could not have been *deceived*, having feen it for feveral years together; and was expofed to the full view of every one that paffed by.

From the above facts, Mr. Smeaton inferred, that though the upper cruft was hard gravel, yet a much more foft and yielding material lay underneath it, otherwife the bottom, could not have been turned up in the manner mentioned, and that the foundations would have been more fecure, had they been laid upon the upper cruft without its being broken, by excavating or driving piles: and which is evidently proved by the fubfequent operations that now come in courfe to be mentioned.

That after this accident Sir Walter Blackett, chufing rather to forfeit the penalty of the bond*, in which he was obliged to uphold the bridge for feven years, than to attempt it again, the magiftrates of the county employed Mr. Wooler, then an eminent engineer, to begin another bridge near the fame place: and that in confequence he built a land-

* Penalty 3000l.

breaft

breaft on the north fide; and many preparations being made, and every thing ready for founding a pier, upon the fame fide; on digging the foundation put into the bed of the river, the whole work was put a ftop to for fome reafon or other; and in this Mr. Smeaton could not be deceived; becaufe at that time he was often at Hexham, was well acquainted with Mr. Wooler, and faw the operations going on, and afterwards difcontinued: and it will be further proved by living evidence, that the reafon for difcontinuing the work, was the *foftnefs* of the bottom of the foundation pit; and to afcertain whether this foftnefs was particular to *that place*, a pit within the folid bank was dug upon the oppofite fide of the river, in the bottom of which the fame foft ftratum prevailed; and to this fome of the magiftrates were themfelves witneffes, as will alfo be proved: it can alfo be proved from a letter of the faid Mr. Wooler, now deceafed, that in his opinion *a permanent bridge could not be eftablifhed here without building a folid wall of fix feet high, and 42 broad, quite acrofs the river;* which he obferved was an expedient that has fucceeded where *expenfe was not regarded.* He further fays, " that the attempting to fet a bridge " upon fuch an enormous depth of quickfand, over a river fo fubject to floods as the " Tyne, may be deemed fo hazardous as to be next to imprudence itfelf."

Mr. Smeaton therefore could not be *deceived* as to the exiftence of a hard ftratum of gravel at the top, with a quickfand under it, at either of the places juft defcribed*: as Mr. Wooler's operations fully prove what was the caufe of the failure of the firft bridge.

Mr. Jonathan Pickernell, who was brought from London to be employed in this bufinefs under Mr. Wooler, and recommended by him, will further prove, that he was employed to bore the river, which he did in various places, and verbally reported to Mr. Smeaton, that wherever he had tried the river he conftantly found a quickfand to take place underneath the upper bed of gravel; and that he had reported the fame to *the clerk of the peace.*

In this Pickernell might have deceived Mr. Smeaton; in his never having made fuch borings, nor made fuch report: but Smeaton did not reft it here; for in the very place where he built the bridge, he tried the bottom of the river, not by boring, but by driving down a pointed iron bar, in various places acrofs the river, to the depth of nine or ten feet, and conftantly found, that though the entry of the bar was in all places difficult, yet at every blow it went more and more eafy; however at this depth the gravel felt fo tolerably compact as

* N. B. The Dwarf wall, the place recommended by Mr. Mylne, is near the original bridge.

to be deemed by him fufficient for fupporting a bridge, *provided* the upper cruft was not *broken by digging or piling;* and provided, the weight upon each pier was not *too great,* which would be the confequence of *wide arches;* and as was the cafe with Sir Walter Blackett's bridge: the middle arch being 70 feet.

Now, if in the conclufion I drew from thofe trials of the river's bottom by the bar, I have been deceived; I deceived :*myfelf,* as I was not beholden to the *patience and perfe-verance* of Mr. Pickernell only, for as I not only directed but affifted in the operation myfelf, I faw with my *own eyes,* and felt with my *own.hands.*

Thefe are the facts on which Mr. Smeaton built his plan of operations, which facts, if valid, he challenges Mr. Mylne to fhew, that the methods he grounded them on, were not the moft likely, *as far as was then known,* to produce a permanent bridge, that if, at all, muft be erected at fuch a moderate expenfe as the county of Northumberland was likely to raife: for to build a folid wall acrofs the river whereon to fet it, he judged with Mr. Wooler to be improper, except where *expenfe was not to be regarded.*

Let us now fee what Mr. Mylne has on the other hand eftablifhed to invalidate thefe facts. He fays, page 3d, (297) " I have bored the river at the bridge to the depth of 23 feet " below the latter water level, in a place where I might not be led aftray by any altera-" tion formed by the faid flood in its milder velocity; and I have found under the " teftimony and perfeverance of Mr. Wake, that the foil and texture of the bed of the " river at this place is uniformly a compofition or congeries of roundifh and flat ftones, " gravel and fand, of equal quality and confiftence in the whole of that depth."

Mr. Mylne then caufed a hole to be bored to the depth of 23 feet into a bed of gravel: but was this the only hole? Mr. Mylne makes no mention of any other. Where did he bore it? Mr. Mylne does not fay; but contents himfelf with telling us, he did not bore it in a place where the gravel was liable to be difturbed by the impetuofity of the river. But does Mr. Mylne think this was fufficient to determine what the foil and texture of that part of the bed of the river was which *was liable* to be fo difturbed? Mr. Mylne fays, page 4th, (297) that the bed of the river " though hard to the touch of boring, and compact " to the eye, and feeling of inftruments, is wonderfully loofe, and unconnected in its " parts, infomuch that the bed of the river Tyne feems to fhift and alter its form, extent, " and fituation with every flood more or lefs; and tearing up at one time to a great " depth that fair moulded and well-laid hollow, which the ftream had laid for itfelf

" upon

" upon fome former occafion." This was not the part, therefore, that Mr. Mylne
chofe to examine otherwife than in his mind's eye ; he chofe a place not fo liable to
alter, although the greateft part of the bridge, and particularly that part that firft gave
way, was founded, and obliged to be founded, upon *this fair moulded and well laid hollow*,
which the ftream had *laid for itfelf upon fome former occafion.* Now Mr. Smeaton cannot
be deceived in this, that it was in this very part where he drove down his bar: it was
in the deepeft part of the river; and all his trials were made in a *boat*, though the
river was then in its low ftate. Whereas it will be proved in evidence, that the part
where Mr. Wake bored was upon a dry gravel bed, nearly abreaft of the place where
Mr. Smeaton put down his fecond coffer dam pier: and where the foundation was
actually funk into the bed of the river, as deep as it poffibly could be for water * ; in
confidence that the gravel bed there was quite as deep as Mr. Wake's boring has fince
proved it to be. The fact was, that Mr. Smeaton's original *plan* and propofition was
to build the land-breaft and one pier, on each fide, with coffer dams, finking the
foundation thereof into the bed of gravel, piling underneath and round thefe works,
to as great a depth as fhould appear neceffary on opening the ground ; and which accordingly were done: but finding, on founding the north land-breaft, and alfo the north
pier, that the gravel bed extended ftill further into the bed of the river than was at
firft expected, he directed the fecond pier (oppofite which Mr. Wake bored), to be
laid in the fame way ; leaving then only five of the eight piers to be built in the *fair
moulded hollow* to be fubject to the *vaft powers* of the *floods* of this river ; and this he
was under a neceffity of doing, as he found it impracticable to drain off the water fo
as properly to eftablifh the piers upon a foundation of piling: judging it, therefore, the
fafeft expedient to preferve the upper furface whole, which he found by far the hardeft,
and to found the reft of the piers by caiffons upon the bottom of the river, defending
the bafis thereof by fuch outworks as then were in contemplation.

How, therefore, it happens that becaufe at the place where Mr. Wake bored " the
" foil and texture of the bed of the river is uniformly a compofition or congeries of
" roundifh and flat ftones, gravel, and fand, of equal quality and confiftency throughout the whole of that depth ;" it is and muft be fo everywhere elfe, as well where
he did *not* bore, as where he did ! or how Mr. Mylne becomes entitled to report that
Mr. Smeaton's inveftigation of a fand (or foft matter) below, and a greater hardnefs or
concretion of the matter above, as appeared to Mr. Smeaton, from driving the bar as
already defcribed, is an unfounded idea, Mr. Smeaton cannot fee to arife otherwife

* It rofe fo plentifully from the bottom as to employ near 40 men at the pumps.

than from an intuitive knowledge in Mr. Mylne that Mr. Smeaton does not pretend to : inafmuch as that Mr. Mylne does not adduce the trial of any experiment whatever from whence that inference can fairly be made.

Mr. Smeaton, therefore, refers it to any impartial perfon who fhall perufe Mr. Mylne's reports referred to, and thefe obfervations upon them, whether it is Mr. Smeaton that has been *deceived* in the collection of facts and materials on which he eftablifhed his plan of operations ; or that Mr. Mylne, after hearing every thing that Mr. Smeaton had to offer, has undertaken to judge of works in a condition from whence their original ftate cannot now properly be judged of ; and by examining many perfons concerned and employed in the detail of its execution ; and relying on thofe whofe judgments were incompetent to the bufinefs ; that Mr. Mylne himfelf, in the forming of thefe reports, has been *deceived* and drawn into a *precipitate* opinion.

I fhall now conclude with obferving, that had Sir Walter Blackett's bridge fallen in the day-time, when it could have been obferved that the rapidity of the ftream formed a fall of five feet, or even half of it, I never would have advifed either Mr. Errington or the county to have fet about the erection of another bridge at Hexham, either upon piles, or in any other method, unlefs the fum of money to be expended upon it was unlimited. This fingle fact would therefore, in the prefent cafe, have deterred Mr. Smeaton from propofing the building at all, neither was, nor in its own nature could be known, till the flood that proved the fudden deftruction of the laft bridge ; though it had, without the leaft damage, withftood a flood when the fall was three feet nine inches, in confequence whereof the water would acquire a velocity of about 900 feet in a minute.

And now, being mafter of thefe facts by the fatal deftruction of this work alfo, he is unwilling to injure his reputation by the difappointment of others, in attempting to erect another bridge at Hexham upon a limited fum : that poft of honor he leaves to be occupied by fome more hardy or fortunate adventurer, remembering that a burnt child dreads the fire.

*Grays Inn,*
28th *June* 1788.

To Mr. Donkin.

Sir,                                   Aufthorpe 7th Feb. 1784.

As I heard nothing for fo long after the feffions, I flattered myfelf that Mr. Errington's offer had been accepted.

It is, as you obferved, now reduced to this, that Mr. Errington muft either rebuild or acquit himfelf by law; but it is well worth very ferioufly weighing, whether, to avoid the difagreeable uncertainty of a law-fuit by rebuilding, Mr. Errington will not as certainly incur the difagreeablenefs of both rebuilding and a law-fuit into the bargain; for it appears to me, from the temper and difpofition of his opponents, that they will difpute every ftep that he takes, and if they fo conceive it not to be *ftrictly* done in conformity with the contract, he will be equally liable to be haraffed by an action for non-conformity in part, as for non-conformity in the whole. Let us now, therefore, review the ground on which he will ftand upon each of thefe points.

If he lets the matter go to a jury upon a non-compliance to rebuild, I wifh Mr. Errington would be advifed by the moft able genius in the law upon the point; as to me, it very ftrongly appears, that were I a juryman in the caufe, I never could confent to a greater fum in damages to the county (notwithftanding the penalty), than the fum he actually received from the county, becaufe, as it will undoubtedly appear fufficiently in evidence, that for the purpofe of reinftating the bridge, the materials now upon the place and in their place, are of confiderably more value than thofe Mr. Errington received from the county, notwithftanding their valuation; and if it be true that the law of England is founded in reafon, which the law-men ftrongly contend, that is, in natural juftice, it never can be conformable to either juftice or reafon, that a fum in damage fhould be given to the county beyond the fum in which the county is really damnified; becaufe in that cafe the penalty of the bond would be made to operate as a punifhment upon a perfon who has honeftly endeavoured to fulfill his contract, and who in fact has once fulfilled it to the full content of the other contracting party, and becomes obnoxious to a penalty for no other reafon but becaufe he has been fo unfortunate as to have his work deftroyed by an act of Providence, which no human forefight could inveftigate or guard againft *.

* In other cafes the law does not extend the penalty beyond the damage fuftained, for in the cafe of the common obligation for money borrowed, though the penalty is for double the fum, yet no more is recovered than the fum borrowed and intereft. And why fo? Becaufe it would be contrary to reafon and natural juftice to make a man pay more than he borrowed, merely becaufe he failed in doing it at the day. And I am very apprehenfive, that there will not be found a determined cafe *ftrictly parallel* to the prefent, where it has been otherwife determined; and if not, this cafe being a new one, muft ftand upon the natural ground of the common law, that is, reafon and natural juftice.

U u 2                                       Mr.

Mr. Mylne may talk of the works not having been founded *deep enough*, and that it is uniformly a found gravel, becaufe he found it fo on a trial in a place where nobody that had a knowledge of the fituation would have doubted it; but though for argument fake, it fhould be admitted for *the prefent*, that the gravel was equally compact to an unfathomable depth *quite acrofs* the river, as in the place where he bored, yet we have now the authority of Mr. Mylne to fay, that this gravel, though " hard to the touch of " boring, compact to the eye and feeling of inftruments, is wonderfully loofe and " unconnected in its parts." — If fo, is certainly unable to refift the violence of the floods of the Tyne, which Mr. Mylne very emphatically confirms further, by adding, " infomuch that the bed of the river Tyne feems to fhift and alter its form, extent, " and fituation with every flood, more or lefs, and tearing up at one time to a great " depth that fair molded and well-laid hollow that the ftream laid for itfelf on fome " former occafion." Now this being fo, to what purpofe to drive piles into a body of gravel, if that body is itfelf liable to be taken away, into which the piles are driven? It is neceffary then that this body of gravel fhould be prevented from moving by the depofition of fome matter more compact than the gravel itfelf, to be able to refift the violence of the floods, and protect the ground underneath.

This was indeed done by the depofition of rubble ftones from Oakwood bank quarry, which depofition of rubble is not only in general the moft effectual method my expe-rience has furnifhed, of protecting a gravel foil from the wafh of a violent current, but I efteem the rubble of Oakwood bank quarry of the beft quality for this purpofe that I have feen. The works were amply guarded with this ftone; and fo fatisfactorily fo that I did not think any flood of the Tyne would or could have moved them; and I thought them ftill the more fecure after having refifted the velocity of 900 feet per minute, without the leaft derangement, but on the contrary, rendered the whole more even and compacted by the interftices being filled up fmooth and even with gravel and matter of a leffer fize. Had this defence of Oakwood bank quarry rubble laid ftill there was nothing perifhable in the bridge to be any caufe of decay; but a flood comes fo rapid and fudden that its velocity, amounting to 1000 feet per minute, was fufficient to move the defence of the Oakwood rubble, and fo even and equally was it diftributed that the failure of all the parts dependent thereon was in a manner together.

Mr. Mylne entertains an idea, that if our works had been founded deeper they would have fared better; and fo far I agree with him in opinion, that had they been as deep again, they might have ftood an hour longer: but when we know that the river Tyne

is

is capable on every flood, more or lefs, of tearing up that *fair moulded hollow* (I fuppofe he means the channel of the river) to a *great depth*, it perhaps may be very difficult to afcertain what depth will be out of the reach of its action; nor is this difficulty rendered the lefs by his fhifting the refolution of this very difficult queftion from himfelf to us. I know a part of this river where the water has gulled away its bed (as well as I remember it, on meafure) to the depth of 24 feet below the low ftate of the water's furface.

I have thought it proper to fay thus much by way of obviating thofe hints and furmifes which tend to make it feem as if every thing had not been done that could be done in the fituation.

I come now to confider what ground Mr. Errington ftands upon in cafe of rebuilding; and here it ftrikes me, that not only Mr. Mylne, but every one elfe alfo, has confidered this matter in the fame light as if it had never been done, that is, confidering how Mr. Errington was to acquit himfelf of the whole agreement: whereas the bridge has been built complete, and to the entire fatisfaction of all the contracting parties, according to the agreement, and fo certified by two magiftrates as therein appointed: had, therefore, the laft covenant been omitted, which obliges Mr. Errington to *maintain* and *uphold* for feven years after the magiftrates' certificate, he would moft certainly have ftood acquitted as having fully performed his contract. It would, therefore, feem to me, that what remains obligatory in the contract is what remains fo in force of the laft claufe, and it would further feem to me, that in an unfortunate cafe, like the prefent, where the thing has once been fatisfactorily done, that a liberality of conftruction of this laft claufe cannot be extended in disfavor of Mr. Errington, but rather, as far as the nature of the thing will reafonably admit, in his favour.

I fhould therefore wifh that the following queries were fubmitted to fome genius in the law.

1ft, As a confiderable part of the bridge is left ftanding and unhurt, whether, if more favourable in point of conftruction to Mr. Errington, this may not be placed under the idea of an obligation to *repair* in contradiftinction to that of *rebuilding?*

2dly, Whether the words of the contract *uphold, fupport, repair, maintain,* and *keep in good and fubftantial repair* the faid bridge, will oblige Mr. Errington to make the bridge
exactly

exactly what it was before, or strictly conformable to the design annexed to the contract, and to which it refers, notwithstanding that from the nature of the accident, the ground of the site is so altered that an exact restitution is now impracticable?

3dly, Whether Mr. Errington (having once acquitted himself in point of satisfactory construction) is obliged (further than he may think necessary for his own security) to make any additional defences, or to be at greater expenses in rendering the foundation more secure than it was before, but may do the work in the same way it was before, where that can be done; and where change of circumstances has rendered that impracticable, in any other equivalent way, in the judgment of Mr. Smeaton, to whom the whole matter by the agreement has been referred, and confirmed by act of parliament?

4thly, Whether Mr. Errington may not, for his own security, take any other method of rebuilding any of the particular parts deranged, that from change of circumstances shall be, in the judgment of Mr. Smeaton, more likely than the former to render the work permanent?

5thly, Whether the words *good and sufficient repair* are not qualifying words which will enable Mr. Errington to make use of the old materials in the places where they are now at rest though not in the same position in which they were originally built; insomuch that had the flood turned some of the pillars topsy turvy, whether they may not be made use of in part of the re-erection, provided that, in the judgment of Mr. Smeaton, they are or can be rendered as secure as in the original construction; though in point of position they will not be strictly conformable to the original design referred to, but will be made as safe and effectual for the use of His Majesty's subjects as a bridge, as if every thing were restored to what it was at first?

6thly, Whether, in virtue of the words " good and effectual repair," Mr. Errington may not in this repair omit every expense that is merely ornamental, in case he so chuses?

7thly, As the risk of the bridge's standing after the repair, lies with Mr. Errington for a certain term, in which, if it should fail, he will be obliged to repair *toties quoties*, whether in case he again employs Mr. Smeaton, and Mr. Smeaton will again engage in the concern, whose judgment, by the agreement, is made the *dernier resort*, whether this will not bar all interference and controul of the magistrates as to the modes to be pursued of repair and re-erection of the particular parts that require it?

8thly,

8thly, Whether the certificate of the magiftrates, and the acquiefcence of the whole body, by paying Mr. Errington the laft payment due on fuch certificate, be not a full bar to the magiftrates to any cavil that may be raifed now, whether Mr. Errington fulfilled his contract in the erection, which, in reality, he did to the fulleft extent, in the judgment of Mr. Smeaton, and which nobody thought of controverting till after the fatal difafter had happened ?

9thly, Whether the time that has lapfed by the procefs of a treaty of compofition, which the magiftrates have admitted from Mr. Errington, and to which they have never given a final anfwer till the Chriftmas quarter feffions held in January 1784, be not to be reckoned in part of the feven years from the time of the certificate, which was at the fame time three years before ?

10thly, Whether if Mr. Errington, affifted by the judgment of Mr. Smeaton, repair the bridge fo as to be in every refpect as effectual for ufe, and as fufficient in point of ftability, as the bridge was before the accident, and is fo maintained by him to the end of the term to which he is obligated, he will not then be legally difcharged from the penalty of the obligation into which he entered ?

I beg leave, before I conclude, to declare once more, that, though I never thought it a difficulty to reinftate the bridge upon a principle of permanency equal to what it was at firft, yet from the experience of this accident I do not know of any method at any expenfe within the bounds of Mr. Errington's obligation to reinftate it, fo that it fhall, in my opinion, have the fame profpect of permanency that I thought it had before the accident ; yet fuch a principle of ftability as it had before the accident may enable it to ftand longer than the lifetime of any man in being, or it may be thrown down again the next year ; for as it is out of my power to calculate the uttermoft powers that Nature can collect, fo it is out of my power to fay what will abfolutely ftand againft every poffible violence.

Before the accident I thought this bridge had this permanent kind of fecurity ; in this, the accident has convinced me that I was miftaken, and that I may be fo again : but this I do apprehend, that had there been but one fingle hour more between the downfall of the fnow and rain in the night and the flood's acquiring its greateft violence by the middle of the next day, it would in that time have fo filled the reaches of the river below the bridge, as to have moderated its rapidity in paffing the bridge from 1000 to 900 feet per minute ; fo that the bridge would have been ftanding at this inftant, and

all

all thofe concerned in its erection receiving the praife that the public was pleafed to attribute to them for fo noble an erection, inftead of the difgrace attending the being unfuccefsful generals.

I remain, Sir,

Your moft humble fervant,

J. SMEATON.

---

A DISSERTATION upon the peculiar Hardfhip of the Cafe of HENRY ERRINGTON, Efq. in regard to his Bond for the Maintenance of Hexham Bridge for Seven Years.

" GIVE me my Bond," fays Shylock ; " I will have my Bond." This, though an ancient legendary tale, has been feized upon by that immortal genius Shakefpear, and wrought up in a ftriking degree, to fhew to what manifeft injuftice human laws are capable of, in particular inftances, when carried to a rigorous execution in thofe cafes to which, as unforefeen, they never have been intended to be applied. Had the laws of Venice been rigidly carried into execution in the prefent cafe, they would have been looked upon with abhorrence by all fucceeding ages ; but in the way it was determined, we cannot lefs ad-mire the ingenuity of the pleader, in finding out a circumftance by which the keen edge of the law was taken off, and ftrict and equal natural juftice rendered to all the parties, than the renowned decifion of *Solomon* between the two harlots ; or the celebrated decifion of *Sancho Pancha* between the cook and the defendant.

What, in this cafe, do the magiftrates of the county of Northumberland purfue Mr. Er-rington in an Englifh court of juftice to obtain ? Why, to obtain the payment of 9000l. from a perfon who never received from them more than 5700l. Do the magiftrates then mean to make money of Mr. Errington for the benefit of the county's purfe, merely becaufe they have caugh thim upon the hip ? No ; their own honour, jointly as well as feverally, will not prompt them to avow this. No ; they fay that they fue Mr. Errington for the penalty of the bond of 9000l. to force him to re-erect the bridge, and maintain it for feven years in the terms of the contract. But fuppofing Mr. Errington to pay the 9000l., will any man undertake to erect a bridge for that fum, to maintain it for feven years, giving a bond for the performance, of 9000l. ? (It fhould be 11,000l. to be a ftrict parallel), without

2

which

which they will not ftand upon the ground Mr. Errington does, nor the county be any better affured of their having a bridge. Certainly no man will, unlefs he is mad, or in no
degree inftructed by the leffon the late fatal experiment has taught. Why then, what
is the refult, but that the magiftrates of Northumberland, taking advantage of a particular turn of the law of this kingdom, mean to force a fum out of the pocket of
Mr. Errington that he never received, in order to put it into the county's purfe, to enable
the county, by a further addition of their own, to lay out a fum of money upon a further
and more extenfive experiment, far greater than Mr. Errington's contract was to
receive.

" But," fays Shylock, " it was your bufinefs to have confidered the confequences before
you entered upon the bond. You executed the bond with your *eyes open*, and you muft
pay the penalty."

But did Mr. Errington enter into this bond with his *eyes open?* Why, no; he certainly did not. Were any other perfon *now* to enter into a fimilar bond, it may be faid,
that he really and truly enters upon it with his eyes open; that is to fay, with the neceffary degree of information to give him fome idea of the extent of the difficulties and
hazard that were likely to attend it. The late erection may be confidered as a proper
experiment to prove the degree of rapidity and violence that the river Tyne is fubject
to fo high up in its courfe as the parts oppofite to Hexham; but, previous to
this, there was nothing to furnifh an adequate idea, much lefs a pofitive proof, of
the degree of violence of which this river is at times, under certain circumftances,
capable.

A bridge new built oppofite Hexham, at the upper or weftern end of the town, of a
conftruction fomewhat fimilar, though (according to the doctrine of fome) more fecure,
as having piles and ftrong platforms under all the piers, in the compafs of a fingle night,
in the inundation that happened in Nov. 1771, was totally taken down and deftroyed,
nothing remaining the next morning but the north abutment. This will naturally fuggeft great violence in the river, or great weaknefs in the conftruction of the bridge; but
to which the cataftrophe was principally to be attributed, does not pofitively appear: for
the bridge being apparently right at darkening, and totally demolifhed at break of day
the next morning, nobody happened to be witnefs of its deftruction, or of the fall that
the water had in paffing the bridge from the up ftream fide to the down ftream fide thereof. All that could be feen next morning (the water being then a good deal fubfided)
was, that from the marks it had left it had been very uncommonly high, and that it had

not only taken the fouth abutment clean away but widened that fide of the river by 60 or 70 feet; all which indicate marks of great rapidity and violence, but *what degree* of it by no means appears: for if the water had rofe to the fame height, and had been ftagnant like a mill pond or tide river at high water, no degree of height or depth of water ought to hurt a bridge that is exprefsly built not to take any damage from mere wet; nor did any thing appear by which a fall even of two feet could be inferred: its failure therefore muft appear to be owing not fo much to the weaknefs of the conftructed matter of the bridge as to the weaknefs of the ftratum whereon it was founded.

The height of the flood that occafioned the demolition of the laft bridge, was in the middle of the day; and the beginning and progrefs of its fall witneffed by many perfons; and before any derangement had happened, it had been remarked what member of the bridge the water was even with on the up ftream fide, and what member from the down ftream fide, which from its known dimenfions, amounted to near upon five feet of differ-ence of level; fo that the water came down in this flood with fo much rapidity and fud-dennefs, that not being able to fill the reaches of the river and vallies below fo faft as it came down, it formed there a *breaft* of the aftonifhing height or fall of near five feet perpendicular: and from a fall of lefs than five feet, that is, of four feet five inches, there neceffarily refults a rapidity of the torrent, amounting to one thoufand feet in a minute; a quantity of fall and rapidity, that *could it have been known* from the deftruction of the former bridge, or even had there been found a fall of *half this quantity*, Mr. Smea-ton can take upon him to fay, it would have deterred him from encouraging Mr. Erring-ton to have had any thing to do with undertaking the propofed bridge: fo that it may be fairly faid, that neither Mr. Errington, nor any of his advifers, either had or could have that degree of information as to warrant its being faid, that Mr. Errington executed the bond with his *eyes open*: but any one now that enters upon a fimilar obligation muft enter into it with his eyes open; becaufe from the late fatal experiment, he will know this *capital and leading maxim*,—that by the fudden melting of fnow, accompanied with a violent downfall of rain, pouring from the fteep fides of thofe hills extending to the very fources of two large rivers that join a little above Hexham, there is a capability of the waters coming down with that fudden violence, as that in the fituation of the late bridge, the torrent is capable of forming a breaft of near five feet, and confequently, of acting with a certain rapidity of at leaft a thoufand feet per minute; and knowing this for a cer-tainty alfo, drawn from the fame experience, that this velocity is capaple, not only of tearing up the bed of the river, but of removing all fuch rough materials of ftone as may be depofited for the defence of the regularly conftructed works.

Whoever

Whoever will therefore now undertake to build a permanent bridge, muſt be provided with ſuch a deſign as not only will be proof againſt all the violence and cauſes of derangement already aſcertained and deſcribed, but, as it cannot be known for a certainty that the violence already experienced is the uttermoſt that Nature is capable of, in this place, he ought to be ſtill more firmly fortified on that account, to reſiſt ſuch further violences as may poſſibly happen : all which, in Mr. Smeaton's judgment, cannot be expected to be done for a much larger ſum than the penalty of Mr. Errington's bond ; much leſs for the ſum of money and value in materials, that Mr. Errington actually received.

Had Mr. Errington drawn the magiſtrates of Northumberland into the ſcheme and idea of building a bridge at Hexham, merely to ſerve his own purpoſes, and after a very conſiderable expenſe to the county, it had ended in the ill-fated cataſtrophe that has happened ; had Mr. Errington had a view to make a profit of this buſineſs ; had he been ſparing of any apparently neceſſary expenſe for the accompliſhment thereof ; had he let it by the great, to be executed by under workmen, and thereby eaſed himſelf of the trouble and attention neceſſary to ſuch a work, and withal put a round ſum of money into his own pocket ; had he employed incompetent artiſts to direct and ſuperintend the work, or adviſed with ſuch, as to the practicability and mode of accompliſhing it, who were not of eſtabliſhed reputation in the country for works of the kind ;—in ſhort, had Mr. Errington practiſed or attempted to practiſe, any fraud in the conduct of this affair, or acted with any ſiniſter views, in any of theſe caſes it would have been natural for the magiſtrates, finding themſelves cheated, deluded, and diſappointed, to have purſued Mr. Errington with the rancour and vindictive ſpirit they are now doing : but if it ſhall appear that the very reverſe of all theſe things is the truth ; if it ſhall appear that the magiſtrates had entertained this ſcheme and idea from the ſuggeſtions of the late worthy and univerſally eſteemed and reſpected Sir Walter Blackett, the upſhot of which, as reſpecting him, was the total demolition of the bridge, as already deſcribed, in 1771 ; if the diſappointment ariſing from this fatal cataſtrophe was ſo great to the magiſtrates, that, on Sir Walter Blackett's refuſal to be concerned with the bridge any further, chooſing rather to pay the penalty of a bond that he had entered into with the County, than further embark in ſo ill fated, in ſo apparently uncertain and hazardous an undertaking * : I ſay, if on this refuſal of Sir Walter, the magiſtrates ſtill remained ſo eager for a bridge that they actually began to erect another near the ſame place, and for this purpoſe engaged an eminent engineer to direct and

* Sir Walter Blackett entered into a bond, the penalty whereof was 3000l. the identical ſum that he received from the County, towards building a bridge at this place. The bridge certainly coſt a far larger ſum ; but the voluntary ſubſcriptions of ſeveral gentlemen whoſe eſtates lay in that neighbourhood, made a part of the extra ſum expended, no part of which was refunded by Sir Walter.

ſuperintend

superintend the work †, with a refident furveyor of his own choice and recommendation ‡; and if after making fome progrefs in the work, this fame engineer gave it up as impracticable, on account of the infufficiency of the foundation; if the magiftrates notwithstanding ftill remained unfatisfied, and eager to that degree as even to attempt to draw in the poor working mafons of the county to take this great hazard upon them, (after a declared impracticability by an eminent engineer) to the probable ruin of themfelves and families; I fay, if, after all this ardour for a bridge on the part of the magiftrates, Mr. Errington was the unhappy facrifice delufively taken in by their offers §; then it may be fairly faid, that the magiftrates have drawn Mr. Errington into a fcrape, and not Mr. Errington the magiftrates; and that in taking the thorn out of their fide he put it into his own; for which act they are now fully bent to punifh his ill fuccefs by a rigorous exaction of a fum of money that he never received.

True it is that Mr. Errington had a view, he had a *motive* in this bufinefs; but if that motive was a laudable one, why punifh his want of fuccefs with fo much rigour? We do not read of any age where men undertook works for the public fervice, merely for the fake of having the trouble of performing them; mankind in every age had a *moving caufe* of action. Did Sir Walter Blackett engage in building a bridge oppofite Hexham without any other moving caufe than the mere good of the public? Certainly not. Sir Walter Blackett had a very confiderable eftate oppofite the upper or weftern part of the town of Hexham, and was lord of the manor of the whole. Had Sir Walter Blackett had no moving caufe but the mere *public utility*, to fix the deftination of the good he intended to mankind, he might have found many other places to promote the building of a bridge in England, and fome in Northumberland; but it muft be allowed, that the above cir-

---

† Mr. John Wooler, who was then rebuilding the bridge of Newcaftle, overthrown by the fame flood, and which was afterwards finifhed by him.

‡ Mr. Jonathan Pickernell, from London; afterwards made Surveyor of the County Bridges of Northumberland, and recommended by fome of the magiftrates to Mr. Errington and Mr. Smeaton, to build Hexham Bridge under his direction.

§ Whoever was the leaft converfant with the public tranfactions of this county, in the years 1775 and 1776, will well remember an advertifement coming from the magiftrates of this county, and appearing in all the three weekly papers of Newcaftle, *importing* their readinefs to treat with any perfon that would engage to build a bridge at Hexham, according to a defign or plan for a *fuperftructure*, lodged for their infpection with the clerk of the peace. The offerer to be at liberty to purfue *his own plan or method* of conftructing the *foundations* (under water), but to give fecurity for the permanency thereof for feven years. This was the purport, as it occurs to memory; but the mafons and working mechanical artificers were too wife to take the rifque of fuch impending ruin: and the confequence was, that after continuing the faid advertifement weekly for the greateft part, if not the whole of a year, without any adequate offer, it was Mr. Errington's ill fate to be feduced by thofe colours fo hung out.

cumftances

cumftances ought not to take from the merit of Sir Walter's fixing his efforts at Hexham, and building the bridge fo as to land, on the north fide, upon his own eftate.

In like manner, Mr. Errington having an eftate oppofite to the town of Hexham, on the eaftern or lower part of the town, being alfo defirous to have the bridge to land at the north end upon his eftate, the benefit that was likely to accrue thereto was a fufficient inducement to him to engage in that very great fcene of trouble and attention that muft neceffarily arife in the building of a bridge in fuch a fituation, and of fuch a magnitude, without the leaft profpe&t of any profit, or other advantage by the building; on the contrary, it was Mr. Errington's profeffed declaration to Mr. Smeaton, when he applied to engage his advice, opinion, and affiftance, that fo far from meaning to be a gainer by the undertaking, he fhould not be difappointed if he were 2 or 300l. out of pocket; and which, it will be proved, a&tually turned out to be the cafe.

In regard to the competency of Mr. Smeaton, to form a judgment of and properly execute fuch undertaking, this may be feparately difcuffed: but it will certainly exculpate Mr. Errington in having employed him, when it is recolle&ted, that at *that time* Mr. Smeaton had had *no bridge fallen*, had been employed in the full fcope of bufinefs in his profeffion no lefs than 25 years, and at that time had given proofs of his abilities by a conftant fuccefs in executing the moft difficult works, in almoft every part of Great Britain, without *then* the failure of a *fingle fubje&t*.

If therefore Mr. Errington a&ted the part of a wife, a prudent, an upright, and a difinterefted man, willing to promote the public good, along with his own; if in a laudable attempt towards this end he has failed, from caufes that could not in their *own nature* be *forefeen;* it muft be allowed to be *hard*, nay *very hard* indeed, that Mr. Errington not only cannot be excufed in his failure, but he muft be a&tually pofitively *punifhed*. Does the law admit no remedy? The law-men fay of *none;* nay it is not even in the power of the magiftrates to remit any part or tittle of the *penalty of the bond!* fo fays the *written* law of England!

But what fays *natural juftice* in the cafe? that law which is written in the breaft of every rational man, and is properly deemed the law of *equity* and of *good confcience;* does this agree in the fame determination? No, it revolts from the idea, and fays, that to exa&t the full penalty of the bond in the prefent cafe, would be the moft fhocking injuftice to Mr. Errington. Is no one then in a legiflative fenfe, in poffeffion of the executive part of this law of equity, and good confcience? Yes, reafon anfwers, every man in quality of
a *Bri-*

a *Britifh juryman* poffeffes it, and has the adminiftration of it in his power; he alone can draw the line of right and wrong, in the *particular* cafe before him, where the written general law of the kingdom can (from the nature of the thing) draw none; and it is a wife provifion of the general law, to leave the decifion of particular cafes to the *breaft* of a jury; where the law itfelf has not, nor can fix the limits of right with an adequate degree of precifion. For this reafon, a juryman's determination is *upon oath ;* the purport of which oath is to the effect, that he will ufe his judgment according to his confcience; for there is no need of an oath, if it is to be fuppofed that a juryman is previoufly and externally fixed to a point; that is, that he is to judge, without having the ufe of his judgment.

When a juryman has heard all parties and perfons who have any thing to fay in the queftion; he is afked by the judge, whether he finds for the plaintiff or the defendant? that is to fay, if the right is for the plaintiff or the defendant? if for the plaintiff, what are the damages? Now is it poffible that any juryman can lay his hand upon his breaft and fay the plaintiffs are damaged to the amount of 9000l., when it is fully proved, that the value of every thing that the defendant received from the plaintiffs, inclufive of the road that ftill remains, was but 6100l., and in this 6100l. they are damaged no more, than upon a fair eftimate it would coft to reinftate the building as agreed for *; for the magiftrates cannot be allowed to fay, that to reinftate the building as it was agreed for, will not now be likely to anfwer our end in point of permanency; he fhall either build a *better bridge*, or *give us money* wherewith to do it; this certainly cannot now be admitted from the plaintiffs *in foro confcientiæ ;* becaufe it is the light drawn from the experiment made by Mr. Errington, that alone makes the magiftrates now to fee the neceffity of building upon a more extenfive plan; and fince it was they themfelves that practifed the delufion, in order to draw any unwary perfon they could into the fcheme (which unfortunately for himfelf, happened to be Mr. Errington), they certainly ought not to be *gainers* by this delufion. If Mr. Errington pays them back what upon a fair eftimate it is likely to coft, to reftore it to what it was; even this puts the magiftrates in a *better* fituation, than they would have been, if by the combination of natural caufes, this flood had been no more violent than common, and which this bridge had withftood before; in which

---

* The contract was, to be paid 5700l. in money by inftalments as the work advanced, for the bridge, and 400l. to Mr. Donkin for making a road, in the whole 6100l. ; alfo Mr. Errington was to receive all fuch materials of timber, ftone and iron, as the magiftrates themfelves had provided during their efforts to get a contract; — thefe materials were valued by the magiftrates to Mr. Errington at 2000l., though in reality they never were of 1000l. value to him; and as the road ftill remains, and there is much more value now upon the premifes towards a bridge, than Mr. Errington received, the magiftrates in receiving the prefent materials from Mr. Errington towards a bridge, will be gainers upon what he received from them.

cafe

cafe it would have now been left upftanding; becaufe they are aware, in attempting to reinftate it, that fuch powers of refiftance are *now* required, of which before its down-fall they could not poffibly have formed any conception; which knowledge will be of moft *important ufe* to any one who finds courage to become a future builder and guarantee for the upftanding thereof.

Nay, fuch is the utility of the knowledge gained by this experiment, that in the eye of reafon, the whole charge thereof certainly fhould not be folely upon Mr. Errington. The magiftrates, if *nolens volens* they muft have a bridge at Hexham, fhould contribute to this experiment as well as he; and therefore only a moiety of the fum eftimated for rein-ftating the building, fhould be reimburfed by Mr. Errington, as his proper fhare of an unfortunate adventure; and the other half fhould be allowed by the magiftrates for the experience gained: and then upon this fund of money and knowledge, and an adequate fum in addition for its more extenfive conftruction, they may with fome degree of profpect hope to poffefs a permanent bridge at the public expenfe, without forcing it out of a private man's pocket.

That juries do in other cafes exercife this kind of judgment, is manifeft. I become bound in the penal fum of 2000l., that *A.B.* fhall duly fulfill the truft repofed in him by a certain company, deliver a true account, and all moneys, papers, and writings committed to his care, upon being thereto required. It happens that he abfconds, carries off every thing that he had in his charge, and 1000l. of the company's cafh. The company fue me with an intent to recover 2000l. upon the bond; but though he has broken every article for which I became bound, yet unlefs the company can prove the intrinfic value of the books and writings, and that by the breach of his honefty, they have in fact fuffered a real pecuniary lofs of more than the real fum of money deficient; will any jury punifh me by the payment of 2000l. for the delinquency of another, when the whole of their intrinfic lofs is no more than 1000l.? furely they will not.

Again, in the common cafe of a bond for repayment of money borrowed; the common condition of the obligation is, that if the obligee pays a certain fum (commonly one half of the penal fum) upon a certain day with lawful intereft for the fame, *without fraud or further delay*, then the obligation to be null and void, otherwife to remain in full force and virtue. We will fuppofe the obligee fails in every article, wherein he has been bound; he neither pays the fum nor the intereft at the time; he makes many fraudulent promifes of payment at a future day, merely to gain time till he can abfcond; and makes ufe of every fhuffling pretence, at any rate to *delay* the payment: It may happen, and often

does,

does, that for want of the money at the time, the obligor is prevented from fulfilling a purchafe that he has made; by which he would have got a profit equal to the fum, and in failure thereof may himfelf be fubject to an action; and after all this he is put to the neceffity of profecuting an action to recover his own; this is *very hard*, but will a jury punifh this obligee for his fraud, failure and neglect by finding in damages the whole penalty? No, they will calculate what his principal and intereft comes to; allow him that, with expenfes of fuit, and give damages accordingly. But here we are told, that this difpenfing power in the breaft of a jury, is limited in confequence of an *act of parliament*, exprefsly made for this purpofe, fo that let the penalty be what it will, the obligor can recover no more than the real pecuniary damage.

Very well; but does not the very making of fuch an act of parliament infer the judicial power of a jury, that fubfifted in their *breafts*, before the act was made? and therefore, where there is no act to fix the line of law, that it ftill refts in the breafts of a jury? the thing fpeaks for itfelf. Acts of parliament are not made to remedy an evil that has never exifted. It is not to be doubted, but that before this act, it was common for a defigning obligor to trump up a detail of the damages he had fuffered, by the money not being duly paid according to the obligation of the bond, in order to extort the whole penalty; and though this might fometimes be the cafe, that the obligor might fuffer by the neglect of the obligee, yet it doubtlefs would be pretended ten times for once, that it really was the cafe. It might alfo happen to fome juries not to have that accuracy of reafon and difcernment, but to fuppofe themfelves under an abfolute *obligation* to find the full penalty, though an adequate damage might not be fo clearly made out, conceiving the penalty to be the proper punifhment for non-performance. In order fully to fettle thefe matters upon the beft general grounds, an act has been found neceffary to reftrain the jury's difcretion from proceeding beyond a certain line; and nothing can be a ftronger proof of the full extent of a juryman's power of difcretion, than this act of parliament, to reftrain it in this particular cafe. It is not therefore to be doubted, but that every jury-man who is fully fenfible of what is due to his own character, and the full extent of Mr. Errington's cafe, will perceive, that as he has the power, he undoubtedly will have the will, to draw the mercilefs teeth of the lion, and not fuffer them to fix unreafonably deep in the flefh of an innocent man; and this even at the expenfe of his own oath in giving exaggerated damages, where mitigated damages only are due.

## BERWICK BRIDGE.

The REPORT of John Smeaton, Engineer, upon the State of Berwick Bridge, from a View thereof, the middle of September 1771.

### Notes taken on View.

THE key from the upper birth to the bridge, appears to be in a ruinous condition.

N.B. The north weſt abutment wall of the bridge ſwayed out, and overleaning, it ſeems proper to be ſupported by a buttreſs.

The ſplay wall up ſtream wants pointing.

The ſecond courſe of arch ſtones in the firſt arch from Berwick, is decayed in the weſt face near the crown of the arch ; the ſtones that are periſhed ſhould be ſhifted.

The Piers to the fourth incluſive being in hand to be pointed and repaired, and in great part done, it is needleſs to ſay any thing as to the reparation thereof.

Fifth Pier—Some ſtones deficient, and the lower point of the ſtarlings frame beginning to decay, theſe matters ſhould be attended to.

Seventh Pier—Several ſtones out of the ſides of the pier, and alſo ſome ſtones out of the old ſtarling jetty or frame, as well as the new, which laſt is ſettled and conſiderably deranged.

Ninth Pier—The north ſide of the ſtarling wants planking, and the upper peach is in bad condition; it may be rebuilt, or with repair, may be expected to laſt a few years longer.

Tenth Pier—The ſheeting on both the north and ſouth ſide of the ſtarling wants repairing, as alſo the peak.

Eleventh Pier—The down ſtream point of the ſtarling is decayed, the north ſide wants planking half round, and the ſouth ſide wants the looſe ſtones righting up againſt it.

Twelfth Pier—The down ſtream point of the ſtarling is out of repair, the ſides want planking all round, and the ſouth ſide more particularly wants repairing.

Thirteenth Pier—The ſtarling wants planking all round, and the looſe ſtones righting up in the paſſages, and laying regularly againſt the ſides of the ſtarling; the peak of the pier has ſettled, but with repair may probably laſt a few years longer.

Fourteenth Pier—The ſtarling wants planking on the north ſide, and its down ſtream point is deficient. In the face of this arch and the next ſome ſtones are wanted on the weſt ſide, but as they are found in the ſoffite underneath, they are not in imminent danger, and by ſhifting the decayed ſtones may be properly repaired.

Several lengths of the parapet of the bridge are overleaning, ſome of the worſt places ſhould be taken down and rebuilt, and the terminating pillars of the abutment parapets on the ſouth end of the bridge being in a ſtate of diſrepair, ſhould be made up again to prevent further derangements to the parapets.

N. B. Reference has alſo been had to the report of Meſſrs. Wilſon, Dods, Buglaſs, and Steel, dated 21ſt June 1771.

This bridge having been built ſeveral centuries, has felt the effects of time, and beſides the decay of the ſtones in ſome particular places, the mortar is in general periſhed from the joints, as commonly happens to buildings of the ſame age, and alike expoſed. Nothing would ſo much contribute to an appearance of good repair as to have the joints new pointed, and could this be done in an effectual manner it would in reality contribute to its duration; but the imperfect union of new mortar with old, and with ſtones in a ſtate of waſte, is ſuch, that the froſts and rains in the courſe of a very few years generally undo whatever of this kind is done, even when executed with great care: however, it being very ſatisfactory to the public eye to ſee matters of this kind look well, I cannot but recommend it to thoſe who have the management of the repairs of the bridge, to do as much of this as they can, without neglecting what is really eſſential.

This bridge, like moſt of the ſame age built in tide rivers, has been founded on piles ſawn off above low water, for the defence of which from the injuries of the weather, and alſo to ſecure the ground round each aſſemblage of piles for the forming of a pier, each aſſemblage has been ſurrounded with another work called a ſtarling or frame, which is

2                                                                                    terminated

terminated fo much above low water as to cover the wood work immediately under the pier; and as thefe ftarlings are confidered only as a work of defence, whofe furface is in general wood, they muft of courfe be fubject to decay from the injuries of the weather and the wear of the water; the piles in the fides of the ftarlings in time therefore want renewing, and when that is the cafe, as they can eafier be done by furrounding the former work with a frefh row of piles than drawing out the ftumps of the old ones, it happens that the *eafieft way* is often taken: but as this increafes the bulk of the ftarlings, and of courfe contracts the water ways, the rapidity of the flood tends in a greater degree to fap and deftroy thofe outworks, and then recourfe is had to ftones to fecure the ftarlings, and the intermediate bed of the river, which ftill increafing the obftruction, form alto-gether a kind of dam, over which the water falling pools a great hole below the bridge, which often threatens the deftruction of the whole. For thefe reafons I cannot commend the firft laying of the ftones round the ftarlings, and lefs the extending the ftarlings them-felves fo far from the fides of the piers as to require it; but when done, as one evil is frequently applied to cure another, perhaps it may not be fo perfectly fafe juft now to take away the furrounding ftones already placed; but it is very material that the repair of the ftarlings fhould be duly attended to, as the fafety of the whole bridge depends upon them; and when any repair is made to do it in fuch a manner as not to contract the prefent water way, but whenever it is poffible by contracting the ftarlings fideways to increafe it, and whenever the down ftream points of the ftarlings become dangerous, by the pooling of the ground below the bridge, and taking it too much away from the tail of the ftarlings, it will be the beft and cheapeft way to fecure the bottom by throwing in rough rubble ftones from the quarry; and though this may in fome fenfe be confidered as a contraction, yet being chiefly in the eddy of the pier, will not have the effect of fuch, and is to be applied only in fuch a degree as the matter there has become defective in quantity: it is poffible that after fuch application fome of thefe ftones, by being laid too fteep againft the ftarlings, may be carried down by the fucceeding floods; but as this will only happen by their being too fteep, by fliding down they enlarge their bafe, and by filling up the place from whence they have flided, they will in the end remain immove-able like a rock.

By attempting to put the ftones into a frame at the tail of the feventh ftarling for the fecurity of the down ftream part of that ftarling, in the cafe above mentioned, and for want of fomething to keep the ground from being carried away round this new jetty, it has fettled and the ftones therein have become deranged; but if the wood work could be drawn out, and the ftones fuffered to form their own flope, and frefh

ftones

ſtones added where they may be defective, I apprehend this ſtarling would thus be effectually ſecured.

In regard to the up ſtream peaks of the piers which have ſettled and cracked, as the arches do not immediately depend upon them, the building thereof is a matter leſs preſſing, eſpecially when it is conſidered how difficult it is to unite the new work with the old, ſo as to prevent a ſeparation even after they are rebuilt.

J. SMEATON.

*Auſthorpe*,
16th *June* 1772

ELEVATION

Red 170.

Scale of Feet

DESIGN for a BRIDGE over the RIVER DOVERAN at BANFF

ELEVATION of the CENTERINGS.

PLAN

20.0

30.0

17.0 19.6 14.6 40.0 14.6 17.0

PLAN of the FOUNDATION.

SECTION of ONE
of the PIERS

13 by 1
36 feet 6 long

30 Feet

9 Feet

12 Feet

17 Feet

12.0

Common Water

line when the tide is out

6.0

14.0

12.0 2.0

2.0 2.0

13.0

Scale of Feet

35 Feet.

J. Farey Jun. del.

J. Smeaton 1773.

# BANFF BRIDGE.

## (See Plate 14.)

ESTIMATE for building a Bridge over the River Doveran near Banff, to be of 410 Feet between the Abutments, 20 Feet wide over all, and to confift of Seven Arches; the Aifler of the Piers, Arches, and Facings of Freeftone, and the Core of the Piers, Spandrell Walls, and Parapets of Rubble.

| | £. s. d. | £. s. d. |
|---|---|---|
| **PREPARATION for founding the PIERS.** | | |
| To timber in coffer dams for the fix piers and two abutments, exclufive of what is to remain connected therewith, 1278 cube feet at 3s. 6d. | 223 13 0 | |
| To iron work to do. 12 cwt. 2 quarters, at 5d. per lb. | 29 3 4 | |
| To pumps, piling machines, and utenfils | 100 0 0 | |
| | | 352 16 4 |
| **The PIERS and ABUTMENTS to the Spring of the ARCHES.** | | |
| To excavations and pumping for eight foundations at 15l. | | 120 0 0 |
| To timber in plank piles, bearing piles, ribbands, &c., for the eight foundations, 2788 cube feet, at 4s. | 557 12 0 | |
| To iron work for the fix piers and two abutments, 22 cwt., at 5d. per lb. | 51 6 8 | |
| To lead 10 cwt., at 2d. per lb. | 9 6 8 | |
| | | 618 5 4 |
| To aifler in 14568 cube feet, at 6d. | 364 4 0 | |
| To rubble work in do. 82 roods Scots, at 4l. 16s. | 393 12 0 | |
| To 2d. per foot extra upon 4024 cube feet of rubble blocks to ferve in the infide for the bottom, tie and bond courfes as aifler | 33 10 8 | |
| To 50 tons of Pozzelana, at 3l. per ton, ground and fifted ready for ufe | 150 0 0 | |
| | | 941 6 8 |
| To 450 cube yards of rubble to be depofited round the piers for fecuring the bottom, at 1s. | | 22 10 0 |
| Total of piers and abutments, | | 2054 18 4 |

The

|  |  |  | £ | s. | d. | £ | s. | d. |
|---|---|---|---|---|---|---|---|---|

## The SUPERSTRUCTURE.

To centering for the arches which contain at a medium 1220 feet superficial in the soffite, and for seven arches 8540 feet, at 10d. - | | | | | | 355 | 16 | 8

To aisler in the arches, which at two feet thick contain 17080 cube feet, at 6d. - - - - - - - 427 0 0

To aisler in the capings and facings 4792 feet at 6d. - - 119 16 0

To rubble work in the whole superstructure, 104 roods, at 4l. 16s. including scaffolding - - - - - 499 4 0

To extra work 16787 feet, face measure of rubble walling in the spandrells, abutment walls, and parapets, 1d. - - - 69 18 11

To cramps and lead for the caping - - - - 15 8 0

To filling the spandrells and abutments with quarry scraps and levelling the whole length of the bridge with gravel, 2417 cube yards, at 6d. - 60 8 6

|  |  | 1191 | 15 | 5 |
|---|---|---|---|---|

Total of the superstructure, 1547 12 1

## The TERMINATING SLOPES by way of Access to the Bridge.

To dry rubble building wharf walls for confining the forced earth at the north-west end of the bridge, containing 25½ roods, at 3l. per rood - 76 10 0

To rubble parapets in mortar to do. at 4l. 16s. - - - 21 12 0

To extra work on 400 feet running of rubble caping and base of the parapets, at 3d. - - - - - - 5 0 0

To gravel, &c. for levelling the south east end of the bridge, and for forming the slope on the north west end, so as not to be steeper than one in twelve, containing 3400 cube yards, at 6d. - - 85 0 0

188 2 0

## ABSTRACT.

The piers and abutments - - - - 2054 18 4

The superstructure - - - - 1547 12 1

The terminating slopes - - - - 188 2 0

Neat estimate ——— 3790 12 5

Contingencies on the above at 10 per cent. - - - 379 1 3

The above being supposed to be performed by those who are expert in this kind of works, we may add for surveyors and profit of undertakers or inexperience of workmen 10 per cent. more - - 379 1 3

Total, 4548 14 11

N. B.

N. B. The above estimate is on suppofition that every thing is to be provided, and nothing to remain; but a confiderable quantity of old materials of the late bridge has been preserved and depofited near the place; I suppofe to the amount of 200l., and the timber of the centers and coffer dams will remain at last to be difpofed of.

A quarry very near the bridge, which affords excellent rubble, greatly contributes to the fmallnefs of expenfe for fo large a building.

J. SMEATON.

*Aufthorpe,*
19th *February* 1772.

---

EXPLANATION of the manner of laying the Foundations for the Bridge of Banff.

A—Plank piles encafing the pier and ferving as a coffer dam for laying the foundation.

B—String piece firft laid for regulating and connecting the plank piles above the common furface of the water.

C—Another ftring piece for connecting the tops of

D—Planking for keeping out the tide water.

The infide is then to be excavated to the depth required.

E—Bearing piles to be increafed or diminifhed in number, length, or fize, according as the ground is found more or lefs firm: thofe drawn will do for a middling gravel.

F—A ribband or ftring piece which is to be adapted to the infide of the piles, and with fcrewed bolts, (whofe threads lay hold of the planks on the outfide, but go freely through the ribband) to be fixed to every other plank, and as many of them as are here reprefented like.

G—Are

G—Are bolts with flat fhoulders, and an eye in the head, by which they are tied with cramps to

H—Blocks of rough ftone holding the cramps for tying in the outfide cafing firmly to the pier.

I—Flat rubble blocks for the ground courfe to tie the bafe together.

When the pier is raifed above low water mark, then the upper part of the coffer dam with its ftring piece, crofs beams, &c., are to be removed, and made to ferve for another pier, the plank piles are to be cut off with chizzels at the dotted line K, five or fix inches under water, and then the ftring piece B, comes away and ferves for the next pier.

The outfide muft be guarded with rubble round the cafing.

N. B. By way of eafing the driving of the plank piles, the ground fhould be excavated fomewhat wider than the pier, and as deep as the water will permit.

The plank piles (which are four inches thick) will be beft made of beach, which can be had in planks of proper dimenfions from Suffex.

J. SMEATON.

*Aufthorpe,*
19th *February* 1772.

---

## EXPLANATION of the Center for the Bridge of Banff.

THIS defign fhows how it will ftand under the middle arch of 50 feet fpan.

A—One of the tie beams for framing one of feven fimilar ribs.

B—One of the curbed fellies, there being fix in each rib.

C. C—

CC—Struts or props of Scots firs about fix inches diameter in the middle, and of their natural taper.

DD—Two braces in each rib for throwing part of the weight of the middle upon the two adjacent fupports.

E—Shews the ends of two-inch planks for covering the center.

F—The vaufoirs or penftones upon the centers two feet in height.

GG—Pillars or props of Scots firs of their natural taper, ten inches diameter in the middle: as there are fhewn five under each rib, they will be feven deep, and each row of feven framed into head and fell pieces, whofe ends are fhewn at

HI—Ten inches fquare, and 20 feet long.

KK—Denote fhort blocks of about five inches in height laid perpendicularly over each of the pillars above mentioned, fo as to level up the tie beams to their proper height, which blocks are fplit away from under the tie beams on ftriking the centers.

The pillars G G are of fuch a height that the frames H G I ftanding upon the ground will fupport the center when the two arches next the town are built (the length of the tie beam A being adapted to the fpan of the fmalleft arches), in which cafe large flat ftones well bedded upon the gravel, one correfpondent to each pillar G, will fufficiently fupport the fell pieces I; but as the arches will rife higher, and efpecially thofe which are over the channel of the river at low water will be more fafely fupported by

LL—Piles of Scots fir about 10 inches in the middle, driven into the gravel five or fix feet more or lefs, till they make a confiderable refiftance to the ram; they are to be driven in rows as nearly correfpondent to the prefent places of the pillars G as may be, and then their heads levelled down fo as to fuffer each fell piece I to lay true upon each row of feven piles.

M—Curbed pieces anfwerable to each rib, put in fo as to make out the center from the fmalleft to the greateft arches; thefe reft upon a plate piece whofe end is reprefented at N, which are each of them fupported by feven ftanchions or props O, ftanding upon the loweſt offſett of the baſe of the piers, which may be of Scots fir five or ſix inches diameter in the middle.

To prevent confuſion, the ſtays that will be neceſſary to keep the work in place while putting together, I have omitted, as theſe will be obvious to every carpenter, but it is to be noted that every thing of this kind below the tye beams muſt be ſtruck away after the center is got together to prevent reſiſtance to ſpeats, in caſe theſe ſhould happen while the center is up.

J. SMEATON.

*Auſthorpe,*
31ſt *December* 1772.

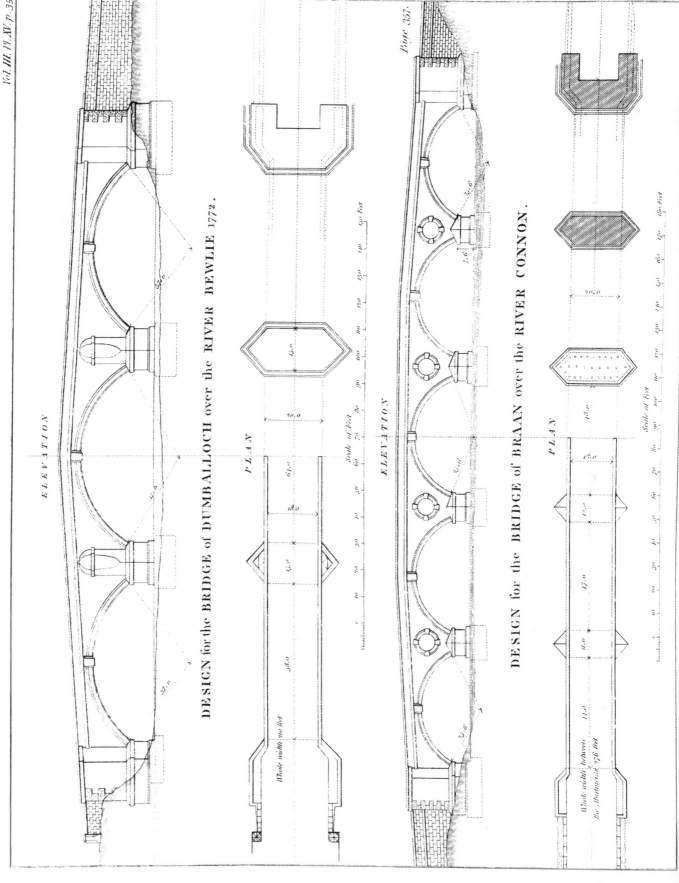

Page 357.

ELEVATION

DESIGN for the BRIDGE of DUMBALLOCH over the RIVER BEWLIE 1772.

PLAN

Scale of Feet

Whole width 20 feet

ELEVATION

DESIGN for the BRIDGE of BRAAN over the RIVER CONNON.

PLAN

Scale of Feet

Whole width between 13.0
the Abutments 27.6 feet

# DUMBALLOCH BRIDGE.

## (See Plate 15.)

ESTIMATE for building a Bridge over the River Bewlie at Dumballoch, being 210 Feet between the Abutments, 20 Feet wide over all, and to consist of Three Arches to be built with Freestone.

| | £. s. d. | £. s. d. |
|---|---|---|
| **PREPARATION for founding the PIERS.** | | |
| To timber for coffer dams for two piers and two abutments (exclusive of what is to remain connected therewith), 783 cube feet, at 3s. 6d. | 137 0 6 | |
| To iron work for do. 7 3, at 5d. per lb. - - | 18 1 8 | |
| To pumps, piling machines, and utensils - - | 120 0 0 | |
| Total of the Preparations | | 275 2 2 |
| **The PIERS and ABUTMENTS to the Spring of the ARCHES.** | | |
| To excavations and pumping for the four foundations, at 20l. each - | | 80 0 0 |
| To timber in plank piles, bearing piles, ribband, &c. for the four foundations, 2568 cube feet, at 4s. - - | 513 12 0 | |
| To iron work for two piers and two abutments, 12 cwt. 3 qrs., at 5d. per lb. - - - | 29 15 0 | |
| To lead, 6 cwt., at 2d. per lb. - - | 5 12 0 | 548 19 0 |
| To aisler in do. 12,304 feet (the leading being supposed not to exceed one mile) at 6d. - - - | 307 12 0 | |
| To rubble work in do. 64 roods Scots, at 5l. 10s. - | 352 0 0 | |
| To 2d. per foot extra upon 3600 cube feet of rubble blocks, to serve in the inside, for the bottom, top, and bond courses - | 30 0 0 | |
| To 35 tons of Pozzelana, ground and sifted ready for use, at 3l. 3s. - | 110 5 0 | |
| To 1432 cube yards of rubble, to be deposited round the piers for securing the bottom, at 1s. 6d. - | 107 8 0 | 907 5 0 |
| Total of the Foundations | | 1811 6 2 |
| **The SUPERSTRUCTURE.** | | |
| To centering for the arches, which contain in the whole 4200 feet of soffite, at 1s. - - - | | 210 0 0 |
| To aisler in the arches, which at 2 4 thick contain 9800 cube feet, at 6d. - - - | 245 0 0 | |
| To aisler in the capings and facings, 2650 feet, at 6d. - | 66 5 0 | |
| Carried forward £ | | |

| | £. | s. | d. | £. | s. | d. |
|---|---|---|---|---|---|---|
| Brought forward | | | | | | |
| To rubble work in the whole superstructure 71 roods, at 5l. 10s. | 390 | 10 | 0 | | | |
| To extra work upon the face of the spandrells, abutments, walls, and parapets, 9606 feet, at 3d. | 120 | 1 | 6 | | | |
| To lead and iron for cramping the caping | 8 | 8 | 0 | | | |
| Fitting the spandrells and abutments with quarry scraps, and levelling the whole with gravel, containing 1921 yards, at 1s. | 96 | 1 | 0 | | | |
| | | | | 926 | 5 | 6 |
| Total of the superstructure | | | | 1136 | 5 | 6 |

## The TERMINATING SLOPE at the South End of the Bridge, and Access on the North.

| | £. | s. | d. | £. | s. | d. |
|---|---|---|---|---|---|---|
| To dry rubble building in the wharf walls for confining the forced earth at the south end, containing 13¼ roods, at 3l. 10s. | 47 | 5 | 0 | | | |
| To rubble parapet in mortar 1½ roods, at 5l. 10s. | 8 | 5 | 0 | | | |
| To extra work on 270 feet running of caping and base, at 3d. | 3 | 7 | 6 | | | |
| To 1500 cube yards of gravel for levelling the north end, and forming the slope on the south end so as not to be steeper than 1 in 12, at 6d. per yard | 37 | 10 | 0 | | | |
| | | | | 96 | 7 | 6 |

## ABSTRACT.

| | £. | s. | d. |
|---|---|---|---|
| The pier and abutments | 1811 | 6 | 2 |
| The superstructure | 1136 | 5 | 6 |
| The terminating slopes | 96 | 7 | 6 |
| Neat estimate | 3043 | 19 | 2 |
| Contingencies on the above, at 10 per cent | 304 | 7 | 10 |
| The above being supposed to be performed by those who are expert in this kind of works, we may add, for surveyors and profit of undertakers, or inexperience of workmen, 10 per cent. more | 304 | 7 | 10 |
| Total | 3652 | 14 | 10 |

N. B. The above estimate supposes that rubble stone can be procured at 1s. 6d. per rood of wall laid down in place, and lime at 10d. per boll: if those materials cannot be procured at those prices, allowance must be made in the estimate upon them for the difference.

J. SMEATON.

*Austhorpe,*
19th *Feb.* 1772.

N. B. The foundations for this bridge are to be laid in the same manner as for Banff bridge.

# BRAAN BRIDGE.

## (See Plate 15.)

ESTIMATE for building a Bridge over the River Conon at the Boat of Braan, to be 276 Feet between the Abutments, 20 Feet wide over all, and to confift of five Arches; the Aifler of the Piers, Arches, and Facings to be of Freeftone, and the Core of the Piers, Spandrell Walls, and Parapets of Rubble.

| | £. | s. | d. | £. | s. | d. |
|---|---|---|---|---|---|---|
| **PREPARATION for founding the PIERS.** | | | | | | |
| To timber in coffer dams for the four piers and two abutments, exclufive of what is to remain conneƈted therewith, 340 cube feet, at 3s. 6d. | 59 | 10 | 0 | | | |
| To iron-work to do. 3 cwt. 2 qrs., at 5d. per lb. | 8 | 3 | 4 | | | |
| To pumps, piling machines, and utenfils | 120 | 0 | 0 | | | |
| | | | | 187 | 13 | 4 |
| **The PIERS and ABUTMENTS to the Spring of the ARCHES.** | | | | | | |
| To excavations and pumping for fix foundations, at 20l. each | | | | 120 | 0 | 0 |
| To timber in plank piling, bearing piles, ribbands, &c. for the foundations, 3162 cube feet, at 4s. | 632 | 8 | 0 | | | |
| To iron work for the four piers and two abutments, 16 cwt. 2 qrs., at 5d. per lb. | 38 | 10 | 0 | | | |
| To lead, 7 cwt. 2 qrs., at 2d. per lb. | 7 | 0 | 0 | | | |
| | | | | 677 | 18 | 0 |
| To aifler in do. 15,120 cube feet (the leading being fuppofed not to exceed one mile), at 6d. | 378 | 0 | 0 | | | |
| To rubble work in do., 65 roods, at 5l. 10s. | 357 | 10 | 0 | | | |
| To 2d. per foot extra upon 4602 cube feet of rubble blocks to ferve in the infide, for the bottom, tie, and bond courfes, as aifler | 38 | 7 | 0 | | | |
| To 40 tons of Pozzelana, at 3l. 3s. per ton, ground and fifted ready for ufe | 126 | 0 | 0 | | | |
| To 1152 cube yards of rubble, to be depofited round the piers for fecuring the bottom, at 1s. 6d. | 86 | 8 | 0 | | | |
| | | | | 986 | 5 | 0 |
| Total to the piers and abutments | | | | 1971 | 16 | 4 |
| **The SUPERSTRUCTURE.** | | | | | | |
| To centering for the arches, which contain 5780 fuperficial in the foffite, at 8d. | | | | 192 | 13 | 4 |
| To aifler in the arches, which at two feet thick contain 11,560 cube feet, at 6d. | 289 | 0 | 0 | | | |
| Carried forward £ | | | | | | |

|  | £. | s. | d. | £. | s. | d. |
|---|---|---|---|---|---|---|
| Brought forward | | | | | | |
| To aifler in the capings and facings, 3060 feet, at 6d. - | 76 | 10 | 0 | | | |
| To rubble work in the whole fuperftructure, 92 roods, at 5l. 10s., including fcaffolding - - - | 506 | 0 | 0 | | | |
| To extra work upon 11,664 feet, face meafure, of rubble walling in the fpandrells and abutments, walls and parapets, at 3d. - | 145 | 16 | 0 | | | |
| To cramps and lead for the caping - - | 11 | 4 | 0 | | | |
| To filling the fpandrells and abutments with quarry fcraps, and levelling the whole length of the bridge with gravel, 2384 yards, at 1s. - | 119 | 4 | 0 | | | |
| | | | | 1147 | 14 | 0 |
| Total of the fuperftructure | | | | 1340 | 7 | 4 |

The TERMINATING SLOPES, by way of Accefs to the Bridge.

|  | £. | s. | d. | £. | s. | d. |
|---|---|---|---|---|---|---|
| To dry rubble building in wharf walls for confining the forced earth at the ends of the bridge, containing 14 roods, at 3l. 10s. - | 49 | 0 | 0 | | | |
| To rubble parapets in mortar, to do. 2¾ roods, at 5l. 10s. - | 15 | 2 | 6 | | | |
| To extra work on 384 feet running of rubble caping and bafe of the parapets, at 3d. - - - | 4 | 16 | 2 | | | |
| To gravel, &c. for making the accefs to the bridge, and for forming the flopes fo as not to be fteeper than 1 in 12, viz. 2048 yards, at 6d. - - - | 51 | 4 | 0 | | | |
| | | | | 120 | 2 | 6 |

### ABSTRACT.

|  | £. | s. | d. |
|---|---|---|---|
| The piers and abutments - - | 1971 | 16 | 4 |
| The fuperftructure - - | 1340 | 7 | 4 |
| The terminating flopes - - | 120 | 2 | 6 |
| Neat eftimate | 3432 | 6 | 2 |
| Contingencies on the above, at 10 per cent. - - | 343 | 4 | 7 |
| The above being fuppofed to be performed by thofe who are expert in this kind of works, we may add, for furveyors and profit of undertakers, or inexperience of workmen, 10 per cent. more - | 343 | 4 | 7 |
| Total | 4118 | 15 | 4 |

N. B. The above eftimate fuppofes that rubble ftone can be procured at 1s. 6d. per rood of wall laid down in place, and lime at 10d. per boll : if thofe materials cannot be procured at thofe prices, allowance muft be made in the eftimate for them upon the difference.

J. SMEATON.

*Aufthorpe,*
19th *Feb.* 1772.

N. B. The foundations for this bridge are to be laid in the fame manner as Banff bridge.

## ALTGRAN BRIDGE.

ESTIMATE for building a Bridge over the River Altgran, near above the prefent Wood Bridge, to confift of One Arch of 44 Feet Span.

|  | £. | s. | d. |
|---|---|---|---|
| To excavation for the abutments and laving the water for laying the foundations | 5 | 0 | 0 |
| To aifler in the abutments to the fpring of the arch, 392 feet, at 6d. | 9 | 16 | 0 |
| To do. in the arch, 1800 feet, at 6d. | 45 | 0 | 0 |
| To do. in the facings, 302 feet, at 6d. | 7 | 11 | 0 |
| Rubble work in the whole, 11 roods, at 4l. | 44 | 0 | 0 |
| Centering for the arch, fuppofed to be brought from the bridge at Braan | 15 | 0 | 0 |
| Filling the fpandrells and levelling the bridge with gravel | 3 | 7 | 6 |
|  | 129 | 14 | 6 |
| To 5 loads of dry rubble walling to wharf up the flopes | 12 | 10 | 0 |
| To 548 yards of earth for forming the flopes | 6 | 17 | 0 |
| Total | £.149 | 1 | 6 |

N. B. It is probable the work may be undertaken for lefs money, but the quantities may nearly be depended upon.

The above eftimate fuppofes the rubble ftone can be procured at 1s. 6d. per rood of wall laid down in place, and lime at 10d. per boll : if thofe materials cannot be procured at thofe prices, allowance muft be made in the eftimate upon them for the difference.

J. SMEATON.

*Aufthorpe,*
19th *February* 1772.

## BRIDGES AT BEWLIE, CONON, ALTGRAN, &c.

THE REPORT of John Smeaton, Engineer, upon fundry Matters referred to him by a Minute of the Board of Truftees of annexed Eftates, dated 29th January 1770; as alfo by the Letter of Mr. Secretary Barclay, of 17th July 1771.

The feveral fubjects are as follow:

(Board's Minutes, 29th January 1770.)
Bridge over the River Bewlie.
Do. - - Conon.
Do. - - Altgran.
Harbour or Landing Place at Portleich.
Mills and Watercourfe of Miltoun of New Tarbet.

(Mr. Secretary Barclay's Letter.)
Stone Piers at the Ferry of Inverbreakie.

HAVING carefully viewed the above fubjects in the month of Auguft 1771, I tranfmitted to the Honourable Board plans, directions, and eftimates for conftructing the three bridges comprehended in the three firft mentioned articles, on the 19th of February laft; but being in the courfe of this bufinefs fuddenly called away to London upon affairs relative to the Grand Canal, as I then by letter informed the Board, and having been by a fucceffion of preffing affairs occupied ever fince, it is only now that I am able fully to acquit myfelf of the above articles.

What regard to the three firft articles I have in a great meafure delivered myfelf upon as above mentioned; it only remains that I affign my reafons for preferring the fituations for which thefe plans are refpectively drawn.

Refpecting the River Bewlie.—The river at the prefent ferry is not only much wider and deeper than at the place propofed at Dumballoch, but being much more in the tide's way, would on that account alfo be more expenfive; and furthermore as the river feemed to

3                                                                          threaten

threaten to break through the neck of the loop, at the extreme bend of which is the ferry, it might thereby happen that the bridge would become deferted; and if the bridge were originally built upon this neck, the cutting of the neck and forming a dam acrofs the river at the prefent ferry, with a road upon it of fufficient height to be out of the water's wafh, would be in itfelf a confiderable expenfe. At Dumballock the river has a fair run, is moderately wide and deep, and has a good gravel bottom, and as I am informed that it will not carry the road much about, I think upon the whole that this place is to be preferred to any that was pointed out to me for the fituation of a bridge over the river Bewlie.

The fitteft place for a bridge over the river Conon has been by the country in general judged at the place called the Rock. The reafon why I did not think this place proper I gave in my letter of 19th of February laft, which accompanied the plans and eftimates, and which, to fave trouble of a reference thereto, I here repeat: " That I find the total " length of a bridge there (at the Rock) will be greater by almoft 50 feet than where I " propofe it; and that it will require one arch to be built of 114 feet fpan, and as the " water is very deep underneath, it will require fo great an expenfe to center and con- " ftruct fuch an arch, befides making good the paffage over the remainder of the river, " which is almoft 200 feet more, that I apprehend it is far more eligible to conftruct the " bridge in the place I have fixed upon, where there is a fair and open run of the " river.

" The Rock, as fo called, being a kind of concreted gravel, which appears in a degree " to wear away by the action of the water, it is in no ways certain that the river may not " in part undermine the foundation of fo great and neceffarily heavy an arch, and which " from the depth of the water it will be difficult to guard againft by artificial means."

Befides the place I have propofed, viz. at the Boat of Braan, there is alfo another fituation weftward, at the Turner's Boat of Braan. This is fo nearly equivalent with the former, that the fame defign would fuit either place; but by being further weftward, I apprehend it muft take travellers further round, the paffage at the common ferry over Conon being eaftward of all thofe places.

Refpecting the paffage of Altgran, there is no choice; there is already a weak timber bridge which may occafionally take over a horfe.

## PIER or Landing-place at Portleich.

PORTLEICH is fituated upon the north weft fide of the bay of Cromartie, which is in the whole fo entirely land-locked, that there cannot at any time be any great fea, fuch as places adjoining upon the ocean are fubject to. The bay of Cromartie may therefore in the whole extent of it be juftly confidered as a fafe harbour; however, being fome miles broad, it fometimes happens that there is a greater fwell raifed within itfelf than is agreeable to veffels lying upon the ground at low water to deliver or take in a cargo. The fhore at Portleich lying upon a very gentle flope, and the tides being rather fhort, as not exceeding 11 feet at common fpring tides, it happens that in order to get a very moderate depth of water at high water, it will be neceffary to advance the pier or landing place a confiderable way from the high water mark into the bay. A pier therefore of the length of 188 yards, will but barely reach from high to low water mark at fpring tides ; fo that alongfide this pier head, which I propofe to be the landing place, there will be about 11 feet water at fpring, and about 7 or $7\frac{1}{2}$ feet at neap tides About 24 yards next the pier head is defigned to be of 12 feet broad at top, fo as to be convenient for a landing place for veffels to lie on either fide, the reft of the pier being only 9 feet broad at top, is confidered only as a caufeway for making a communication for carriages between the fhore and the landing place; but will be very ufeful in fheltering fifhing boats, for the fake of which a few large whinftones are roughly thrown together. The new pier is propofed to take its commencement from a ftone that I marked when there, a little to the weftward of the prefent landing-place, and to extend in a direction towards the caftle of Cromartie ; fo that veffels which while unloading want to take the fhelter of the pier may lie on either fide thereof, according to the wind and feafon of the year. As the fhore is a muddy kind of fand, and a very flat flope, a pier run out in this manner, by intercepting the circulation of the water, may gather mud on each fide of it; but to prevent its gathering near the head in the part joining the caufeway to the pier head, two arched perforations are left, through which the feas and current acting will be the moft likely means of preventing the gathering of mud near the pier head, where the water is moft required to be deep. The foundation of the pier is fuppofed to be laid originally one foot under the furface of the fand, fo that being 15 feet in the whole height, it may be allowed to fettle one foot more in the fand, and yet remain two feet above the high water of fpring tides, which I apprehend where the feas are fo moderate as in the bay of Cromartie, will be a fufficient elevation.

In

In regard to the eſtimate of this, as well as for the reſt of the propoſed works, the only thing that I can be certain of is the quantities, for the prices depend ſo very much upon the kind of workmanſhip employed, and carriage, that I cannot be at any certainty about them. Where the carriage does not exceed a mile by land or five or ſix miles by water, I apprehend the prices I have annexed will ſuit, on a ſuppoſition that labour is not dearer in the ſhire of Roſs than in the ſouth of Scotland.

---

ESTIMATE for building a Pier or Landing-place at Portleich, upon the Bay of Cromartie.

|  | £. | s. | d. |
|---|---|---|---|
| THE pier head, comprehending 78 feet in length, to be 15 feet broad at bottom, 12 feet top, and 15 feet high from the foundation, at a mean thickneſs of two feet outſide, will contain 6620 cube feet of rough ſcappeled block freeſtone, which being laid dry, the quarrying, freight or carriage, and building, at 5d per foot - - - | 137 | 18 | 4 |
| To 340 cube yards of quarry rubble or broken whinſtones, for filling the body, winning, freight or carriage, and laying in place, at 1s. 6d. per cube yard | 25 | 10 | 0 |
| To cramps and lead for confining the corner ſtone at the termination of the pier head - - - - | 1 | 0 | 0 |
| The pier head or landing place | 164 | 8 | 4 |
| To continuing the pier or cauſeway from the landing-place above mentioned to the ſhore, being 162 yards in length, and 9 feet broad at top, which at the mean thickneſs of 18 inches, will contain 21,141 cube feet of ſcappeled free ſtone, at 4½d. - - - - | 396 | 7 | 10 |
| To 1062 cube yards of quarry rubble or broken whinſtones for filling the body, at 1s. 6d. per cube yard - - - - | 81 | 13 | 0 |
| To extra work in the two arches - - - | 4 | 0 | 0 |
|  | 482 | 0 | 10 |
| Sum of the above, being the neat eſtimate - - | 646 | 9 | 2 |
| To contingencies at 10 per cent. - - - | 64 | 12 | 11 |
| Total | 711 | 2 | 1 |

The

## The MILLS and Watercourſe to the Mills of Miltown of Newtarbet.

The mills of Milltown of Newtarbet are indeed in very bad repair, and need rebuilding; but ſo far as I can learn, when in repair are ſufficient to anſwer the purpoſes; and as their method of grinding in that part of the world is in a particular way, which their conſtruction of mills is adapted to anſwer, I cannot venture to give any new deſigns, which might require different millwrights and millers to what the country furniſhes, to make them ſucceed, and at laſt probably not give ſatisfaction: I therefore refer the Board to get their mills rebuilt in the moſt ſubſtantial way that is in practice in the country where they are, as ſoon as a proper ſituation can be fixed upon. The matter here, which is moſt preſſing, and about which ſome advice is the moſt neceſſary, is the dam head, by which the water of the river is turned into the watercourſe or lead from thence to the mills. The dam head and watercourſes, as relative to the ſituation of the mills, muſt be looked upon as a very remarkable piece of projection, for after raiſing a dam head to about three feet, to an extent of 250 feet in length, acroſs a river that in general is not above 50 feet wide, and after conducting the water along the face of the rocks, in an aqueduct built up with ſods, which are perpetually failing and tumbling down, it is in a courſe of about half a mile in which there is a fall of about 20 feet, ultimately conducted to the troughs of the mills, from whence to the bottom of the wheels there is *only about* 11 *feet of fall* made uſe of after loſing twice that quantity by the road; ſo that it would ſeem that the original projector had extended his watercourſe upon the river in order to meet with difficulties: had the work been permanent there had been no error, but a loſs of ſo much labour in the firſt erection; but the misfortune is, that the dam head being only built with looſe ſtones, it is broken by every ſpeat that comes down the river, and as certain tenants are obliged to build it up as often as it comes down, and ſometimes obliged to work up to their middles in water, even in winter, it becomes not only a heavy burthen upon them, but a great interruption to doing the buſineſs of the country, that is thirled to thoſe mills.

Two ways offer themſelves to avoid theſe inconveniences, one is, by building a new dam head near the place where it now is, and removing the mills; the other, to build a new dam head at a new place, and conducting the water to the place near thereto, that may be more convenient. By reference to the plan made by Mr. David Aitken the ſurveyor, it appears, that if inſtead of the dam head, which turns the water into the

preſent

present lead, a dam head is erected higher up, and a new lead or watercourse cut on the other side of the river, in the direction that the mills can have a fall into the river there, which he makes to be no less than 18 feet 3 inches, which indeed is more than sufficient for the purpose of the mills, they having now in effect only 11. The objection that occurs to me is as follows: A part of the pleasure-ground, or policy of Balganoun, lies contiguous to this dam's head, and the ground on that side being elevated scarce 18 inches above the level of the water in the dam, would be very liable to be overflown by the pen of the present dam in speats, were it not for the great length of the dam which takes off the water in a sheet of no less breadth than 250 feet, and generally in the greater speats breaks down. But were this water pent up by a dam, whose length will be no more than 84 feet, according to Mr. Aitken, the water will be confined in so narrow a compass, in proportion to what it occupies at present, that it would undoubtedly be more liable to overflow the policy above mentioned: whatever damage may happen to the policy by the present dam, it may, as I apprehend, be prescribed for; but in case the situation of the dam is changed, or the dam itself altered, if the laws of Scotland are in such cases parallel to the laws of England, such new dam cannot be prescribed for, but must be liable to all damages that may arise from it. In order, therefore, that the dam and watercourses may be clearly freed from the possibility of litigation, I observed that there was no kind of necessity to bring the water from the old dam, but that the level would admit of the water's being taken up even below the bridge where the Cromartie estate extends on both sides, and can be conveyed into some part of the present lead before it arrives at the mills, and consequently, that the situation and power of the mills will remain the same as now. The surveyor reports, that from the supposed new dam head to the surface of the water in the mill lead, is a fall of 8 feet $2\frac{1}{4}$ inches; he does not clearly distinguish whether he supposes the dam head raised up to the level of a marked joint of the middle pier of the bridge, which he says is 3 feet $\frac{1}{4}$ inch higher than the surface of the ground at the place proposed by me for the dam head; but supposing the level taken from the said joint of the pier (as the most unfavourable supposition), if we cast off the 3 feet $\frac{1}{4}$ inch for the height of the said joint above the said ground, and (if need be) two feet more to bring the dam's crown within the land, there will still remain three feet of descent from the dam head, through the new part of the watercourse into the old one, which in a course of 550 yards is more than sufficient; it therefore follows, that the new dam head need not be raised more than is sufficient to divert the water into the new watercouse, and there appears to be a fall of full four feet to the head of the mill troughs, if the mills were rebuilt, where there seems to have been formerly a walk mill, this four feet might be added to the present fall of 11 feet, and make 15, which would cause the mills to work still more briskly; but

but if the prefent fituation is found more commodious, the fall at the mills may be confiderably increafed by raifing one of the banks fome part of its length.

The defign for a dam head, that I have given, amounting to 180l., fuppofes the work to be done in the moft fubftantial manner, and fo as to be likely to be permanent; but as it appears to me, that a few flat ftones fet upon the edge, fo as to form a line acrofs the river, might be fufficient to divert the water into the new watercoufe, and the work may be done in this way, in cafe the Board is willing to fave a good part of the outlay, by fubjecting the work to fmall derangements by floods.

ESTIMATE for a new Dam Head and Watercourfe for the Mills of Miltown of Newtarbet.

|  | Cubit feet. | £. | s. | d. |
|---|---|---|---|---|
| To fir timber in bearing piles, plank piles, ftring piece and cover, containing | 209 | | | |
| To oak and fir timber in the fluice - - - - | 17 | | | |
| Timber in the whole, at 4s. per foot, wood and workmanfhip, and meafured neat in place - - - - } | 226 | 45 | 4 | 0 |
| To iron work in the whole - - - - | | 3 | 0 | 0 |
| Carpentry total - | | 48 | 4 | 0 |

### MASONRY.

|  | Roods. | £. | s. | d. |
|---|---|---|---|---|
| To cafting foundations - - - | | 3 | 0 | 0 |
| The dam's end wall to be four feet thick, and fuppofed eight feet high from the foundation, being 96 feet long, will contain - - - | 4¾ | | | |
| The flue walls within and without the clough or fluice, being two feet thick and four feet high from the foundation, and in the whole 70 feet long, will contain, fay - - - - - - | 1 | | | |
| Roods of rubble mafonry at 5l. 10s. - - - | 5¾ | 31 | 12 | 6 |
| To allowance at 3d. per foot, face meafure, upon the dam's end wall next the water, and on the top to allow for larger ftones, caping and rough working the fame, 1152 feet - - - - | | 14 | 8 | 0 |
| Carry forward | | 49 | 0 | 6 |

|  | £. | s. | d. |
|---|---|---|---|
| Brought forward | 49 | 0 | 6 |

To freeftone fcappeled for the furface of the dam, 72 feet long, 19 feet broad over all, and at a medium of 18 inches deep, will contain 2052 cube feet, at 5*d*.  42  15  0

To 72 yards cube of quarry rubble, or broken whinftones for filling the body and fecuring the fkirt of the dam, at 1*s*. 6*d*.  -  -  5  8  0

|  |  |  |  |
|---|---|---|---|
| Total of mafonry  - | 97 | 3 | 6 |

## SPADE-WORK.

To cutting a new aqueduct of five feet bottom, fuppofed three feet mean depth, with proper flopes, 555 yards running, at 6*d*.  -  13  17  6

To banking acrofs the flat from the dam's end wall to the high ground, being about 27 yards long, 3 feet top, flopes as 3 to 5, to be covered with fods, containing 234 cube yards, at 4*d*.  -  -  3  18  0

|  |  |  |  |
|---|---|---|---|
| Total of fpade work | 17 | 15 | 6 |

## ABSTRACT.

| | £. | s. | d. |
|---|---|---|---|
| Carpentry and iron work  -  -  - | 48 | 4 | 0 |
| Mafonry  -  -  -  - | 97 | 3 | 6 |
| Spade-work  -  -  - | 17 | 15 | 6 |
| Neat eftimate | 163 | 3 | 0 |
| To contingencies, at 10 per cent.  - | 16 | 6 | 4 |
| Total | 179 | 9 | 4 |

The PIERS, or Landing-places, at the Ferry of Inverbreakie:

On this head I am defired by Mr. Secretary Barclay's letter of 17th of July 1771, " to " examine into the ftate of the above ferry, and confider the propriety and utility of " building the pier propofed." The ferry of Inverbreakie is fituated upon a narrow part of the bay of Cromartie, fomewhat fimilar to Queensferry upon the firth of Forth, but I believe not half the width of the latter; and as the bay of Cromartie is land-locked on every fide, and the tides rife and fall far lefs than at Queensferry, there is accordingly far lefs fwell and ftrength of tide; of confequence the ferry of Inverbreakie is by *nature* attended with fewer inconveniences than Queensferry. The

two

two fhores at Inverbreakie are upon a moderately quick decline from high water till about the low water of neap tides, after that there is a flat ground on both fides, and for a confiderable extent upon the fouth fide, which at low water fpring tides lies dry. The confequence is, that whenever the water over the faid flats becomes too fhallow for the boats to go over them, the ferry is altogether inacceffible to the land except for foot paffengers, who may then be carried over the flats on men's backs; and as there is no fort of piers, landing-places, or projecting rocks, or any other than a floping beach, it is not a very commodious ferry for the boating of *horfes* at any time, and as to carriages, it feems altogether impracticable, without taking the carriages to pieces. As I went to Miltown of New Tarbet from Invernefs by way of Bewlie and Dingwall, and returned by the ferries of Inverbreakie and Keffack to Invernefs, I had a kind of practical proof of the utility of thefe ferries, and beg leave to fay, that were the equipment of the ferries of Keffack and Inverbreakie for taking in horfes and carriages, as complete as they might be made, no travellers into the eaftern parts of the fhire of Rofs, &c. would go the other road, except thofe who would rather go 20 miles round than crofs any water at all; and even there are two fmaller ferries where the inconveniences of embarking and de-barking are almoft as great, greater indeed than at Keffack, where there are already tolerable landing-places made by art on each fide, though not fo good as they might be rendered. The main difficulty therefore at prefent is at Inverbreakie, where if two ftone piers or landing-places were run out from the fhore on each fide, viz. to the length of 80 yards on the fouth fide, and 100 yards on the north fide, this ferry would become practi-cable for travellers at all times, except for about two hours at the low water of a fpring tide. The length of 100 yards on the north fide will carry out the pier far enough to enable the boats to work at low water of fpring tides; but to do the fame on the fouth fide would require that pier to be extended 70 yards further than above propofed. But as thofe flats feem to be compofed of nothing but an oozy kind of fand, I am in hopes that as thofe piers will in a fenfible degree increafe the motion of the water over them, there is a chance that the remainder of the flat fand extending beyond the fouth pier towards the north, may be removed by that increafe of the current. It will however very greatly contribute to the utility of the ferry to have the piers carried to the extent above fpeci-fied; and which may be afterwards extended in cafe the utility thereof to that part of the kingdom demands it.

ESTIMATE

ESTIMATE for the Landing-places at the Ferry of Inverbreakie

|  | £. | s. | d. |
|---|---|---|---|
| The landing-place being at a medium of fix feet high from the foundations, 10 feet broad at bottom, and eight feet at top, will contain of fcappeled freeftone, at a mean thicknefs of 18 inches in one yard, 90 feet cube, at 4½d.   -  | 1 | 13 | 9 |
| To 2¾ yards of rubble for filling the body, at 1s. 6d.   -   -  | 0 | 4 | 0 |
| Neat eftimate per yard running   -  | 1 | 17 | 9 |
| Ten per cent. for contingencies   -   -  | 0 | 3 | 9 |
| Total per yard running   -   -  | 2 | 1 | 6 |
| To 180 yards of pier or landing-places, including both fides of the ferry of Inverbreakie, at 2l. 1s. 6d.   -   -   -  | 373 | 10 | 0 |

N. B. As there is water-carriage from eminent quarries upon the Cromartie bay, it is likely this work may be contracted for to be done confiderably cheaper.

## SUTTON BRIDGE.

THE REPORT of JOHN SMEATON, Engineer, upon the Defign for a new Bridge over the River Hull, near Sutton.

HAVING infpectedthe defign for a Turning or Swing Bridge, intended to be built over the river Hull near Sutton, and which was exhibited to me by the direction of Meffrs. Pool, Watfon, Etherington, and Page; and the queftion being, Whether it will, and in what manner, affect or endanger the flooding of the grounds in Holdernefs, or the navigation of the faid river? I am of opinion as follows:

That every bridge may be confidered in fome degree an impediment to drainage and navigation; yet being a work of public utility, it feems to me, that it cannot be properly objected to where the rifk is fmall and the advantage adequate: refpecting drainage, the principal objection that appears to me will be by ftopping the ice upon the breaking up of a froft, which all timber bridges built on piles are in fome degree apt to do; but the river Hull being a flow running river, will not be fo fubject to form fuch obftructions as thofe that run more rapidly. The bridges from Fulham to Putney, at Kingfton, Staines, Maidenhead, and Henley, over the river Thames, thofe at Beal and Haddlefey over the river Aire, thofe on the Dutch river over the river Dun, and that at Hull over the river Hull, are fo many inftances among numberlefs others of the fame kind; and though it fometimes happens, yet it does not ordinarily happen that the ice is ftopped thereby.

I fee no objection to it in point of navigation, but the trouble of opening it; but I look upon a moveable bridge as a lefs hindrance to veffels than a fixed one; and, I believe, it was never fupported as an objection to the building a bridge, viz. that of the river being navigable. I take it for granted that the width of 22 feet is fufficient for the paffage of all veffels that can go there, otherwife it ought to be made fo; and, I think, it ought to be made fo high that veffels may go under it, in cafe of the turning part happening to be out of order.

J. SMEATON.

*Aufthorpe,*
7th *Feb.* 1772.

## WALTON BRIDGE.

THE REPORT of John Smeaton, Engineer, upon the State of Walton Bridge, as it appeared upon a View taken in July 1778.

THIS bridge having been built with the greateſt care and circumſpection under the care of that eminent artiſt Mr. Etheridge, and of the very primeſt timber that could be procured, and which having been finiſhed not many more than 20 years, before it diſcovered ſigns of failure, is a monument of the fallibility of all timber conſtructions where great weights and preſſures are concerned.

I found the four middlemoſt diviſions of the great arch ſupported by four rows of pillars raiſed upon a ſcaffold, which were ſupported by four rows of piles driven into the bed of the river, by which means the great arch appeared ſufficiently ſecured from any further failure that was likely to happen in a ſhort time, and this at the ſame time furniſhed the means of examining that part of the bridge which, by the decay of ſome particular pieces of timber, was in a likely way to have fallen into ſudden ruin.

I examined every part of the bridge as carefully as I poſſibly could, and as nearly as I could come at them ; and it appeared to me, that the timber in general was as perfectly ſound, and much more hard than it was when the bridge was erected : but in the part where the firſt ſymptoms of complaint had ſhewn themſelves, I found ſome parts of the timber ſo perfectly rotten and decayed, that it was reduced even lighter and leſs connected than touchwood, at the ſame time that the remaining part of the ſame pieces appeared perfectly ſound ; and this decay taking place at a critical point where ſeveral pieces were aſſembled, and thereby looſing the geometrical bond and dependence of one part upon another, at the ſame time that it was in a way inevitably to have brought down the whole of the great arch, it proves that the decay was not owing to the fault of the particular pieces concerned, but to their local ſituations, of which effect we will now enquire the cauſe.

The great arch being 130 feet ſpan, and originally not above 27 feet riſe above the ſpringers, muſt, as an arch, be conſidered as a very flat one, but as there is in the very crown of the arch between 10 and 11 feet in height from the under ſide of the arch

pieces

pieces to the upper fide of the hand-rail, this affords a fpace that in geometrical timber framing would have a very great degree of ftiffnefs independent of its rife as an arch, fuppofing it not to be overpreffed with weight. The main ftrength therefore of this bridge has been intended by its architect to confift in the ftrength of the two geometrical frames, making the two fides or external faces of the bridge ; between and from which frames he has fufpended a floor by ftrong ftraps of iron, at every divifion of the arch ; and as he has propofed to cover this floor with a confiderable thicknefs of gravel, which amounted to upwards of 100 tons burthen, over the opening of the great arch, it feems that he has confidered that the laying plain beams acrofs from fide to fide with about nine tons of gravel upon each beam near the middle of the bridge, exclufive of the weight of the timber work, would be fo great a burthen as to caufe the beams to fag, and totally to fail when they became a little tainted ; he has therefore upon the middle of each of thefe crofs beams erected a king poft, in the way of the king poft of the prin-cipals of a roof, and fupported thefe king pofts by oblique ftrutts or truffes on each fide, footed into the crofs beams near the ends, fo that thefe crofs beams will refemble the tie beams of a roof, the ends of which, inftead of being fupported upon a wall, are fup-ported, as has been faid, by ftrong iron ftraps from the fide frames. Now as the rife in the king pofts is but fmall, the fide truffes or ftrutts will act very obliquely, and therefore to fupport the king poft, and in confequence the middle of the floor, with its fuper-incumbent weight of gravel, thofe ftrutts will require to be very firmly footed into the crofs beams near the ends, and on this account require to be funk a good depth into the crofs beams, in order to make the footing good : again, near the middle of the bridge, the upper furface of the crofs beams will be nearly horizontal, fo that any wet that in time of rain percolates through the gravel and makes its way down thofe oblique ftrutts, to adjacent parts, will run into thofe footings, and there not being able to efcape or fuddenly to dry, cannot get out but by a flow evaporation in dry and warm weather; this producing a fermentation, it is eafily feen that in thofe parts fo fituated a ftrong tendency to rot muft take place, while in thofe crofs beams further from the middle of the arch, their upper furface being more inclined to the horizon, the wet will lodge in a lefs degree, and confequently they will be lefs fubject to rot. It is alfo eafy to fee that fome of the timbers fimilarly fituated by accidental circumftances in the cover, being fome more, fome lefs liable to take the rot, even though the quality of the timber was all alike. It is further neceffary to explain, that befides the intention of ftrengthening the floor by the truffing abovementioned, in the middle of the bridge a fet of ribs are thrown in and abutted one againft another, fo as to form an arch upon the fame principle as the under fet of arch timbers do on the outfide ; but as thefe ribs fo thrown in, cannot be connected with exterior timbers in the manner the outfide ones are con-

nected,

nected, and framed with the hand-rail, the middle rib therefore confisting of single timbers for three or four bays in the middle of the bridge, till they get more cover, can there afford but a feeble additional support in proportion to the strength of the outside frames. This being premised, it will readily be conceived, what in reality has proved to be the case, that the footing of the strutt with the cross beam, being the second from the middle towards the Middlesex side, on the down stream side of the bridge, being totally rotted through both strutt and beam, this end of the beam will cease to be suspended by the iron straps and the weight of the floor, and bed of gravel will become discharged almost immediately upon the middle rib, which being single, though perfectly sound, was sprung in the middle, that is, so far broken as to have lost the greatest part of its strength by large splinters thrown off on the under side. The failure of this rib again in part discharged the weight upon the adjacent cross beams, which being also decayed, though not perfectly rotten, must soon have given way in like manner as the former, had not the application been made in due time by the supports under the most weak and decayed parts, as I found it, which giving time for consideration, the question now is, what to do with it?

On examination I found all the truss footings at both ends of the four middle cross beams to be tainted, and some of them considerably decayed, as well as that which was totally so, and caused the immediate failure. As the scaffold did not extend further, I could only judge of the soundness of the adjacent cross beams from viewing them at the distance of 10 feet; and as I found the rest of the timber work in general very little decayed, the parts particularly specified and examined excepted, I am in hopes the taint in the parts I have mentioned has not extended in any great degree much further. It seems therefore to me, that as the greatest part of the structure remains perfectly sound, and the decayed parts capable of being restored, strengthened, and supported at an adequate expense, by means of the scaffold already erected, it would not be for the interest of the proprietor to make at present any material change in the structure, as a repair judiciously conducted may be very well expected to make it last for a term equal to what it has lasted already; and then it will be time to consider whether it may be worth supporting, by throwing another pier into the middle of the river, and thereby making the great arch into two, covered with wood, brick, or stone, as may be then advised.

I look upon it that the present temporary support, exclusive of the deformity it gives to so noble a structure, might be effectual for some years to come, could it be secured against injury from ice floating down the river at the break of a great continued frost; the

effects

effects of which, on the arrival of such an event, cannot, I think, be ascertained before hand; there appears, however, a reasonable time for consideration.

When this bridge was undertaken, I apprehend it was the boldest design of the kind that had then been executed, at least in this kingdom; and the principles of its construction carried to a greater extent than had before been experienced. It is not therefore to be considered as wonderful, that after the term of years that has elapsed since its construction, errors have shewn themselves that probably would not have been foreseen by those who now can readily see them, but which in this repair I would as much as possible endeavour to relieve.

The first and greatest error of all was in making so large a middle arch. Had this bridge consisted of three arches more nearly equal, or rather of five arches, and equally well performed, in all probability no material derangement would have happened in the course of a century. The great span of the middle arch made it desirable to make it as flat as possible, otherwise it would have been impracticable to have driven carriages over it;—as it is, though in reality too flat for the span, yet it is still too steep an ascent for loaded carriages, because in their descent they are obliged to drag, and this makes it necessary to keep a cover of gravel much thicker, and consequently much heavier than would otherwise be necessary, and this again, together with the flatness of the arch, so very greatly increases the natural pressure, that overcoming those of the side arches, it makes both the piers lean to landwards, and also lays so great a stress upon the bed timbers which the arch pieces rest upon, that their ends are perfectly sunk into the bed timbers, so that from the whole of the pressure sideways, the crown of the arch is come down above two feet in perpendicular height; and I look upon it, that these are consequences of errors in the first construction, independent of what has happened by the rotting of the timbers; again, the weight of the cover has produced the idea of trussing the cross beams, and this again has brought on the rotting of the beams and strutts at their junctures: I would therefore recommend the following mode of repair:

1st, In those cross beams that need shifting, I would recommend to use whole beams without trussing, and rather to fill up the spaces by doubling the beams than the use of strutts; the timbers being thus unwounded will be less subject to retain moisture and rot.

2dly, I would remove the gravel totally from above the nine middlemost bays of the middle arch; and therewith by increasing the height and length of the slopes, reduce

3                                                                                    the

the general declivity, and by bringing a greater weight over the fide arches, enable them more nearly to counterbalance the lateral preffure of the great arch. This middle fpace thus laid bare, I would cover with fea coal cinders fuch as are collected by the duft carts, and fifted near London, for the purpofe of burning bricks; and this cover of cinders being gradually reduced to about fix inches thick at the crown of the arch, by this means, I think, that near about 40 tons weight may be taken off from the middle of the great arch; for the cinders being much lighter than gravel, and fticking together much more clofely, a lefs thicknefs is neceffary, and by eafing the defcent, the dragging over the great arch may probably be prevented, or at leaft rendered lefs detrimental.

3dly, By way of a 'general ftrengthening of the whole of the great arch, I would advife under all the arch timbers between the radius pieces, to chock in balks of fir, and to tighten every piece feparately by oak wedges, and when all are in, to tighten up the whole, nailing lead or copper plates over the joints to keep out the wet, and this I would advife to be done with refpect to the middle ribs, or arch pieces, as well as the outfides; all which having fair footings cut into the ftone of the pier, above the cordon, will alfo greatly relieve the preffure of the prefent arch timbers upon their bed timbers.

J. SMEATON.

*London,*
30th *March* 1779.

ESTIMATE

ESTIMATE for building a brick Superſtructure upon the preſent Piers, with the addition of a new Pier in the middle of the River, for a Bridge over the River Thames at Walton in Surrey.

|  | £. | s. | d. |
|---|---|---|---|
| To clearing and preparing the foundation for laying down a new pier in the middle of the river, ſuppoſe - - - - - - - | 40 | 0 | 0 |
| To conſtructing a caiſſon and bottom for the pier to reſt on - - - | 136 | 9 | 0 |
| To the conſtruction of two centers for turning the arches, and removing each of them into a new place - - - - - - | 339 | 8 | 0 |
| Preparing the foundation, and carpentry, and iron-work | 515 | 17 | 0 |

### THE PIER.

|  | £. | s. | d. |
|---|---|---|---|
| To caſe the ſame with Purbeck aiſler will contain 2396 cube feet, which if ſet in place, at 3s. per foot, will be - - - | 359 | 8 | 0 |
| The inward caſe or filling of ditto with grey ſtock brick, and terras mortar quite ſolid, will contain 2548 cube feet, which at 13d. per foot - - - | 138 | 0 | 4 |
| Maſonry and brick work | 497 | 8 | 4 |

### THE ARCHES.

|  | £. | s. | d. |
|---|---|---|---|
| The two great arches, at 2½ brick thick, will contain 22⅖, ſay 23 rods of brick work, entirely of grey ſtock quite ſolid, in good mortar, at 8l. 15s. per rod, - - | 201 | 5 | 0 |
| The two leſſer arches, at ſame thickneſs, will contain 18 rods of works, at 8l. 15s. | 157 | 10 | 0 |
| The four new arches | 358 | 15 | 0 |

|  | £. | s. | d. |
|---|---|---|---|
| The ſpandrell walls, to be at a medium two bricks thick, completely faced with grey ſtock bricks outſide, will contain in 270 feet length on each ſide, in the whole 37 rods of work, which at 8l. per rod - - - - | 296 | 0 | 0 |
| The walling and vaulting in order to fill up the ſpandrells between the arches and ſupport the gravel road, will contain 42 rods of the beſt place brick work, at 7l. 7s. | 308 | 14 | 0 |
| To ſpandrell walls and vaulting to ſupport the road | 604 | 14 | 0 |

|  | £. | s. | d. |
|---|---|---|---|
| To 540 feet running of Portland ſtone wall caping for the parapets, to be 8 inches thick, will contain 660 cube feet, which laid on and cramped, at 4s. 6d. per cube foot, will come to - - - | 148 | 10 | 0 |
| To flagging a walking path on each ſide with Elland edge flaggs, two feet broad, and 270 feet long, will contain 1080 feet, at 1s. - - - | 54 | 0 | 0 |
| To 550 loads of gravel to form the road, at 1s. - - - | 27 | 10 | 0 |
| To caping, paving, and gravelling | 230 | 0 | 0 |

AB-

## ABSTRACT of the preceding ESTIMATE.

| | £. | s. | d. |
|---|---|---|---|
| Preparations for the foundation, caiſſon, and centers   -  -  - | 515 | 17 | 0 |
| Maſonry and brick work in the pier   -  -  -  - | 497 | 8 | 4 |
| The four new arches   -  -  -  -  -  - | 358 | 15 | 0 |
| The ſpandrell walls and vaulting, to ſupport the road   -  - | 604 | 14 | 0 |
| To caping, paving, and gravelling   -  -  -  - | 230 | 0 | 0 |
| Neat eſtimate | 2206 | 14 | 4 |
| To allow for contingent expenſes 10 per cent. | 220 | 13 | 5 |
| Total | 2427 | 7 | 9 |

N.B. The above ſuppoſes the bottom of the river to be a gravel of ſufficient conſiſtence to lay down the pier without piling, as the former piers appear to have been, and alſo thoſe of Weſtminſter bridge.

No allowance is made for taking down the wooden ſuperſtructure, as the value of the timber when taken down, will be much more than the expenſe.

---

# HARRATON BRIDGE.

**ESTIMATE** for a Scaffold Bridge on Stone Pillars, at Harraton upon the River Wear.

|  | £. | s. | d. |
|---|---|---|---|
| To making coffer dams for keeping off the water for the four foundations | 116 | 11 | 0 |
| To digging and framing the four foundation pits to found upon the stone head | 60 | 0 | 0 |
| To making a horse chain-pump | 60 | 0 | 0 |
| To works of horses and drivers in drawing the four foundation pits | 63 | 0 | 0 |
| To clearance for the masonry | 299 | 11 | 0 |

## MASONRY.

|  | £. | s. | d. |
|---|---|---|---|
| To 1088 cube yards of terras rubble work in the four square foundations to support the pillars, at 9s. per cube yard | 486 | 0 | 0 |
| To 3100 cube feet of terras work cased with aisler at 6d., to bring the work above low water | 77 | 10 | 0 |
| To 627 cube yards of aisler building in the superstructure, at 4s. | 125 | 8 | 0 |
| To 640 cube feet of aisler work in the parapets of the terminating pillars, at 7d. | 18 | 13 | 4 |
| To lead and iron in cramps for the masonry | 20 | 0 | 0 |
|  | 727 | 11 | 4 |

## CARPENTERS.

| | Feet. | | | |
|---|---|---|---|---|
| In the two great truss beams, | 882 | | | |
| In the cross beam, standards, &c. | 485 | | | |
| 1367 at 2s. | | 136 | 14 | 0 |

## SMITH'S WORK.

|  | £. | s. | d. |
|---|---|---|---|
| To 8¼ cwt. in 60 main bolts at 50s. per cwt. | 20 | 12 | 6 |
| To 5½ do. in spike nails, &c. at 42s. per cwt. | 11 | 11 | 0 |
|  | 32 | 3 | 6 |

## PLUMBER'S WORK.

|  | £. | s. | d. |
|---|---|---|---|
| To sheet lead to cover the two main truss beams, 28 cwt. at 1l. 1s. | 29 | 8 | 0 |

AB-

## ABSTRACT.

|  | | | | | | £. | s. | d. |
|---|---|---|---|---|---|---:|---:|---:|
| Preparing foundations | - | | - | | - | 299 | 11 | 0 |
| Masonry | - | - | - | | - | 727 | 11 | 4 |
| Carpentry | - | - | - | - | - | 136 | 14 | 0 |
| Smith's Work | - | - | - | - | - | 32 | 3 | 6 |
| Plumber's work | - | - | - | - | - | 29 | 8 | 0 |
| | | | | Neat estimate | | 1225 | 7 | 10 |
| Add 10 per cent. for contingent expenses | | - | - | - | | 122 | 10 | 9 |
| | | | | | | 1347 | 18 | 7 |

*Lambton,*
26th *June* 1783.

_____

## ESTIMATE of the Stone Arch and additional Pier.

|  | | | | £. | s. | d. |
|---|---|---|---|---:|---:|---:|
| To finking and timbering the foundation pit | - | - | - | 13 | 5 | 0 |
| To drainage of the water | - | - | - | 15 | 0 | 0 |
| To 270 cube yards of rubble in terras | - | - | - | 121 | 10 | 0 |
| To 775 cube feet of aisler in terras, at 6d. per foot | - | - | - | 19 | 7 | 6 |
| To 157 cube yards of aisler building, at 4s. | - | - | - | 31 | 8 | 6 |
| To 810 cube feet in the arching, at 7d. | - | - | - | 23 | 12 | 6 |
| To 1152 cube feet in the spandrells and parapots, at 7d. | - | - | - | 33 | 12 | 0 |
| | | Neat estimate | | 257 | 15 | 0 |
| To 10 per cent. for contingent expenses | - | - | - | 25 | 15 | 6 |
| | | | | 283 | 10 | 6 |

I. SMEATON.

*Lambton,*
27th *June* 1783

METHOD

METHOD of founding and building the Stonework of Harraton Bridge. The Method of founding the two Land-breaſt Pillars of Harraton Bridge upon the Stone Head, as follows:

THE ſtone head at the north pillar lying but about eight feet, and the ſouth pillar ſomewhat exceeding twelve feet below low water-mark, I ſhall confine my remark to the ſouth pillar, becauſe what will do for that will become perfectly eaſy as to what concerns the north pillar.

The ſouth pillar being confiderably within the bank, I would ſet out the pit about 18 inches or two feet wider on each ſide, and longer at each end than the intended foundation; and then going down perpendicularly, I would ſupport the ground by framing in the manner of a common ſquare pit or ſhaft; but as the bearings on the ſides would be 30 feet in length, I would not only make the ribs pretty ſtrong (ſuppoſe half balks,) but ſupport them with two or three ſetts of croſs-ſtretchers, and then get out the matter with corves and baſkets, till you get down to the ſtone head, carrying down the frame, and putting in new ribs as often as you find neceſſary from the looſeneſs of the matter, as you do in ſinking a pit.

The ſtretchers may either be borrowed back or walled in, and the greateſt part of the framing may alſo be borrowed back, the ground tier of ribs excepted, which may alſo be avoided by driving down boards endways as ſoon as they will reach the ſtone head.

The two pillars in the bed of the river I would get down to the ſtone head, by driving a caſe or coffer dam of rebated piles down to the ſtone head; for the deeper pillar which is the ſouth one, the piles would not want driving above 9 feet into the ground before they would come down upon the ſtone, and the piles being 20 feet in length, would reach above high water mark. Theſe too flat ſides when the water was pumped out, would want ſupporting with croſs-ſtretchers, as deſcribed in the foundation pits for the land-breaſts.

I would compoſe each ſide of this caſe of whole timbers well ſhod with iron, to make them enter the ſtone, if of a ſoft quality, and thereby get the better footing. The intermediate ſpaces I would fill up with half timbers rebutted edgeways, and into the main timbers, ſo as to form in the whole one ſolid ſheet; this would greatly eaſe the pit of

water,

water, that would otherwife be produced, and from the infide of this cafe to get out the matter fo as to found upon the ftone head.

What will greatly contribute to facilitate all thefe operations, will be to have pump-machinery quite mafter of the water, and for this purpofe it will be well to have two pumps, one for each end of the pit, but which may each be worked by the fame horfe work, fo that the water may readily drain each way, and each be ferviceable if the other is out of order.

The piles compofing a cafe for the fouth pier will afterwards do for the north, and after all, the timber will be very little the worfe for ufe.

---

## CONSTRUCTION of the Mafonry for Harraton Bridge.

The fquare foundations, fuppofed to bring the folid work up to the bed of the river, comprehended in the 1080 cube yards of terras rubble in the eftimate, I fuppofe to be compofed chiefly of blocks, flatted and bedded with the hammer, fo as to make them lie flat, and being all of a height in the fame courfe, the interfpaces may be carefully filled and packed with common rubble work, fo as to bring the top of the courfe to a level : the blocks compofing the outfide of each courfe to be ftraightened and rough jointed with the hammer.

The mortar for this purpofe to be thus compofed: for every two bufhels of unflaked or clod lime, allow one of terras, and three bufhels of coarfe clean fand, or ditto mixed with fine gravel ; a fufficiency of which mortar being allowed as common backing mortar for common rubble work, it will be worth while to ufe large ftuff, and bring it a little together by hammer jointing, to fave the mortar.

The 3100 cube feet of terras work cafed with aifler, being the whole of the folid of the fhaft of the columns from the above ground folid, for four feet in height, which will bring the work above low water ; I fuppofe the average of two feet breadth from the outfide to be of aifler, and fet in terras mortar ; viz. to a bufhel of clod lime a bufhel of terras, and a bufhel of clean coarfe fand, well beaten, and ufed as foon as brought to a

proper

proper confiftence, the interior part of the rubble work in the fame mortar as prefcribed for the fquare foundations; fo that the ufe of the internal blocks flatted and bedded as before mentioned, will not only fave the backing mortar but render the work ftronger.

The reft of the folid of the pillars computed to amount to 627 cube yards, to be built with aifler in the proportion of the laft article; only as no terras will be neceffary, the terras muft be fupplied by the addition of an equal quantity of fand, that is, to two bufhels of clod lime four bufhels of fand.

J. SMEATON.

*Aufthorpe,*
14th *November* 1783.

## BRIDGE AT CARLTON FERRY.

Mr. SMEATON's Opinion concerning the building of a Bridge over the River Aire at Carlton Ferry.

HAVING confidered two defigns for a bridge over the river Aire at Carlton Ferry, one being for a bridge of ftone, the other for a bridge of wood upon ftone piers, both drawn by Mr. John Carr, Architect, I am of opinion that the defign for a wood bridge being no more than $11\frac{1}{2}$ feet head-way above the level of fpring tides, will not be fufficient for the paffage of veffels under the fame in time of frefhes; fuch, as in other refpects, allow the river to be in a navigable ftate: but that the defign for a ftone bridge, the crown of which arch appears to be intended to be elevated above 21 feet above the level of the fpring tides, will be fufficient for the purpofes of navigation. But that in cafe the wood bridge before mentioned, is furnifhed with draw leaves or falls properly conftructed to anfwer the purpofes of navigation, then I fee no objection to the faid defign on account of navigation.

J. SMEATON.

*London,*
25th *April* 1774.

---

### On Mr. CARR's Defign for Carlton Bridge.

THE hindrance that an additional ftone bridge of any kind muft neceffarily be to the navigation, I apprehend has been fully confidered by the proprietors of the navigation previous to obtaining the act; and as this bridge will, according to the meafures ftated thereon, afford a fufficiency of head room for the veffels in all extremes of floods, I do not fee any material objection thereto that can be removed. But as the arch will be always more or lefs of it under water, and as much as poffible to prevent damage to the arch and to the veffels by ftriking, and the mafts and ropes by rubbing againft it, I would
advife

advife that the angle which the face of the arch makes with the foffite, be rounded off by a radius of about one foot, quite round from one fpringer to the other, on both fides of the bridge.

As a confiderable part of the prefent water-way appears to be interrupted by a folid, it feems worthy of confideration by Mr. Carr, whether a dry arch on each fide may not tend to the fecurity of the bridge in giving more ample paffage to the water; this matter chiefly relates to the proprietor of the bridge; yet if any thing material fhould happen to the bridge from the want thereof, the navigation may in confequence fuffer interruption.

J. SMEATON.

*Aufthorpe,*
18th *January* 1775.

## MONTROSE BRIDGE.

ESTIMATE of a Bridge according to a Sketch propofed for Croffing the South Efk from the Inch Breck to the Fort Hill at Montrofe.

|  | £. | s. | d. | £. | s. | d. |
|---|---|---|---|---|---|---|
| TO mafonry in the bridge, extending from the north abutment towards low water mark, upon an extent of 150 feet from high water mark, at the foot of the brae, towards the main channel, from a computation of particulars - - - | | | | 1697 | 0 | 0 |
| To nine bays of timber fcaffolding, at 45 feet each, extending in length 405 feet, from a computation of particulars, at 233l. each bay - | | | | 2097 | 0 | 0 |
| To the mafonry extending from the fouth end of the faid fcaffolding to the fouth fhore, if taken at the fame price as that in the north (though lefs work), will be - - - | | | | 1697 | 0 | 0 |
| | | | | 5491 | 0 | 0 |

### EXTRA ARTICLES.

|  | £. | s. | d. | £. | s. | d. |
|---|---|---|---|---|---|---|
| To a great piling engine for driving the large bearing piles - | 150 | 0 | 0 | | | |
| To fmall piling engines, boats, punts, various machines, fcaffolding, centers, and utenfils - - - - | 500 | 0 | 0 | | | |
| To fuperintendance, fuppofe for two years - - - | 240 | 0 | 0 | | | |
| To incidental and unforefeen expenfes - - - | 250 | 0 | 0 | | | |
| To expenfes in temporary fcaffolding, and driving the main bearing and fender piles, being 56 in number, at 3l. each - - - | 168 | 0 | 0 | | | |
| To the contingence of quarry rubble for the purpofe of fecuring the ground about the foundations, where it may be found neceffary in the courfe of the work - - | 38 | 0 | 0 | | | |
| | | | | 1346 | 0 | 0 |
| Total of the bridge | | | | 6837 | 0 | 0 |

### INCH BRIDGE.

|  | £. | s. | d. |
|---|---|---|---|
| The bridge over the inch burn being of mafonry, and taken at the fame expenfe as the north end of the main bridge, will come to - | 1697 | 0 | 0 |
| | 8534 | 0 | 0 |

N. B. In the above eſtimation no roads by way of acceſs, nor any thing is comprehended beyond the termination of the parapets of the wings or retaining walls, no charges of law, intereſt of money, or damage of grounds for laying and working materials, or quarry leave, nor any remuneration to the engineer in chief.

The main walls are ſuppoſed to be of white craig rubble ſtone, built in the beſt manner, in the ſtyle of the fronts of the new ſchools : all the building under high water-mark to be done with mortar proper for the ſalt water : the arches, blocking courſe, and caping, and ſuch quoins and courſes of the foundations as are neceſſary to be hewn, are eſtimated of the beſt Lauriſton ſtone. The timber is all ſuppoſed to be of the beſt Riga or Dant-zick fir.

The charge per foot running of the maſonry part is 9l. 14s., that of the ſcaffolding 5l.

# TINMOUTH BARRACKS.

To the Honourable Mr. St. John.

Sir,                                                    Newcaſtle, 22d January 1780.

Mr. WALTON received your letter of the 28th of December laſt, deſiring our opinion upon certain matters therein mentioned, and accompanying a plan of a parcel of ground part of His Grace the Duke of Northumberland's manor of Tinmouth, upon which barracks have been erected, and alſo a particular and valuation of the ſaid parcel of ground, and the coal ſuppoſed to be under the ſame, propoſed to be given in exchange for ſundry parcels of land, which His Grace holds by leaſe from the crown.

We now beg leave to acquaint you, Mr. Smeaton came to this place on Monday the tenth inſtant, and that we proceeded upon the buſineſs the following day, and wrote to the Duke of Northumberland's principal agent William Charlton Eſq., and to Mr. French, His Grace's bailiff for the manor of Tinmouth, and ſeveral perſons whom we thought proper to have the aſſiſtance of on our view. We had no anſwer from Mr. Charlton, either to our joint letter, or to one Mr. Walton wrote to him upon the fifth inſtant, and Mr. French gave us for anſwer that he had no directions, but notwithſtanding that he would have attended us upon our view, if he had not been under the neceſſity, (as deputy clerk of the peace for the county of Northumberland) to attend the ſeſſions; ſo that we were diſ. appointed of any aſſiſtance from the Duke of Northumberland's agents.

We proceeded on our view on Wedneſday the twelfth inſtant to the barracks, and made the neceſſary obſervations there, to enable us to ground our opinion upon; and as a new ſurvey of the barracks and ground ſeemed to us neceſſary, an able ſurveyor was immediately ſet to work, not only to make a ſurvey and plan of the ground upon which the barracks are built, but of the buildings alſo, and of the ſeveral high roads, foot paths, &c., which plan you receive herewith; and alſo one laid down by a ſmaller ſcale, taking in Tinmouth Caſtle, Clifford Fort, &c., accompanied by an eſtimate of the yearly value of the lands, ſuppoſing no buildings to have been made, and of the expenſe which it is computed the barracks might at firſt coſt, on the foundation of prices paid in this country for work of a ſimilar nature.

3 D 2                                                              Having

Having since given this business all the attention in our power, we sit down to inform you, that we find the ground upon which the barracks are built, with the space on the outside of the buildings, does not amount to more than seven acres and 19 perches, according to the survey made by Mr. Fryer, and we apprehend the difference of two roods and 25 perches arises from the waste of ground by the sliding of the banks into the sea; and that the quantity will continue to diminish, and the sea in the course of years entirely destroy the buildings on that side of the square of the barracks joining upon the seashore: under this very material inconvenience, which in our opinion can no ways be prevented but at an immoderate expense, it becomes very difficult to ascertain what may be the present worth of the fee-simple and inheritance of the said parcel of ground, with the way-leave or right of passage over the adjoining lands of the Duke to and from the barracks, proposed to be granted therewith: but having maturely considered the subject, we must observe, that the value of lands principally depends upon local situations; and we do not see a better rule or guide than that which other lands of equal quality in the same situation are let for; and as we are well informed His Grace's lands adjoining upon the barracks are let at fifty shillings per acre, the same rate at which Mr. Fryer values them, we cannot think the barrack grounds deserve more; so that the annual value thereof upon the quantity now in being, according to Mr. Fryer's survey, will amount to seventeen pounds fifteen shillings, which, at twenty five years' purchase, will amount to four hundred and forty-five pounds; but considering the Duke of Northumberland is under no obligation to sell, and it must be supposed he sells merely for the accommodation of the Crown, and considering that such accommodation breaks into His Grace's estate, we think the present worth of the fee-simple and inheritance thereof, with the way-leave or right of passage to and from the barracks proposed to be granted, will deserve thirty years purchase, amounting to five hundred and thirty-four pounds; His Grace reserving to himself the sea banks behind the barracks, and granting a liberty at all times to the servants of the Crown to enter upon and take such methods as they shall think proper for the preservation of the same.

The public highway for carriages from Tinmouth to Shields, passing very near the barracks on the north side thereof, will be very sufficient to answer every purpose of the barracks, and the foot paths mentioned and described will be very sufficient for every communication between the barracks and Clifford's Fort, as well as the Spanish Battery.

We have endeavoured to inform ourselves whether any, and what damages may be done by the soldiers in the adjacent lands, and from thence and our own observation, we think that damage (if any) must be very inconsiderable; and from the commanding officer

on

on duty, when we were there upon our view, we were informed not one complaint had been made during the time of his refidence.

It is very clear to us, that it will be attended with no inconvenience to His Majefty's fervice for His Grace the Duke of Northumberland to referve the banks between the barracks and high water mark, as mentioned in the faid particular, there being no appearance of ufe from thofe banks to the Crown, for the purpofe of bringing ftores by fea to the barracks, or for any other purpofe whatever, the banks or cliffs being fo fteep and rugged as to be very difficult of accefs; nor could any veffel be brought to the foot, or lie there with fafety without digging a harbour, and erecting proper piers for defending the veffels laid therein.

In regard to the coal under the faid feven acres and nineteen perches of land, we had that matter under full confideration, and are clearly of opinion, that no calculation of quantity, quality, or value, can be made thereof, upon that fair ground which we think ought to determine the value to a purchafer. Conjecture of feveral feams of valuable coal being within and under the lands before mentioned, cannot juftify us to found a calculation upon; we have defired His Grace the Duke of Northumberland's agents to acquaint us if any borings have been made in the lands of His Grace, at or near Tinmouth, but to this they have given us no anfwer; and as no judgment can be juftly formed without proper trials to be made by boring, we beg leave to be excufed from making any eftimate with refpect to the coal. Our obfervations with refpect to the coal would have ended here, but as we think ourfelves obliged to make fome remarks on the valuation handed to us, we beg leave to ftate the following particulars :—The quantity and value, as fet down by the Duke of Northumberland's agents, cannot be applied to the value of the coal fuppofed to be under the barrack lands, as being founded on what His Grace lately agreed for at Tinmouth Moor, which was then and now is an open working colliery, as no notice is taken of the great expenfe which attended the winning of it, or of laying waggon ways, building fteaths, and other incidental expenfes of bringing the coals to market, together with other common expenfes attending the vend; we therefore repeat, the circumftance of the Duke's agreement at Tinmouth Moor cannot apply itfelf to the value of the coal fuppofed to be under the barrack grounds, which lies unknown, unafcertained, and fubject to the expenfes before mentioned to obtain it. We have faid unknown and unafcertained, for at prefent there is nothing that appears to us to fix the certainty of the exiftence of any valuable feam or feams of coal under thofe lands; there may be feams of coals that may be very valuable, for any thing that we can fay to the contrary, but whether there are fuch or not feems to us problematical. The colliery now working by

His

His Grace the Duke of Northumberland on Tinmouth Moor, is at $2\frac{1}{4}$ miles diftance, or thereabouts, in a ftraight line ; nor is there any other colliery to our knowledge now worked, according to the neareft line, nearer than half a mile from the barracks, and the feam of coal worked there, according to the general courfe of the rifing of the ftrata towards the fea, fhould be either out of or near the furface of the ground at the barracks ; but the depth and courfe of ftrata may be fo varied, even in the laft mentioned fpace of half a mile, by interruptions called dykes, as to leave no more than that kind of probability or profpect, which may be fufficient to induce a land owner or an adventurer to be at the expenfe of making a trial of the ground by boring ; and after that is afcertained with favourable circumftances, yet the expenfe of making an actual winning, fo as to bring the colliery into actual working, and the clear profit thence arifing, is in its own nature fo uncertain, that the undertaking becomes no other than an adventure, which, if really profitable, by far the greateft part is due to the rifque and hazarded capital of the adventurer, and not to the owner of the ground ; fo that whether the owner himfelf becomes an adventurer, or any other, the land owner's intereft in the coal is no more than what would be given for the privilege of working it, by the general eftimation of thofe who make thefe adventures their object. Having gone thus far, you will excufe us when we add, that when we know the colliery which is fuppofed to be under Tinmouth Barracks is won, and in a working ftate, with the time in which the coal will be turned into money, we will have no difficulty in giving a valuation of the fum it is worth to be purchafed, at any given period. At prefent we have no data to go upon, we have no evidence even of probability from His Grace's agents, and till that kind of evidence is had to reduce the matter to as great a certainty as the nature of the affair will admit of, a doubt muft remain with us, and confequently we cannot think it right for His Majefty's fervice to purchafe the fuppofed property of coal at any rate whatever. But here it may be afked, what will become of the barracks, if His Grace the Duke of Northumberland fhould find the colliery good, and work away the fame? To this we anfwer, that the working of this colliery feems to us to be at a great diftance in point of time, and that though the working of a colliery will occafion the furface of the ground to fall in irregular hollows, and thereby very greatly damage lofty and maffy buildings, yet in the prefent cafe, we cannot confider the barracks as fubject to the fame damage, as being low and light buildings, confifting of a ground floor only, and in that refpect, in our conception, much on a footing with cottages and pitmen's houfes, which are commonly erected on the very ground under which a colliery is working, or to be worked ; and as often as fuch damages happen, reparation is made at a fmall expenfe after the ground is fettled, and the habitations rendered comfortable again to the occupiers. For thefe reafons, therefore, with what follows relative to the inftability of the barracks, from the fliding banks of the fea fhore,

3

we

we have not a doubt but that it will be for His Majefty's fervice to run the rifque of any damage which may be fuftained by working the colliery under the barrack grounds.

Having finifhed what relates to the lands upon which the barracks are built, and the colliery which may be under the fame, we proceed to make our obfervations refpecting the condition of the barracks themfelves, their ftability, propriety, and convenience of fituation.

On this head we muft obferve, that though we cannot look upon ourfelves as judges of the propriety of military ftations or conveniences, yet we muft neceffarily take notice, that the banks lying between the barracks and the fea fhore, of which His Grace the Duke of Northumberland propofes to keep poffeffion, are of a very confiderable perpendicular height, and being compofed wholly of a factitious clay, mixed with ftones and gravel, and irregularly interfperfed with a ftratum of fand, in wet feafons they become fpringy; and as the fea wafhes away the foot of the cliffs, the banks gradually flide down, and the degree of wafte of the banks may be judged of as follows:—

According to the plan accompanying the particular we received from you, a clear fpace is defcribed between the fouth line of the barracks and the banks, fufficient to receive the fouth-eaft and fouth-weft baftions entire, together with the full breadth of feventy-fix feet in the curtain on the fouth fide, agreably to the reft. At the time of our view we found a confiderable part of the fouth weft baftion had flid down into the fea, together with a great part of the original curtain, fo that in one place the verge of the broken ground was not twenty feet from the fouth wall of that fide of the barracks; we have therefore no reafon to doubt but that in a few years a part of the buildings themfelves will flide down the bank, and in time the whole, unlefs fome effectual method be taken for fupporting thofe very irregular inclining banks from the depredations of the fea. We doubt not but that the original ftate of the ground was according to the plan; it is however now fo far changed, that if nothing is done, or the means ufed by the King's fervants for that purpofe fhould prove ineffectual, it becomes, in the next place, a confideration how far thofe banks can or ought to be fupported, which is to be done at the expenfe of the Crown, and confidering this as a matter relative to the good of His Majefty's fervice, we do not conceive, that with refpect to His Majefty's more than any private fervice, any thing is worth more to preferve than it originally coft to produce, provided it can be reproduced with the fame coft, and to the fame intent. In this ftate of the matter, though we do not think it impoffible to fecure the banks from further depredations of the fea, yet if done in an effectual manner, the expenfe would be fo great that we are very clear it would be far lefs coft to

build

build up the barracks anew as they happen to fail, nay even to build the whole new upon frefh ground, than to fecure the banks; and as the foil and fite of Tinmouth Caftle are not only the property of the Crown, but in our idea proper for the reception of barracks, there cannot, as we judge, be a doubt of its being eligible to remove them to that fituation, in cafe they cannot be continued where they are at a moderate expenfe.

And now having concluded our obfervations, and given our opinion on the feveral points referred to us, we truft what we have done will be fatisfactory; we can however fay that we have gone through the whole impartially, and with care and attention; but we cannot difmifs the fubject without making this general obfervation, — that though it appears to be improper to purchafe the coal fuppofed to be under the barrack grounds at any rate whatever, yet we think it would be right to purchafe the lands, not only to give the Crown a power of enjoying the barracks, during the continuance of their being ufeful, but to veft an authority in His Majefty to remove the buildings to the caftle, or to let the remains ftand as fhall be thought expedient. This feems to us to be the beft method of fettling the bufinefs on an equitable footing to the parties, as His Grace will have fatisfaction for the property he parts with, and that property will be vefted fo that it may be made fuch ufe of as fhall be judged beft for His Majefty's fervice.

We are, Sir,

Your moft obedient fervants,

NICHO$^s$. WALTON.
J. SMEATON.

P. S. Since forming our ideas and opinions upon which this Letter was written, we have feen Mr. Charleton, but have not received any information from him to induce us to alter the fentiments we have expreffed.

To the Honourable John St. John,
Surveyor General of the Crown Lands.

## LEEDS INFIRMARY.

TO the Committee of Leeds Infirmary.

Gentlemen,

I HAVE infpected the ftone cifterns made at this hofpital by the late —————— Craven; I find them nearly of the capacity propofed; but as I underftand they never held water from the firft, nor do they feem in a capacity of doing fo now, or of being made to do fo, without taking them all to pieces, and putting them together with the greateft care and firmnefs, which would of itfelf be a confiderable expenfe. Each double ciftern, when full, is charged with above fix tons of water, and relative to this weight they have not only been much too flightly bonded together in their joints with iron, but I apprehend the foundation upon which they have been built has fettled under the weight of the whole; for the leaft fettlement or giving in the joints muft and will necefsarily difunite the cement of whatever kind, and thereby render them leaky. What the nature of the agreement was you beft know; if to make them complete to hold water, it was then the undertaker's bufinefs to look to the foundation and ftrength of his bondages: but yet, Gentlemen, if thefe cifterns can be rendered ufeful to you, it would feem more eligible fo to do, than to return the materials upon the hands of the widow, which, when taken to pieces, will be of little value. I take it for granted you wanted, and ftill want thefe cifterns; and had they not been fo undertaken to be made of ftone, you had ordered them of lead, in which cafe ftrong wooden frames of timber would have been wanted to have fupported the leaden lining, which frames of timber, though oak, would have been liable to need rebuilding in 20 years, in which cafe the workmanfhip upon the lead would have been to renew likewife: had therefore the propofition been how to render thefe cifterns *durable*, like the reft of the buildings, and which every public building ought to be rendered as much as poffible, then you could not have been better advifed than to have put together a cafing of large ftone flags, which the country affords, in order to fupport the leaden lining, and in this refpect thefe cifterns as they now ftand, will completely anfwer that end; and though in putting them together fo as to hold water without fuch lining, there is work upon them that would have been unneceffary; yet if a reafonable allowance be abated from the firft ftipulated price for this fuperfluous work, it will be much lefs lofs to the widow, and much lefs difadvantageous to the truft than either the removal of the materials, or an attempt to render them water-tight in the manner they are. According to my eftimate they may be completely lined, as they now ftand, with lead of 7lb. to the foot, for about or fomewhat under 30l.; of which about 8l. will be workmanfhip, the other will be the neat value of the lead and folder.

According as it has been given in to me, Craven's eftimate was:

|  |  |  | £. | s. | d. |
|---|---|---|---|---|---|
| Stones | - | - | 14 | 14 | 0 |
| Cramps and cement | - | | 2 | 5 | 3 |
| Work | - | - | 13 | 2 | 6 |
| | | | 30 | 1 | 9 |

How far the ftones were properly valued at 14l. 14s. (carriage, I fuppofe, included) I know not, having never had occafion to purchafe any of that magnitude. There are in the whole 20 ftones, fo that they will come to 14s. $8\frac{1}{2}$d. nearly per ftone, which on an average contains about 24 feet fuperficial each; but as the Committee have agreed to take them at that price, on fuppofition they were rendered effectual to the purpofe, fuppofe

|  |  |  | £. | s. | d. |  | £. | s. | d. |
|---|---|---|---|---|---|---|---|---|---|
| Widow Craven paid for the ftones | - | | | | - | | 14 | 14 | 0 |
| Cement and cramps | - | | 2 | 5 | 3 | | | | |
| Work | - | - | 13 | 2 | 6 | | | | |
| | | | 15 | 7 | 9 | | | | |

Half of which is 7 13 $10\frac{1}{2}$

which deducted on account of what would have been fuperfluous, on fuppofition of the ftone frame being made purpofely to be lined with lead, then there will remain for cramps, mortar, and work - 7 13 $10\frac{1}{2}$

And there will be payable to Widow Craven - 22 7 $10\frac{1}{2}$

which, in my opinion, is a high valuation of the work for the purpofe above mentioned; and there will be 7l. 13s. $10\frac{1}{2}$d. retained in the hands of the Committee toward the workmanfhip of the lead lining, computed at 8l.; fo that the truft will have very little more than the value of the lead and folder to pay for, to make the work effectual in the fame manner as if originally intended for lead cifterns in ftone cafings, which, in my opinion, would have been the complete way of doing it; and, at the fame time, the wooden covers,

the

the conveyances for bringing in the water, and conduits for carrying it away will be preferved, which would be loft in cafe the work were totally demolifhed. That this is an equitable way of confidering the matter appears hence: that if the Widow Craven were to line the ciftern with lead, fhe would then undeniably complete the work; and as fhe would add fo much real property to the hofpital as the value of the lead, for this fhe ought to be allowed in addition to the 30l. 1s. 9d., and fhe would have to pay 8l. for the work inftead of allowing 7l. 13s. 9d.

I muft obferve, that a couple of refervoirs capable of holding between 12 and 13 tons of water, which will be filled many times in the year by rains, will not only be a fafety to the hofpital, but an acquifition well worth the expenfe. Rain water is undeniably the beft for wafhing, and though river water is more efteemed for brewing, yet it differs from rain water only in being mixed with a fmall proportion of hard fpring water, of which kind the infirmary having the command by their well, a proper mixture of the two will anfwer the fame as river water for brewing, fo that (in very long droughts excepted) the infirmary will always have a command of water within themfelves; and, as I apprehend, what water will be wanted during fuch dry feafons to be brought will be reduced to fo fmall a quantity, that the carriage had better be paid for than go to the expenfe of any further pump work, refervoirs, or machinery.

J. SMEATON.

*Aufthorpe,*
11th *Oct.* 1776.

P. S. As I have mentioned the want of ftrength of bondages and foundation as the caufe of the failure of the ciftern's holding water, it may perhaps be doubted whether they will be fufficient when lined with lead. But, I muft obferve, that if the amount was the thicknefs of a piece of poft paper, by which the joint was broken, it would be fufficient to deftroy the ufe of holding in the water; but if lined with lead, the lead will comply without breaking; fo that confidered merely as a cafing to fupport they are fufficiently ftrong.

## EARL OF EGREMONT'S COALS.

THE REPORT of John Smeaton, Engineer, refpecting the Practicability of exporting Coals from the Eftates of Branfty, Birkby, and Afpatria, in the County of Cumberland, belonging to the Right Honourable the Earl of Egremont.

HAVING carefully viewed the fituation of the eftates of Branfty, Birkby, and Afpatria, with refpect to the fea, together with the different propofitions that have been pointed out to me as probable means of carrying and fhipping coals from thence refpectively, I am of opinion as follows; and, firft, refpecting

### The Eftate of Branfty.

It feems here perfectly practicable to carry the coals upon a rail road, acrofs Mr. Ellifon's rope-walk from the place where the colliery is expected to be won, to a yard belonging to the faid Branfty eftate, which has been intended for the depofition of timber, or other merchandife, which is not only adjoining to the public ftreet of Whitehaven, near the arch built by Sir James Lowther, but adjoins upon the wafte ground, being the frontage of the fea-fhore immediately behind, or north eaft of the north pier lately erected as part of, and, as I underftand, the boundary of the port of Whitehaven, as defcribed by act of Parliament: coals therefore being brought from the interior parts of the eftate acrofs the faid rope-walk to the ground contiguous to the fea-fhore, as above defcribed, the queftion is then, Whether it will be more eligible to drop the coals in the yard already defcribed, and from thence cart them through the ftreets of Whitehaven to the harbour, which is within Sir Jame's Lowther's property; or to build a harbour upon the Branfty eftate, contiguous to the faid north pier of the harbour of Whitehaven? The proper refolution of this queftion indeed depends upon a great variety of circumftances; that is to fay,

1ft,

1ft, How far it is practicable that a harbour is capable of being erected and kept open upon the Branfty eftate, fo as to admit of the fafety of veffels; and of what burthen; together with the probable coft of erecting the fame?

2dly, After knowing the probable coft of building a harbour upon the Branfty eftate, then comparing the intereft of the fum that is likely to be funk in effecting this work, with the price of the cartage of coals from the faid ground upon the annual vend of coals, that is to be expected to be fhipped from the faid eftate, the difference will appear.

3dly, On fuppofition, that on this comparifon it fhould appear that the more eligible method is by cartage to the prefent harbour, it will be a further confideration how far, according to the acts of Parliament now in being relative to the harbour of Whitehaven, My Lord Egremont can entitle himfelf to, and infure himfelf of thofe advantages, in difpofing, birthing, and laying on fhips, for the reception of His Lord-fhip's coals, that the nature of the place in its prefent ftate admits of.

4thly, Whether on fuppofition of the preceding enquiry turning out favourable to His Lordfhip's views, he will have power of compelling that the ftreet of Whitehaven through which the coals would be carted, fhould be kept in their prefent ftate of repair, on fuppofition that Sir James Lowther fhould think proper to ceafe his cartage through the fame avenues?

As an engineer, the firft article, comprehending the practicability and expenfe of erecting a harbour upon the Branfty eftate, is the only part of the propofition that properly comes before me, and as this is a neceffary foundation for making the other comparifons, to this I fhall at prefent confine myfelf, leaving the other difcuffions to be determined where they more properly may, by the gentlemen already employed as coal viewers, to advife upon the winning of the faid collieries, and His Lordfhip's agents in the country.

The north pier of Whitehaven, which has been erected within thefe few years paft as a boundary to the port, appears to me as an unfinifhed part of a larger and more extenfive defign; but what this defign is, or has been, I had not an opportunity of getting any certain intelligence. From the place of commencement and direction in which it is carried out, it would feem to me, as if Sir James Lowther does intend, or

has

has intended, to carry forwards his coals over his archway, and to have fhipped from this pier thofe of his coals that are now carted from the faid archway to the quays; were this done, and My Lord Egremont fuppofed to have an equal right to do the fame, his purpofe would thereby in whole, or in part, be alfo ferved; at prefent it does not feem fufficiently carried forwards to ferve the purpofe of either: but taking it as it is, it appears to me to be a very fufficient defence or break-water, fo far as it is gone to the fouth weft fide of the harbour, that I would propofe for Branfty, and which will therefore ferve the fame purpofe to the Branfty harbour as if it had been built with that intent, except what relates to the right of laying of railways, and leading the Branfty coals thereupon, of which point I am uncertain, but of which it would behove My Lord to be fully advifed, in cafe that, upon the whole, a harbour at Branfty is deemed the moft eligible.

The actual laying out a harbour requires a more minute confideration than the time I was at Whitehaven would admit; it is fufficient to fay, that it appears to me, that a harbour may be made, by running out a pier upon the Branfty eftate, from or near to the north weft termination of the prefent north pier of Whitehaven, fo as to fkreen about half a dozen fhips at a time from the violence of the fouth weft or weftern feas, and when any greater number fhould arrive or be wanted, they can always lie in the port of Whitehaven, either till they can fail after being loaded; or after arrival till a proper birth can be had within the Branfty pier.

I obferve that the north pier is carried out from the fhore about 160 yards, and extends into eight feet water at a neap tide, which from the flow of the tides on this coaft, fhould make from 14 to 15 feet water at the high water of a fpring tide; and which, I apprehend, is as much, or nearly as much, as is generally maintained at the coal quay in the interior part of Whitehaven harbour; and I am of opinion, that by running out a pier upon the Branfty eftate of about 120 yards in length, in a proper direction and conftruction, as much water may be maintained within the fame as at the prefent termination of the north pier.

For this work there can fcarcely be a quarry more commodious than that at Jack-a-dandy Hill, or point within the Branfty eftate, that is now wrought for fupplying the piers of Whitehaven with ftone. Having, therefore, attentively confidered the fituation of the quarry and the work, and having made a careful eftimate of the expenfe of build-

ing

ing 160 yards running of pier, in this fituation it will, according to my eftimate, coft 3059l., fay 3100l.

Now, if upon the comparifons above mentioned, it fhall appear eligible to go into this expenfe, then I would advife this work to be undertaken, and for which I fhall be ready to enter more minutely into the detail, and produce the proper working plans: but if not, I cannot advife it to be attempted. The above expenfe is, however, independent of the laying of waggon ways from the fhore, and the making of proper fpouts, or hurrys, as they are there called, for putting the coals into the veffels, which may propably fwell the above eftimate to 4000l. I would not, however, be underftood that the building 160 yards of pier, as above ftated, will make the very beft harbour that can be made upon the place; what I mean is, that it will be fufficient to begin with, and to fee how things are likely to anfwer; and if fo, it may be afterwards extended either in length, or by building a counter pier on the north fide of Branfty harbour fimilar to the north pier, now in part made as above mentioned, for Whitehaven; the one or the other to be proceeded with as experience fhall direct or fhew neceffary, after the work firft propofed to be erected is put to a full trial. I further looked upon it, that the outworks that have been erected to break off the violence of the fouth weftern ftorms from the harbour of Whitehaven, will be almoft of the fame advantage to the harbour of Branfty, fo that this harbour would anfwer its end much more advantageoufly than if the works at Whitehaven had never been erected.

## 2dly, Refpecting the Birkby Eftate.

It has been pointed out to me, that by way-leaves through the eftate of Mr. Senhoufe, the coals of Birkby can be carried to the harbour of Maryport, at the mouth of the river Ellen, and there fhipped: upon this propofition the opinion of an engineer is unneceffary; but it is further fuggefted, that as the eftate of Afpatria as well as Birkby is fo circumftanced, that this river if made navigable, will be equally ufeful in carrying down the coals thereof to the harbour of Maryport, and it will alfo carry the coals arifing out of feveral other eftates lying upon the faid river; with this idea I have viewed the courfe of the river Ellen, from a point within the manor of Atterby, where a canal has been propofed to be taken off to go through Afpatria towards the fea, at a place called the Blue Dials; from this point to the harbour of Maryport, according to the courfe of the river's valley, is full five miles, and as by a level taken by Mr. Johnfon, from the

faid

ſaid point of the river Ellen, in the manor of Atterby, to the ſea at the **Blue Dial**, there is a fall of 60 feet, we muſt account the ſame deſcent in the courſe of the river Ellen, from the ſaid point to the ſea at Maryport.

From the ſaid point, taken above the head of the aqueduct of Atterby mill, for about a mile downward, the ground lies very well for a canal out of the bed of the river, but then the valley begins to grow confined, and the borders of the high ground to lie very ſteep, and grows very embarraſſed about the corn mill of Roſe Gill, which is a new erected corn mill belonging to Henry Fletcher, Eſquire, and ſeems to have been built at a very conſiderable expenſe ; from thence downward the river runs very rapid, and in general in a very ſhallow and narrow bed for about 2¼ miles, till we are almoſt through the Birkby eſtate ; and the bottom of the valley in this ſpace is ſo narrow, the borders in many places perpendicular rocks, and almoſt every where ſo ſteep and embarraſſed that it would be almoſt impracticable to make the navigation otherwiſe than in the bed of the river. Below Birkby the valley is more open through the eſtate of Mr. Senhouſe, yet not without embarraſſments, ariſing from the houſe, and gardens, and mills belonging to this gentleman, and from the aqueduct leading through the ſame to the iron furnace at Maryport : in ſhort, conſidering the number of locks that muſt neceſſarily be required in a fall of 60 feet, the great rapidity and quantity of water in time of floods, the diſcharge of which muſt in many places be provided for, I cannot lay the expenſe of making the ſame navigable through the above diſtrict at leſs than 3000l. per mile, that is, for five miles 15,000l., and this excluſive of any compenſations to the mills and iron works above mentioned, for loſs of water ; or to two other mills, though ſeemingly of much leſs account ; that is, the mill of Atterby and Sir James Lowther's mill, in the lordſhip of Dereham ; for the river Ellen in dry ſeaſons would not be capable of ſupplying the mills, as well as the navigation, if carried on to any great extent.

Were this river made navigable it would, beſides the manors of Birkby and Aſpatria, belonging to Lord Egremont, be the means of bringing to market the coals of Mr. Senhouſe, as alſo thoſe of the manors of Dereham and Croſby, belonging to Sir James Lowther and ſeveral others ; and were the ſeveral proprietors of theſe lordſhips deſirous, aiding, and aſſiſting in bringing the thing to bear, ſo that there might be no difficulties to remove but thoſe that ariſe out of the nature of the ſubject, then it would be a propoſition well worthy of attention ; but if to the natural expenſe attending the execution of it, as above mentioned, be added thoſe extra expenſes that would attend it, were it to be proſecuted by Lord Egremont upon an adverſe principle with reſpect to the other

4                                                                                                        adjoining

adjoining lords; if carried at all, I am very clearly of opinion the amount would be such as by no means to become an object to Lord Egremont to drive through solely. If this then is the cafe, it will therefore remain, that should a fea fale colliery be won in Birkby, the coals muft go to Maryport by way-leaves, through the eftate of Mr. Senhoufe, who, if he would not grant way-leaves, would, it muft be imagined, oppofe a navigation alfo.

It now remains that I give my opinion with refpect to

### The Eftate of Afpatria.

Exclufive of the river Ellen, what has been propofed for the fhipping of coals from hence, is to take a canal out of the river Ellen, at fome convenient point above the damhead of Atterby mill, and carrying the canal for about a mile upon a dead level, there to terminate the fame, and then to defcend by a waggon way the 60 feet that the faid point of the river has been found to be above the fea, which occurs in about the fpace of ¾ of a mile more, to a place on the fea fhore in the bottom of Atterby bay, called the Blue Dial. To any part of this propofition I fee no *natural* difficulty; the whole queftion as to execution, feems to reft upon the procuring of way-leave or licence to make that part of the canal that lies out of the lordfhip of Afpatria, and the expenfe of building an harbour at Blue Dial.

In point of original eftimation, I would, however, advife that the eftimate of the expenfes attending the winning of the colliery, be made upon the fuppofition of a waggon way altogether, from the propofed place of winning to the Blue Dial, without any canal at all.

Firft, becaufe in order to procure water from the river Ellen, it will be neceffary to cut through a property that lies between Afpatria and the river Ellen, as well as through a property that lies between Afpatria and the fea, whereas, if the whole is done by a waggon way, the difficulty of procuring way leave will be confined to one property only.

2dly, Though an ingenious idea of the ready fhipping and unfhipping of the waggons, containing their loading, from the canal to the waggon ways, and vice verfâ, was communicated to me by Mr. Johnson, yet it is no ways clear to me, that in so short a length of navigation any thing will be saved; whereas, if upon the whole the colliery is likely to bear the expense of winning by a waggon way, and the building of a harbour at Blue Dial, if it shall, after the thing is resolved upon, be found upon entering into a minute detail of the particulars both ways, that any thing can be saved by the execution of a canal, then, by that saving, the thing will turn out just so much the better: the main point, therefore, that seems to rest upon my opinion, as an engineer, respecting Aspatria, is the practicability and expense of erecting a harbour at the Blue Dial.

The principal disadvantage of this place, as to the making of a harbour, is, that as the shore is flat, a pier must be run out to a considerable length before a sufficiency of depth of water can be obtained. Another disadvantage is, that here is no outlet of any river or burn to keep the same clean, which last is indeed a disadvantage common with Whitehaven, but then the sand which is driven coastwise, and a good deal affects the harbour of Whitehaven as well as Maryport, seems there a good deal dispersed, or driven further out to sea; on the other hand, as the seas here are narrow, no considerable swell can come from the northern or north eastern points, nor indeed in the bottom of Atterby bay, where the Blue Dial is situated, do there from the appearance of the shore seem to be very raging seas at any time, so that a single pier, with the turn of an elbow, to screen off the south westerly, westerly, and north westerly winds, will, as I think, be sufficient; and being of a proper construction, I apprehend will not be liable to choak up with sand in any obnoxious degree; so that by running out a pier of 300 yards in length, I apprehend that 14 feet water may be gained in spring tides, with a safe birth for four or five ships, and several more of smaller size; this is much about the same depth of water as at Maryport, and as, like Maryport, the flow of the neap tides is considerably less than the spring tides, no vessel of any consequence will be able to stir at neap tides, and which indeed is a good deal the case at Whitehaven: above a week in each spring tide time, there may be expected from 10 to 11 feet water, and upwards, so that vessels not exceeding that draft of water may move every tide, when wind and weather admit, and for six or seven days, about the neap tide time, vessels under 10 feet water only can move; the dead of neap making only about seven feet, which last will yet float vessels of about 50 tons burthen, or about 75 Cumberland tons of coals.

N. B. The largest vessels belonging to Maryport are said to carry about 300 Cumberland tons of coals, that is, they are vessels of about 200 tons burthen, which kind of vessels generally draw about 11 feet water.

10

Having

Having carefully computed the expenfe of erecting a harbour at Blue Dial, from an excellent free ftone quarry at the diftance of about ¾ of a mile ; the eftimate thereof I make to be 5700l.

From thofe general data, the expediency of the feveral propofitions may be judged of, and the execution of fuch as fhall appear advantageous determined ; and when that is done, it will then be proper to review the premifes, and to enter more minutely into the detail thereof than this general view, which was comprehended in two days only, would enable me to do.

J. SMEATON.

*Auſthorpe,*
19th *October* 1775.

# ROSEVEERN AND OWERS LIGHTHOUSES.

To the Mafter, Wardens, and Affiftants of the Honourable Corporation of Trinity Houfe, Deptford Stroud.

My Lords and Gentlemen,

HAVING at your defire, under the conduct of Captain Bromfield, of your brotherhood, furveyed the ifland of Rofeveern, one of the iflands of Scilly, and alfo the Owers Rock off Selfea, in the county of Suffex, with refpect to the building a lighthoufe on each of thefe fituations, I beg leave to report feparately my opinion on both of them refpectively; and firft,

## Concerning the Ifland of Rofeveern.

We vifited this ifland the 14th of June laft, during the time of fpring tides. The day being perfectly fine, we landed from a boat without the leaft difficulty. The whole of the rock compofing this ifland appears to be granite, which, on trial with a pick, I found to be fufficiently workable. The ifland is of an oval fhape, and nearly a flat upon the top. Its length is pretty near eaft and weft, and it gently rifes towards the weft end, where it is terminated with large granite rocks, naturally piled on each other, higher than any other part. The length of the flat part of the ifland, including the gentle rife, I meafured to 315 feet; and the breadth 132 feet. Without this flat area, the furface declines faft towards the water; and the whole border is compofed of large roundifh granite rocks, difpofed in a floping form, down to low water, and under it.

The flat area is about feven yards above the half tide-mark, and is about four yards above high water-mark of fpring tides. The high rocks at the weft end form a kind of breafting; the fummit whereof rifes near upon 30 feet above the flat area; fo that I apprehend the floping figure, together with the rough breafting defcribed, fo far curbs the fury of the weftern ftorms that they never break bodily over the flat of the ifland, which I alfo infer from this circumftance, that the whole of the flat is covered with a fpecies of foft woolly mofs, that, as I apprehend, will not grow where the fea wafhes very frequently

over,

over, nor does it appear to be attached to the rock fufficiently firm to prevent its being wafhed away and diffipated, if a heavy fea were frequently to fall upon it : however, I judge from the formation of thefe rocks, that in time of ftorms a heavy fpray goes over the whole, and which is alfo corroborated by there being no grafs growing upon them.

From this defcription it would clearly be no difficult matter to erect a lighthoufe upon this ifland, of any height required, if it were eafy to land and unload heavy materials upon it ; but as there is no cove or upright breaft of rocks fo as to afford either fhelter or a fufficiency of water for the veffels to keep afloat long together at any time of tide, it is evident that the ifland can be acceffible to veffels of any fize whatever, only in times of the fmootheft water. It therefore appears to me, that a fufficiency of time upon the ifland cannot be got within a moderate number of feafons, without making an eftablifhment upon it ; wherein a competent number of workmen and an overfeer could be lodged, and victualled with fecurity, during the fummer feafon. And for this purpofe, upon the flat area I have defcribed, there is a fufficiency of room not only for placing the lighthoufe but for the erection of cabins for fheltering the workmen, and for ftores, and cooking their victuals. And as the whole ifland confifts of granite rock, as already mentioned, the principal part of the ftone materials will be found upon the ifland itfelf, or its near neighbour, which is alfo of granite, and can with a fmall boat be readily vifited therefrom.

As a lighthoufe may be placed upon this ifland without being fubject to the ftroke of the fea, its conftruction will not need to be of that degree of firmnefs as if it were expofed to the waves. It therefore may be built fufficiently compact with rubble, or unformed ftone, efpecially if cafed with aifler of fo moderate a fize as to be readily landed, and which may be wrought with granite or moorftone of the beft quality in Cornwall, which, from what I have experienced, is from the parifh of Conftantin near Falmouth ; the cement that I have ufed on former occafions, being of fo compact a nature as to unite the whole building into one mafs of rock ; here being time intervening betwixt the ftorms for the cement to harden before it is attacked by the wafh of the fea.

It may be proper, as a matter on which the whole depends, to fuggeft the manner in which I would propofe to erect the temporary cabins for the lodgment of the workmen. Thofe I would conftruct wholly of wood, to be made and fitted together at a convenient harbour on fhore ; and being carried in veffels at a proper opportunity a fufficient number of them can be landed and fet up in a few tides to contain as many workmen as can go on with the erection and completion of the reft. I would propofe to make

them

them all of fir plank framed crofs and crofs, fo as to compofe fo many large boxes or chefts, when fcrewed together. I would fuppofe them about 10 feet fquare area, and feven or eight feet high; each cabin to have one bed to hold two perfons. Two of thefe to be appropriated to the kitchen and its conveniencies; one having a fire-place of brick, and four more for ftorehoufes, and the overfeer. The whole to be joined together, fo as to form a ftreet, or narrow lane; the doors and windows to be all next the ftreet and facing each other. I fuppofe about 12 of thefe cabins will be fufficient. The whole muft be water tight, every where except the doors, and thefe as light as pof-fible, with air holes to be ufed at pleafure.

I fuppofe what I have faid will enable the honorable corporation to judge of the practicability of the fcheme: when the expediency thereof is determined upon, it will then be in courfe to make out the working defigns; but as very much will depend upon contingence, no draughts that can be made will enable me to form an eftimate on proper grounds. However, from the beft view I can make of the fubject, I cannot judge that this work fhould be commenced upon a lefs capital than 6000l., exclufive of the furniture and machinery required for exhibiting a light therein.

### Concerning the Erection of a Lighthoufe upon the Owers.

The Owers appears to be a hard funken rock, of confiderable extent in the eaft and weft direction; the fhalloweft water being near the eaft end, whence it very gently and gradually deepens towards the weft.

On Saturday the 26th of June, we rowed in a boat from the floating light; and upon the eaftern part had foundings upon an extenfive furface at feven feet, and much more regular than could have been expected. In fome places, more towards the eaftern end, we had fix feet and under; but thofe places feemed more irregular. The wind was frefh at weft, and obliged us to depart rather before the computed time of low water: and as this was the day of full moon, and confequently the fpring ebbs to be expected fomewhat lower, we determined to take another trial, if thofe fprings offered it: and the next day,

Sunday the 27th June, proving very fine, the wind moderate and at N. W. we took another trial, and in a fix-oared row-boat went directly over the rock. We had pretty extenfive foundings at fix feet, fome places 5½, and fome of 5 feet; but the fhalloweft ground feemed the moft irregular. I don't look upon any of thefe foundings to have

been

been made at fo low an ebb as fometimes happens, fo that I have no doubt of the truth of what has been reported to us, that a perfon has ftood upon the bottom in a pair of fifherman's boots; but as thefe times can happen but feldom, I muft lay it down, that a middling fpring ebb does not give a lefs depth than fix feet water; and that to allow fufficient time for bufinefs, we muft fuppofe the foundation to be laid in feven feet water at the leaft. The breadth of the rock in a north and fouth direction could not well be afcertained for want of fixed marks; but fuppofe it where we took our foundings to be 100 yards. It feems fomewhat narrower near the eaftern end, where the leaft foundings are, but there is not a doubt but that there is an ample fufficiency of room for the bafe of any lighthoufe building.

In time of great ftorms the fhallownefs of the water will undoubtedly occafion a great furf to break upon thofe rocks at fome time of tide; in confequence, I cannot apprehend that ftones of any magnitude that it is practicable to bring and depofit here, will lie quietly without being fubject to be removed, unlefs they are firmly detained in their places. I look upon it as practicable to bring hither and retain fuch ftones as may be expected to lie; and confequently, ftone after ftone being laid and others upon them, till a regular platform of ftone can be formed of regular mafonry, above common low water-mark, the building can then be proceeded with upon principles and in methods that have already been put in practice.

There will however occur great natural difficulties to the proceeding with this work, efpecially in its early ftages. Here is too much water at all times to fee to do the bufinefs in any current method, and here is too little water to float veffels of any burthen, to bring and deliver their materials in fize and in quantity: for as they muft neceffarily be of a flat conftruction, and muft not thump the bottom in a troubled fea, the tonnage of ftone that can be expected to be carried out in one cargo cannot be any great matter.

The diftance of the Owers from the land, fuppofed to be at leaft eight miles, and without any good harbour or quay to retreat to, muft confiderably add to the difficulty: and the many fruitlefs voyages that muft neceffarily occur in purfuing a work in which nothing can be attempted but in the calmeft weather, will inevitably render the eftablifhment of a foundation (till it is got above low water) both tedious and expenfive. However, to fhew that it is a work within the bounds of practicability, I will now take the liberty briefly to fuggeft to this honourable Board the method in which I would propofe to proceed.

Befides

Befides conftructing proper veffels, I would, in the firft place, conftruct a diving bell, or rather diving cheft, fuch as was made ufe of about two years fince at Ramfgate harbour, by means whereof large ftones that interrupted the clearing a foundation, were got up in 12 feet water, and which it was found practicable to ufe at a much greater depth had it been neceffary. With this apparatus it is very practicable to drill holes in the rock; and, in the firft place, to eftablifh a center pin which may be afterwards found, by putting a fmall buoy upon it; and to which the craft may be occafionally moored while employed there. At a competent diftance from this center it is equally practicable to drill any number of holes at pleafure, fo as to form a circle, as regular as it could be done upon the plain ground, into which ftrong iron fpindles or pins, being driven to ftand up (fuppofe a foot) above the furface of the rock; large ftones of fix or eight tons let down within this circular area will be detained from flipping without it till the outward circle of pins is filled with ftones of this fize, which may be roughly hewn, but fo as fomewhat nearly to fit each other; laftly, the internal area within the circle of ftones, being chocked in with large rough blocks, put down promifcuoufly, will eftablifh a bafement courfe, that, as I apprehend, will fuftain a winter's ftorm, without material derangement.

The upper fide of the ftones of the outward circle can in their formation have fuch indentures cut in them, as fhall retain in their places another circle of ftones laid upon them, on the fame principle as the firft courfe is retained by the iron pins. The interior part of this courfe being alfo chocked in with rough blocks, thefe two courfes, which may bring the work a foot above low water, and can be further fteadied and bonded with iron cramps, I expect will prove fufficient to fuftain another winter.

In this fituation the work may be brought to a level, the interftices rammed full, and prepared for laying on a cap courfe in water mortar, begun folid from the middle, and dove-tailed together in the fame manner as the Edyftone lighthoufe is. This folid courfe being all united as one ftone, will, both by its weight and connection, firmly bond together the whole of the under courfes, fo that nothing can get out of place though the iron fpindles, or fteady pins, were entirely to wafte and decay with ruft.

After this is done a fuperftructure may be raifed upon this compound bafe on the fame principles, and in a fimilar manner with that of the Edyftone.

A work

A work of this kind is still less a subject of computation, in regard to expense, than that proposed at Roseveern. It will, indeed, very much depend upon good or bad luck of seasons and circumstances; but I think it ought not to be begun upon a less capital than 20,000l.

> I remain with the utmost respect
> to this Honourable Corporation,
> their most obliged servant,
> J. SMEATON.

*Grays Inn,*
*22d July* 1790.

---

## COAL MEASURES.

AN ACCOUNT of the Meafures of Coals at Newcaftle and London, and of the cuftomary Meafures in the Parifh of Whitkirk, and the adjacent Parifhes.

THE Newcaftle chaldron, by which coals are delivered to the fhips, is fixed by act of Parliament to be 53 hundred weight; and by the ftated proportions ufed at the cuftom-houfe, &c., eight chaldrons Newcaftle, are to make out 15 chaldrons London meafure: the weight of the London chaldron will therefore be 28 cwt. 1 qr. 2 lbs.; and the vats by which coals are meafured at London by upheaped meafure, are, or fhould be contrived, of fuch dimenfions as to anfwer the above quantity.

The bufinefs of Newcaftle in the coal way, not only in regard to the delivery of coals for home confumption but for the collieries, is not done by weight but by meafure; the foundation whereof is the coal bowl, which confifts of 36 corn gallons Winchefter meafure ftriked; a veffel therefore of the capacity of 36 corn gallons corn meafure, contains the meafure of a coal bowl, and the coals are meafured therein by filling them level with the top.

The fpecific gravity of clean coals (that is, when free from ftony or braffy matter, which is heavier) does not greatly differ; fo that at a medium it is found and computed that 21 bowls will make a chaldron of 53 cwt., and hence it will follow that the coal bowl will be 276 lbs. neat weight, that is, 2 cwt. 1 qr. 24 lbs., and each of thefe bowls is a fodder, which is looked upon to be a full load for a two-horfe cart, and a fodder is in general eftimation looked upon to be two thirds of a good three-horfe cart load (which is not however the mode of carriage of that country, they generally carrying by fodders); 12 bowls will therefore weigh 29 cwt. 2 qrs. 8 lbs., exceeding the London chaldron by 1 cwt. 1 qr. 6 lbs.

Mr. Smeaton has remembered the collieries of Whitkirk parifh and the neighbourhood above 45 years; he always underftood, that when the owners of a colliery had a mind to recommend their coals to fale, they did it by giving good meafure; that is, by giving fomewhat more than the ufual quantity by making their corves larger, or upheaping

them

them as much as poffible, and when eftablifhed, and having a greater fale than they could readily fupply, they then were apt to fhorten their meafure; that is, to make the corves lefs or upheap them lefs than ufual. Mr. Smeaton never heard nor could learn any ftated dimenfions for a corf; but he always underftood that a corf of coals was, or ought to be a good horfe load: fo that in winter when the roads were but indifferent, the neighbouring farmers feldom carried above 10 corves at once, and fometimes not fo much. In Mr. Smeaton's memory coals, as well as all other goods, were in a much greater proportion carried upon horfes' backs than they have been within the courfe of the laft 20 years. The ftated weight of a horfe pack is 18 ftone, that is, (eight ftone to the hundred) equal to 2 cwt. 1 qr., and this being underftood to be the weight of a corf of coals, a dozen at this rate will weigh 27 cwt., that is, 1 cwt. 1 qr. 2lbs. lefs than the London chaldron; and in confirmation that the corf of coals is not lefs than the horfe pack weight, about eight years fince, Mr. Smeaton being upon fome experiments on the fubject, weighed fix corves of coals, brought home in his own cart from Walton, which were not quite but yet not much fhort of half the weight of a London chaldron, which made him conclude at that time, that the Halton dozen, and the London chaldron, were nearly the fame thing.

From the near coincidence of the dozen of coals confifting of 12 horfe pack weight, the London chaldron, and 12 Newcaftle coal bowls, he is inclined to think they have all originated from the fame thing; videlicet, the coal bowl of Newcaftle, which itfelf has been adapted to the weight they found it practicable to carry on horfes' backs, there being few, if any, wheel carriages in England at that time; and therefore before the invention of the Newcaftle rail or waggon roads, all the coals that were carried down to the fhips muft have been conveyed on horfes' backs, and though the whole coal trade of that country muft have been only a trifle in proportion to modern times, yet the whole of the trade being carried in this way, it muft have been very confiderable as applicable to horfe labour; we may therefore well expect, that in contracting for the carriage of coals to the fhipping, it muft have been of great confequence in thofe times to have the quantity afcertained that each horfe could and was expected to carry. Accordingly the bowl for wheat, and other grain, in that country being moft generally 4 Winchefter bufhels of eight corn gallons each, that is, 32 gallons; it has been found on repeated trials that a horfe was capable of carrying fomewhat more than this bowl of coals, and therefore that inftead of carrying four eight-gallon bufhels, it has been univerfally agreed that it fhould confift of nine gallons, and in confequence the coal bowl of 36 gallons, has ever fince proved an invariable meafure at Newcaftle.

He

He is further inclined to think that originally they fent away three fodders, or 24 bowls, to London for a Newcaftle chaldron (as it ftill continues for home confumption), which at London was fold for two chaldrons, of 12 bowls each; but that as the demand for coals increafed, they gradually fhortened their meafure both at Newcaftle and London, and this abufe and uncertainty increafing called for the interpofition of the legiflature, when, having after a courfe of years forgot the grounds upon which this trade had originally commenced, they fettled it as well as they could to what they found a medium at that time, ordering the Newcaftle chaldron, not to be a meafure, but a weight of 53 cwt.

He has never heard it doubted but that the firft collieries in England were opened at Newcaftle, from whence they would naturally feek inftruction in opening and working collieries in other parts of the kingdom, and finding that a bowl of coals was drawn at a time, which then would be moft convenient, as being a proper meafure for putting upon a horfe, the imitators would naturally do the fame thing for the fame reafon; and as in the manufacturing part of the Weft Riding of Yorkfhire, the neceffity of a confiderable quantity of horfe carriage would fubfift as far back as the firft getting of coals at Newcaftle (fome time, if he recollects his information right, after the year 1200), the load of coals and the horfe pack would eafily be fubftituted for one another; and indeed without fuppofing the dimenfions of a bowl corf to be brought from New-caftle, the reafon of the thing, that is, the natural ability of horfes to carry burthens, would bring the matter fomewhat near, there being only 24 lbs. difference between the pack weight of the Weft Riding, and the coal bowl of Newcaftle, the latter being as before ftated 24 lbs. heavier.

Since the invention of the coal waggon roads and other carriages, as well as powerful machines, or gins for drawing coals from great depths, it has been found convenient to draw more than a bowl at once, and it is now in common practice to draw a 20-peck corf, that is, $2\frac{1}{2}$ bowls at a time, $\frac{1}{8}$ of a coal bowl being denominated a coal peck; and fince the laying of waggon ways in Yorkfhire (the firft of which is in Mr. Smeaton's memory), to carry coals to the navigable rivers, they have delivered as many as they can carry in a waggon for a dozen, or chaldron, now amounting to about from 40 to 45 cwt., taking pay for the coals in proportion; but firft calculated with intent to avoid the payment of the river's toll, which ufed to be taken by the chaldron or dozen before the laft act of Parliament.

J. SMEATON.

*Aufthorpe*,
27th *January* 1779.

N. B.

N. B. The act now paffing for Mr. Brandling ferving the town of Leeds will, as by agreement, probably fix the weight or meafure of the chaldron or dozen at Leeds; but Mr. S. has not yet heard what it is.

To Lady Irwin.

Mr. Smeaton prefents his moft refpectful compliments to Lady Irwin, and inclofed herewith begs leave to prefent her with the beft account he is able to give of the weights and meafures ufed in the coal trade, fo far as they have come to his knowledge.

*Aufthorpe,*
27th *January* 1779.

From Lady Irwin.

Lady Irwin is more obliged to Mr. Smeaton for the paper than fhe can exprefs, and can't, without being infinitely afhamed, reflect upon the trouble fo very juft a ftate of an intricate and difficult bufinefs muft have given him; begs leave to return him her moft grateful thanks, accompanied by every good wifh for himfelf and family.

*Wednefday Evening.*

POZZELANA

## POZZELANA MORTAR.

DIRECTIONS for preparing, making, and uſing Pozzelana Mortar, by JOHN SMEATON, Engineer.

THE firſt thing that ſhould be done, is to ſift it through a coarſe wire ſieve, ſeparating what will paſs through the ſieve from what will not, and then to ſift what has paſſed through the firſt ſieve through one of a finer ſort. A wire ſieve having about 7 or 8 maſhes per inch running, will be of ſufficient fineneſs, and all that will paſs the ſecond ſieve will be fit for uſe, what will not paſs the ſecond ſieve muſt be reſerved for grinding, and what would not paſs the firſt ſieve muſt be broken to a ſize conformable to what would not paſs the ſecond, and then all ground together ; but in breaking the large that would not paſs the firſt ſieve, it will be proper to pick out a kind of grey ſtony matter, as well as other heterogeneous ſubſtances that get accidentally mixed therewith ; and which will readily diſcover themſelves from the true pozzelana, and which have no cementing quality, and render it more difficult to grind. The true pozzelana is of a dark brown or dirty red colour, and the larger pieces being broken will readily diſcover themſelves, eſpecially with an ordinary magnifying glaſs, to be of a ſpongy ſubſtance with innumerable little cavities like a cinder, and not much harder.

That part of it requiring grinding muſt firſt be got perfectly dry, either by the ſun or by a drying kiln, otherwiſe it is apt to clog the mill ſtones, and it is done by far the moſt completely by grinding it upon a pair of corn mill ſtones, which will at one operation reduce it to a proper fineneſs without need of further ſifting ; French ſtones anſwer the purpoſe beſt, for though it may be done by other kinds of mill ſtones, yet being mixed with flinty matter, which cannot readily be picked out, no other kind of ſtones will ſtand the ſervice, if wanted in any conſiderable quantity. The millers however are not very deſirous of meddling with it, on account of its ſpoiling the colour of their ſtones. I have therefore in the larger kind of works that I have been concerned in, found it worth while to conſtruct a mill on purpoſe, to go by water, wind, or horſe, according to convenience.

In making mortar of it, it muſt be mixed with lime in much the ſame manner and proportion that terras mortar is made ; it muſt be obſerved that the better and ſtronger the lime is, the better and ſtronger the cement will be, but like terras it may be uſed with

any

any lime, and in making comparative trials with terras, the fame fort of lime fhould be ufed with both.

The beft kind of lime for water works that I know of, is from Watchet in Somerfet-fhire, Aberthaw in South Wales, and Barrow in Leicefterfhire, and the ftrongeft compo-fition I know is made by an equal quantity of lime, ftriked meafure in the dry powder, after being flaked and fifted, and of pozzelana ground and prepared as above, and if put together with as little water as may be, and beaten till it comes to a tough confiftence like pafte, it then may be immediately ufed; but if fuffered to fet, and it be afterwards beaten up a fecond time to a confiderable degree of toughnefs as before, ufing a little moifture, if neceffary, it will fet harder but not fo quick.

This compofition is of excellent ufe in jointing the ftones that form the lodgement for the heels of dock gates and fluices, with their threfholds, &c. when of ftone.

A fecond kind of mortar is made by ufing the fame proportion of ingredients as terras mortar, that is, two meafures of lime to one of pozzelana beaten up in the fame manner, and which if ufed with common lime, will fully anfwer for the faces of walls either ftone or brick that are expofed to water, either continually or fubject to be wet and dry, in which laft cafe the pozzelana greatly exceeds the terras, as alfo in its lying quiet in the joints as the trowel has left them, without growing as terras does.

As a piece of œconomy, I have found that if the mortar laft mentioned is beaten up with a quantity of good fharp fand, it no ways impairs its durability, and increafes the quantity. The quantity of fand to be added depends upon the quality of the lime, and is thus deter-mined: if to the pozzelana confidered as mortar, you add as much real fand as will make out the whole quantity, fuch as an experienced workman would allow to his lime to make good common mortar, this will fhew the quantity to be added, that is, may be originally beaten up together; thus, if the lime is of fuch quality as to take two meafures of fand to one of lime, then one meafure of pozzelana and three meafures of fand will fatisfy two meafures of lime.

The compofitions above mentioned are feldom ufed further than for 6 inches within the face of the ftone, or at moft, for fetting the ftones and the bricks forming the face of the work, while the backing is wholly done with common mortar, and which under water never comes to the hardnefs and confiftence of ftone, or forms that bond of union which would arife from a ftony hardnefs; I have therefore found it preferable, where

poz-

pozzelana can be had in plenty, to allow one bufhel of pozzelana to eight bufhels of the lime compofing the mortar for backing.

The firft compofition will affuredly acquire the hardnefs of ftone under water, and in 12 months will be as hard as Portland.

The hardening of the fecond and third depends greatly upon the quality of the lime, as alfo that of the fourth, yet there is fcarcely any lime with which the materials well beaten up, in the proportion fpecified, will not acquire a very competent degree of hardnefs under water.

J. SMEATON.

*Aufthorpe,*
23d *September* 1775.

INDEX.

# INDEX.

*Knowch*

*Sandwich*

THE END.

Strahan and Prefton,
Printers-Street, London.

Printed in the United States
By Bookmasters